PARTICLES AND FIELDS

Previous Proceedings in the Series of Mexican Workshops on Particles and Fields

	Year	Held in	Publisher	ISBN
6th	1997	Morelia, Michoacán	AIP Conf. Proceedings vol. 445	1-56396-791-X
5th	1995	Puebla, Puebla	AIP Conf. Proceedings vol. 359	1-56396-548-8
4th	1993	Merida, Yucatan	World Scientific (1994)	981-02-1709-9
3rd	1991	Morelia, Michoacán	Rev. Mex. Phys. 39 Suppl. 1 (1993) 1	
2nd	1989	Puebla, Puebla	Rev. Mex. Phys. 36 Suppl. 1 (1990) 1	
1st	1987	Leon, Guanajuato	unpublished	

Other Related Titles from AIP Conference Proceedings and the Subseries on High Energy Physics

533 Next Generation Nucleon Decay and Neutrino Detector: NNN99
Edited by Milind V. Diwan and Chang Kee Jung, August 2000, 1-56396-956-4

490 Particles and Fields: Eighth Mexican School
Edited by Juan Carlos D'Olivo, Gabriel López Castro, and Myriam Mondragón, November 1999, 1-56396-895-9

484 Trends in Theoretical Physics II
Edited by Horacio Falomir, Ricardo E. Gamboa Saraví, and Fidel A. Schaposnik, July 1999, 1-56396-894-0

478 COSMO-98: Second International Workshop on Particle Physics and the Early Universe
Edited by David O. Caldwell, May 1999, 1-56396-853-3

453 Particles, Fields, and Gravitation
Edited by Jakub Rembieliński, December 1998, 1-56396-837-1

444 Particle Physics and Cosmology: First Tropical Workshop/High Energy Physics: Second Latin American Symposium
Edited by José F. Nieves, September 1998, 1-56396-775-8

To learn more about these titles, or the AIP Conference Proceedings Series, please visit the webpage **http://www.aip.org/catalog/aboutconf.html**

PARTICLES AND FIELDS

Seventh Mexican Workshop

Mérida, Yucatán, México   10–17 November 1999

EDITORS
Alejandro Ayala
ICN-UNAM, México

Guillermo Contreras
CINVESTAV, Mérida

Gerardo Herrera
CINVESTAV, México

Melville, New York, 2000
AIP CONFERENCE PROCEEDINGS ■ VOLUME 531

Editors:

Alejandro Ayala
Instituto de Ciencias Nucleares, UNAM
Circuito Exterior, Cd. Universitaria
Apdo. Postal 70-543
04510 México, D. F.
MEXICO
E-mail: ayala@nuclecu.unam.mx

Guillermo Contreras
Departamento de Física Aplicada
CINVESTAV-Mérida
Apdo. Postal 73 "Cordemex"
97310 Mérida, Yucatán
MEXICO
E-mail: jgcn@moni.cieamer.contacyt.mx

Gerardo Herrera
Departamento de Física
CINVESTAV
Apdo. Postal 14-740
07000 México, D. F.
MEXICO
E-mail: gherrera@fis.cinvestav.mx

Authorization to photocopy items for internal or personal use, beyond the free copying permitted under the 1978 U.S. Copyright Law (see statement below), is granted by the American Institute of Physics for users registered with the Copyright Clearance Center (CCC) Transactional Reporting Service, provided that the base fee of $17.00 per copy is paid directly to CCC, 222 Rosewood Drive, Danvers, MA 01923. For those organizations that have been granted a photocopy license by CCC, a separate system of payment has been arranged. The fee code for users of the Transactional Reporting Service is: 1-56396-954-8/00/$17.00.

© 2000 American Institute of Physics

Individual readers of this volume and nonprofit libraries, acting for them, are permitted to make fair use of the material in it, such as copying an article for use in teaching or research. Permission is granted to quote from this volume in scientific work with the customary acknowledgment of the source. To reprint a figure, table, or other excerpt requires the consent of one of the original authors and notification to AIP. Republication or systematic or multiple reproduction of any material in this volume is permitted only under license from AIP. Address inquiries to Office of Rights and Permissions, Suite 1NO1, 2 Huntington Quadrangle, Melville, N.Y. 11747-4502; phone: 516-576-2268; fax: 516-576-2450; e-mail: rights@aip.org.

L.C. Catalog Card No. 00-106083
ISBN 1-56396-954-8
ISSN 0094-243X
Printed in the United States of America

CONTENTS

Preface ix
Acknowledgments xi

COURSES AND LECTURES

Little Bang at Big Accelerators: Heavy Ion Physics from AGS to LHC 3
    J. Schukraft
Spontaneous Symmetry Breaking and Chiral Symmetry 16
    L.-F. Li
CP Violation 45
    G. Valencia
Top and Higgs at the Tevatron: Measurements, Searches, Prospects 69
    J. Konigsberg
Trilinear Gauge Boson Couplings and Vector Boson Pair Production 76
    A. Sánchez-Hernández
A Short Course in Effective Lagrangians 81
    J. Wudka
Experimental Techniques 102
    J. Engelfried
Physics at the Electron-Proton Collider HERA 122
    A. De Roeck
Charm Hadroproduction 152
    R. Vogt
Recent Results on Charm Physics from Fermilab 172
    J. C. Anjos and E. Cuautle
Deep Inelastic Scattering, Diffraction and All That 199
    C. A. García Canal and R. Sassot

SESSION IN HONOR OF LEON M. LEDERMAN

Welcome address 247
    R. Asomoza
Leon M. Lederman and the High Energy Physics in Mexico 250
    G. Herrera Corral
Origin of Experimental High-Energy Physics at UNAM 252
    J. Flores
SELEX 255
    A. Morelos
The University of Guanajuato Institute of Physics (IFUG).
Leon Lederman, the Big Boss 259
    J. Félix and G. Moreno
The Latin American Collaboration in DØ 263
    A. Sánchez-Hernández

Leon Lederman and 15 Years of Fermilab-CBPF Collaboration 267
    J. C. Anjos
A Challenge to Join the CDF Experiment 271
    J. Konigsberg
**Mexican Participation in the H1 Experiment, A Bit of History,
a Bit of Physics** 276
    J. G. Contreras
The Pierre Auger Observatory 280
    A. Zepeda
A Brief Interview with Leon Lederman 285
    Sistema Tele Yucatán Canal 13

SEMINARS

**Heuristic Derivation of Weinberg's Angle from Space-Time
and Gauge Symmetries Unification** 289
    J. Besprosvany
Zero Momentum Gluons and Perturbative QCD 294
    M. Rigol and A. Cabo
**FCNC and Non-standard Soft-Breaking Terms in Weak-Scale
Supersymmetry** 299
    J. L. Diaz-Cruz
**Bloch-Wilson Hamiltonian and a Generalization
of the Gell-Mann−Low Theorem** 305
    A. Weber
Breaking of Flavor Permutational Symmetry and the CKM Matrix 310
    A. Mondragón and E. Rodríguez-Jáuregui
The Possibility of Discovering New Boson in e^-e^-, $\mu^-\mu^-$, $e^-\mu^-$ Colliders 315
    J. C. Montero, V. Pleitez, and M. C. Rodriguez
The Cosmological Constant and Quintessence 320
    A. de la Macorra and G. Piccinelli
Z Physics Effects of an Additional Non-sequential Bottom Quark 326
    R. Martinez, J.-A. Rodriguez, M. Vargas, and I. D. Zuluaga
Horizontal Interactions and Gauge Theories 332
    W. A. Ponce
**Bounds on Neutrino Mixing Angles within the Context
of $SU(6)_L \otimes U(1)_Y$ Model** 337
    R. Gaitán, E. García, A. Hernández-Galeana,
    and J. M. Rivera-Rebolledo
Flavor Changing Neutral Current Decays of the Top Quark 342
    M. A. Pérez and G. Tavares-Velasco
Itemization of Trilinear Couplings for Neutral Gauge Bosons 346
    F. Larios, M. A. Pérez, G. Tavares-Velasco, and J. J. Toscano
The Neutrino Telescope ANTARES 351
    A. Rostovtsev

Supernova Neutrino Oscillation in the Presence of Random Magnetic Field 355
    S. Sahu

Silicon Drift Detectors in the ALICE Experiment 360
    V. Bonvicini, P. Cerello, E. Crescio, P. Giubellino,
    R. Hernández-Montoya, A. Kolojvari, G. Mazza,
    L. Montano, J. Nissinen, D. Nouais, A. Rashevsky,
    A. Rivetti, F. Tosello, and A. Vacchi

On the Quark Structure of the $f_0(980)$ Meson and the VEPP-2M Experimental Results for the $\phi \to \pi^0 \pi^0 \gamma$ Decay 365
    M. Napsuciale and J. L. Lucio

Observation of the Centrally Produced $\phi\phi$ System at 800 GeV/c 370
    M. A. Reyes, M. C. Berisso, D. C. Christian, J. Félix,
    A. Gara, E. E. Gottschalk, G. Gutiérrez, E. P. Hartouni,
    B. C. Knapp, M. N. Kreisler, S. Lee, K. Markianos,
    G. Moreno, M. H. L. S. Wang, A. Wehman, and D. Wesson

Final State Interference Effects in Hadronic Charm Meson Decays 375
    G. Herrera and M. I. Martínez

Λ^0 Polarization in $pp \to p\{\Lambda^0 K^+\}$ at 800 GeV/c 381
    J. Félix, M. C. Berisso, D. C. Christian, A. Gara,
    E. E. Gottschalk, G. Gutierrez, E. P. Hartouni, B. C. Knapp,
    M. N. Kreisler, S. Lee, K. Markianos, G. Moreno,
    M. A. Reyes, M. Sosa, M. H. L. S. Wang, A. Wehman,
    and D. Wesson

Strong Interaction Corrections to the Neutron Beta Decay and High Precision 386
    A. Garcí and J. L. García-Luna

Single Spin Asymmetries and the Thomas Precession Mechanism 391
    G. Domínguez-Zacarías and G. Herrera

List of Participants 396
Author Index 399

PREFACE

Physics advances through experiments but the evolution of concepts occurs outside the laboratory, where theoreticians undertake the interpretation of observations. The Mexican Workshop on Particles and Fields was the seventh in a highly successful series started in 1987. Since the beginning, the workshop has brought together experimentalists and theoreticians to discuss recent developments in high energy physics. It consists of lecture courses in which students have the opportunity to learn about new trends in the field from experts in Mexico and abroad. There are also parallel sessions of short seminars in which participants present their research work.

The Seventh Mexican Workshop was held in Mérida, capital of Mexico's southeastern state of Yucatán from November 10-17, 1999. It was organized by CINVESTAV Unidad Mérida and the Division of Particles and Fields of the Mexican Physical Society.

During the workshop we had a special session dedicated to Leon M. Lederman. The Division of Particles and Fields of the Mexican Physical Society awarded him a medal in recognition of his support in the creation of experimental high energy physics groups in Latin America.

The contributions gathered in these proceedings also constitute a sampling of the many facets shown by physicists in our country at its present stage of development. A remarkable aspect of this development in recent years is the incursion of our high energy physics community in experimental research. The experimental lectures in this workshop reflect to a good extent the interest of several groups in Mexico that are now starting activities in the most important laboratories of the world.

Another interesting aspect of particle physics development in Mexico is the increasing attendance at the Workshop. In 1989, the number of participants in the Second Mexican Workshop on Particles and Fields was 70. We had approximately 150 participants in this, the Seventh Workshop. In other words, attendance has doubled in the past ten years.

Gerardo Herrera
President of the Division of Particles and Fields
Mexican Physical Society

ACKNOWLEDGMENTS

The Seventh Mexican Workshop on Particles and Fields has been made possible by the support of the following institutions:
Centro de Investigación y de Estudios Avanzados in Mérida and Mexico City,
Centro Latinoamericano de Física (CLAF),
Centro Latinoamericano de Física in Mexico,
Consejo Nacional de Ciencia y Tecnología (CONACyT) and its agreement with the European Center for Nuclear Research (CERN),
Instituto de Ciencias Nucleares and Instituto de Física from UNAM,
International Center of Theoretical Physics (ICTP),
Instituto de Física of the University of San Luis Potosí
University of Puebla, and
the Division of Particles and Fields of the Mexican Physical Society.
We thank all of them for their generous support.

We would like to thank our collegues in the organizing committee:
Lorenzo Díaz Cruz (BUAP),
Juan Carlos D'Olivo (ICN-UNAM),
Rodrigo Huerta (CINVESTAV-Mérida),
Francisco Larios (CINVESTAV-Mérida),
Miguel A. Pérez (CINVESTAV-México).
The success of the event is due to their effort and hard work. We wish to give special thanks to Miguel A. Pérez and Guillermo Contreras for their inexhaustible energy before and during the Workshop as well as to Monika Kullová, Guadalupe Aguilar, Sofía Alonso and Maru Rodríguez for their invaluable secretarial assistance and dedication.
We would like to thank the Academic Secretary of Cinvestav-Mérida Rodrigo Huerta for his hospitality and warm welcome;thanks also to Gerardo Goldman (Director of CINVESTAV-Mérida) and René Asomoza (Academic Secretary at CINVESTAV-Mexico) for their participation in the special session dedicated to Leon M. Lederman. Thanks to the lecturers for the careful preparation of their notes and their excellent talks.

Alejandro Ayala
ICN-UNAM
Guillermo Contreras
CINVESTAV, Unidad Mérida
Gerardo Herrera
CINVESTAV, Unidad Zacatenco

COURSES AND LECTURES

Little bang at big accelerators: Heavy ion physics from AGS to LHC

J. Schukraft

CERN Div. PPE, CH-1211 Geneva 23

Abstract. The field of ultra-relativistic heavy ion physics, which started some 10 years ago at the Brookhaven AGS and the CERN SPS with fixed target experiments, is entering today a new era with the imminent start-up of the Relativistic Heavy Ion Collider RHIC and preparations well under way for a new large heavy ion experiment at the Large Hadron Collider LHC. This overview will sketch a rough picture of the heavy ion program at current and future machines and concentrate on a few important topics, in particular the question if current results show any of the signs predicted for the phase transition between normal hadronic matter and the Quark-Gluon Plasma.

I INTRODUCTION

The aim of high-energy heavy-ion physics is the study of strongly interacting matter at extreme energy densities. Statistical QCD predicts that, at sufficiently high density, there will be a transition from hadronic matter to a plasma of deconfined quarks and gluons — a transition which in the early universe took place in the inverse direction some 10^{-5} s after the Big Bang and which might play a role still today in the core of collapsing neutron stars. The study of the phase diagram of nuclear matter (see Fig. 1), utilising methods and concepts from both nuclear and high-energy physics, constitutes a new and interdisciplinary approach in investigating matter and its interactions. In high-energy physics, interactions are derived from first principles (gauge theories), and the matter concerned consists mostly of single particles (hadrons/quarks). In contrast, on nuclear physics scales the strong interaction is shielded and can, therefore, to date only be described in effective theories, whereas matter consists of extended systems with collective features. Combining the *elementary-interaction* aspect of high-energy physics with the *macroscopic-matter* aspect of nuclear physics, the subject of heavy-ion collisions is QCD thermodynamics, i.e. the study of bulk matter consisting of strongly interacting particles (hadrons/partons). The formalism to be used would ideally be the one of thermodynamics, where complex multi-particle states are described in terms of a few macroscopic variables.

The study of the QGP is of interest to explore and test QCD on its natural

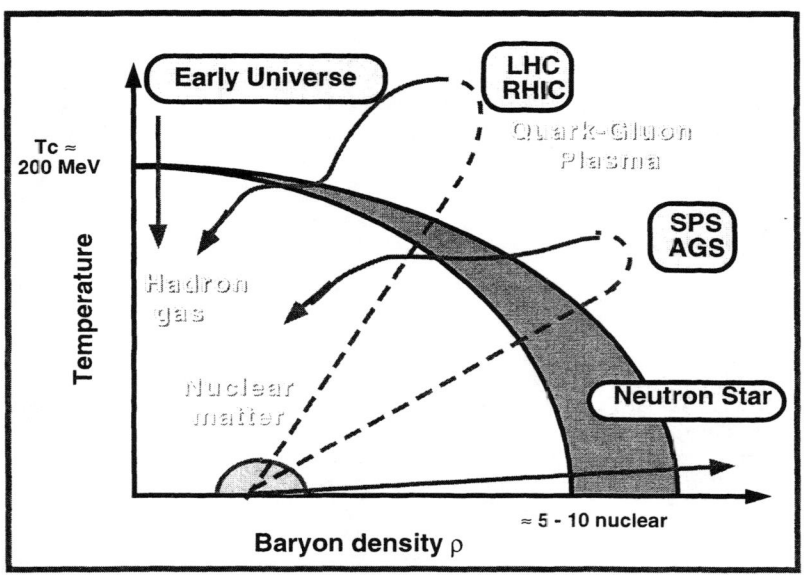

FIGURE 1. The phase diagram of hadronic matter and the hadron gas - quark-gluon plasma phase transition.

scale (Λ_{QCD}) and addresses the fundamental questions of confinement and chiral-symmetry breaking, which are connected to the existence and properties of the quark-gluon plasma. Moreover, it is of general relevance in understanding the dynamical nature of phase transitions involving elementary quantum fields, as the QCD phase transition is the only one accessible to laboratory experiments.

II CURRENT STATUS AND RESULTS

A Initial conditions and global features

The predictions of lattice QCD are rather firm in that a transition to the QGP should exist in the vicinity of a critical temperature T_c of $\approx$ 150 – 200 MeV (whether the transition is of first order, second order, or only 'rapid' is still a matter of debate). However, whether the QGP is actually created in heavy-ion collisions at current energies is a different question and will depend on the dynamics of the reactions and in particular on the initial conditions of the system shortly after the collision. In order to reach the QGP, or even only to use macroscopic concepts (such as 'phase transition') and the language and variables of thermodynamics (such as 'temperature' or 'density'), the system has to be *extended* — i.e. its dimensions ought to be much larger than the typical scale of strong interactions — it has to be in (or near) *equilibrium* — i.e. its lifetime has to be larger than the relevant relaxation times — and the *energy density* ϵ has to exceed the critical threshold for QGP formation. This threshold is predicted by lattice QCD to be of the order of 1 – 3 GeV/fm^3, equivalent to a temperature T_c of 150 – 200 MeV or a baryon density ρ_c of 5 to 10 times normal nuclear matter density (see Fig. 1).

Present results from the ongoing fixed-target program indicate that the initial conditions realized in these reactions could indeed be favourable for QGP formation. In head-on central collisions, hundreds of particles are produced per unit of rapidity, the system expands to a size of the order of 1000 fm^3 (as measured by particle interferometry), and initial energy densities are estimated to exceed 2 GeV/fm^3. However, the expansion is also extremely fast, with an estimated total lifetime of only a few fm/c from the first instance of the collision until the final freeze-out of hadrons.

While these results show that we are certainly *close* to the requirements listed above for QGP formation, they are by no means *sufficient*. In particular the energy density estimates are inversely proportional to the assumed 'formation time', i.e. the time needed to reach thermal equilibrium, and might well be smaller (or bigger?) by a factor of the order of two. Also, the lifetime of the system seems marginal, and even if a QGP is formed it might simply not live long enough for its signals to clearly stand out from the background created in later, hadronic phases of the evolution. The existence of a QGP phase can only be settled experimentally by searching for direct and specific signals.

B Recent experimental highlights

The following sections will concentrate on three main topics which are at the heart of the quest for the QGP, and in which significant progress has been achieved over the last years, i.e. are there experimental indications for *equilibrated hadronic matter, chiral symmetry restoration*, and *deconfinement?*

Equilibrium hadronic matter? While in principle the study of non-equilibrium hadronic matter might be of considerable interest, in practice the huge number of largely unknown dynamical parameters governing the evolution of heavy ion reactions would make the analysis of such a complex system very difficult. The powerful laws of thermodynamics can reduce this complexity and make definite and testable predictions, largely independent of the microscopic dynamics, for those degrees of freedom which evolve in equilibrium. The price to pay is a loss of information concerning events preceding the equilibrium, as the memory of earlier (and possibly more interesting) stages of the evolution is largely lost.

In reality, we will have to deal with a hierarchy of processes and scales, some of which have large cross-sections and correspondingly small relaxation times and therefore might evolve close to equilibrium, and others which decouple early from a thermal evolution and are sensitive to the hot initial phase of the reaction. Prime candidates for the former are hadronic observables, like momentum spectra and particle ratios, and for the latter hard probes and electromagnetic signals.

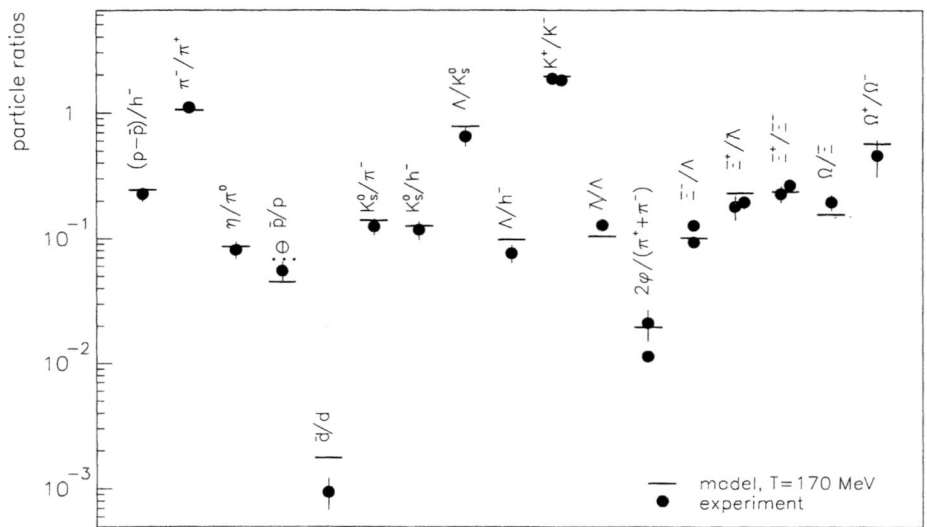

FIGURE 2. Hadrochemical equilibrium model calculation of hadron yields (full lines, calculated for a temperature T of 168 MeV and a baryochemical potential μ_B of 266 MeV) compared to data from CERN SPS.

In a purely thermal system of hadrons, the momentum distributions, when expressed as a function of the transverse mass m_T ($m_T = \sqrt{m^2 + p_T^2}$), will be independent of the particle mass with a slope inversely proportional to the temperature T. In an expanding system, an additional collective flow component can develop which blue-shifts the momentum spectra with a common transverse velocity β_T leading to a mass dependent component. Likewise, the abundance of particle species in equilibrium hadronic matter is given by two independent parameters, i.e. the temperature T and a baryochemical potential μ_B (which reflects the baryon asymmetry in the initial state). A hadronic system in both 'thermal' (momentum) and 'chemical' (particle abundance) equilibrium is therefore fully determined by only three independent parameters: T, β_T and μ_B.

Such a simple prescription seems to be indeed borne out by the data. This is illustrated in Fig. 2, which shows a comparison of measured particle ratios with predictions based on chemical equilibrium [1]. These ratios are in rather good agreement with the equilibrium calculations for a temperature of about 170 MeV and a baryon chemical potential of $\approx$ 250 MeV, corresponding to 1/3 nuclear matter density (at the AGS the corresponding values are T $\approx$ 140 MeV and $\mu_b \approx$ 500 MeV).

Momentum spectra of different particles in Pb-Pb reactions [2] are also well described by a thermal distribution, if, in addition, a common (to all particle types) flow velocity of $\beta_T \approx 0.4c - 0.6c$ is introduced (see Fig.3). On the right part of Fig.3, the inverse m_T-slopes are shown for pp, S+S and Pb+Pb reactions at comparable energies ($\sqrt{s} \approx$ 20 GeV). While the slopes in pp reactions are independent of

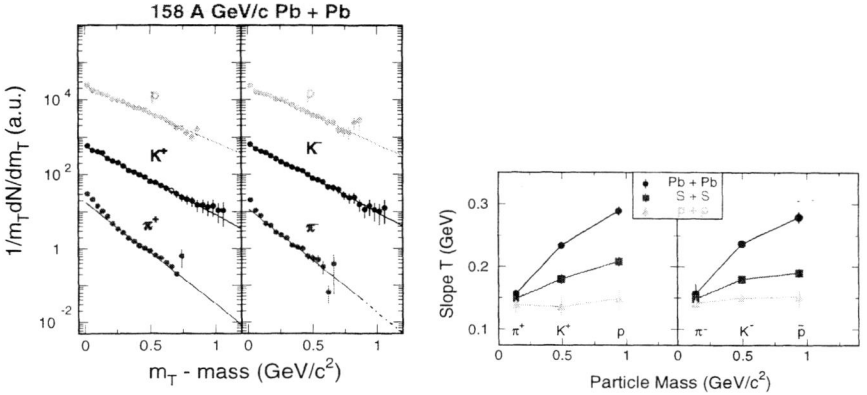

FIGURE 3. Transverse mass spectra (left) and inverse slope parameters (right) of pions, kaons and protons near midrapidity from NA44.

particle type, i.e. exhibit 'm_T-scaling', the slope parameter increases proportional to the particle mass for heavier reaction systems as expected for collective flow. The temperature extracted from momentum spectra at SPS is of the order of 120 MeV, i.e. significantly lower than the one extracted from the particle ratios mentioned above.

A large set of independent hadronic observables, i.e. momentum spectra, particle ratios and HBT correlation results (which are also sensitive to T and β_T [3]), seems to be consistent with a surprisingly simple picture of the late stages of heavy ion reactions: Different particle species are created in relative abundance consistent with chemical equilibrium ratios at a 'temperature' of T $\approx$ 170 MeV; this dense hadronic system then expands and cools to a temperature of about 120 MeV, converting random 'thermal' motion into ordered collective flow until the final freeze-out, when the system is so dilute that all interactions cease. This experimental phase diagram, with both chemical and thermal freeze-out locations as determined from data ranging from very low energies (SIS) up to SPS, is shown in Fig. 4 (taken from ref. [4]). The location of the particle ratio freeze-out point in the temperature-density plane is located very close to the expected phase boundary between hadronic matter and the QGP, and the distance between chemical and thermal freeze-out increases with the beam energy, indicating an increasing dynamical path in the hadronic phase for larger systems (more final state particles).

However, before this intuitively appealing scenario can be taken as fully established, a number of experimental and conceptual questions will have to be clarified. On the experimental side, resolving some inconsistencies between different experiments and better statistics over a large range of impact parameters (in particular for Hyperons) will be needed to come to a more quantitative test of predictions. On the conceptual side, the most puzzling observation is that already very elementary reactions look practically as 'thermal' as heavy ion reactions. While it has

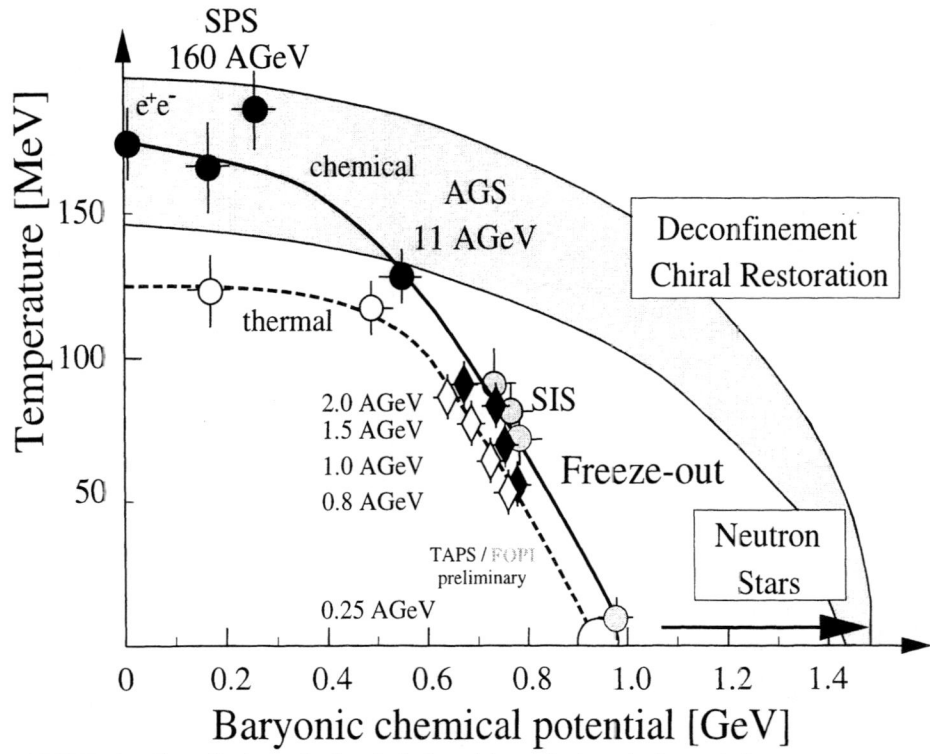

FIGURE 4. Compilation of chemical (particle ratios) and thermal (momentum spectra) freeze-out points from SIS to SPS energies. Reprinted from Ref. [4], with permission from IOP Publishing Limited.

been known, but never 'understood', that momentum spectra in hadronic reactions look 'thermal' (obey m_T-scaling, see Fig.3), a recent re-analysis of pp and e^+e^- reactions has shown that also the particle ratios can be extremely well described with thermodynamics [5]. How can this be possible in systems containing only a few hadrons where a dynamical path from arbitrary initial conditions to thermal distributions via interactions (rescattering) is very unlikely? The success of thermal models to describe particle ratios in reactions ranging from e^+e^- at LEP to Pb-Pb at the SPS could be a hint for a universal feature of the parton-to-hadron (phase?) transition, which might be governed by statistics and phase space at the time of particle creation rather than by dynamical features [4,6].

Assuming that some satisfactory answers to these questions can eventually be found, we could then go on to analyse the hadronic data in more detail to look for information on the dynamics preceding the freeze-out. Relaxation times in a partonic and a hadronic medium are likely to be different, and therefore the questions *how* and *how fast* did the system reach equilibrium in different channels are of interest, particularly in the Hyperon sector, where hadronic relaxation times are

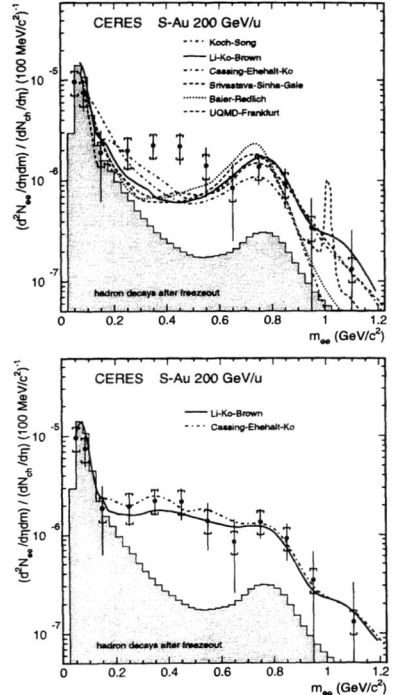
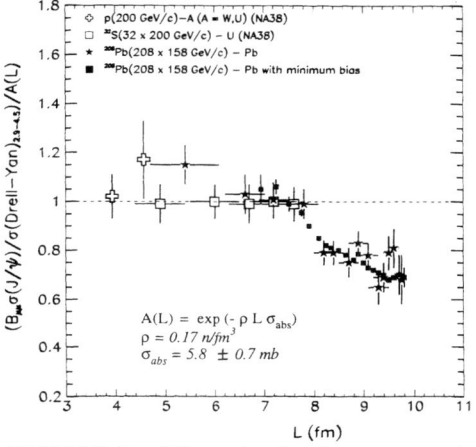

FIGURE 5. Di-electron invariant mass distribution measured by the NA45 experiment in central S+Au collisions, compared to calculations including hadronic decays and effects expected for high pion densities (top) and calculations incorporating in addition a density dependent mass shift of the ϱ meson (bottom).

FIGURE 6. J/Ψ production for proton, sulphur and lead induced collisions relative to the Drell-Yan yield as a function of the thickness L of matter traversed on average. The data are divided by the normal nuclear absorption (exponential in L) which is consistent with the results up to peripheral Pb+Pb reactions. A sudden onset of an additional 'anomalous' suppression is observed for central Pb+Pb.

estimated to be extremely long. Flow patterns should be sensitive to the equation of state of matter and therefore contain indirect evidence for a QGP phase transition preceding freeze-out.

Chiral symmetry restoration? Weakly interacting electromagnetic probes (photons or leptons) are a direct means of gaining information on the early dense and hot stages of the collision, as they leave the interaction volume without being altered by final state effects. While, so far, only upper limits exist for direct (thermal) photon production, recent data on lepton pairs show an unexpectedly large yield at low masses, below the ϱ meson.

Figure 5 shows the electron pair mass spectrum observed in central S+Au collisions by NA45 [7]. The upper part summarizes model calculations which include contributions from hadronic decays (shaded area) and from in-medium pion annihilation and bremsstrahlung. An excess at $0.2 < m(e^+e^-) < 0.6$ remains unexplained. The lower panel exhibits perfect agreement with the data obtained in models which include in addition an in-medium modification of the ϱ and ω masses. A similar excess, consistent with the same model calculations, has been found in the $\mu^+\mu^-$ mass spectrum by NA34/3 [7].

In-medium modification of vector mesons, if experimentally confirmed by more conclusive data, could be a direct consequence of the chiral symmetry transition at the phase boundary between hadronic matter and the QGP. The rapidly varying quark condensate should lead to changes in the properties of hadrons (masses, width) in the vicinity of the phase transition, which will be observable in the lepton mass spectrum for mesons decaying in the dense transition regime. This would indeed be a spectacular verification of the concept underlying the generation of light hadron masses in QCD.

An excess in the intermediate mass range (1.5 - 2.5 GeV) observed in muon pairs by NA50 has so far not found any convincing interpretation [8]. Speculations concerning its origin range from enhanced open charm production and final state rescattering of produced charm quarks to thermal radiation of virtual photons. An experiment is currently proposed at the SPS to address this question.

Deconfinement? Signals originating from hard-scattering processes at the very beginning of the reaction are an ideal tool to probe the state of the surrounding QCD matter. The original idea [9] that J/Ψ production should be suppressed in a QGP relies on a Debye screening mechanism which renders colour interactions short ranged in a dense medium ('deconfinement') and therefore prevents the formation of bound resonances.

J/Ψ suppression with similar characteristics as predicted for a QGP was indeed one of the first results reported from heavy ion experiments in 1987. Its subsequent interpretation, alternating repeatedly between 'trivial' and 'exciting', is probably the best example on how our understanding of nuclear collisions has progressed in a constant interplay between theory and experiment, new explanations and new data. An up-to-date compilation of J/Ψ production relative to the Drell-Yan continuum in pA and AB reactions is shown in Fig. 6 versus the average path length L traversed by the $c\bar{c}$ pair after its creation inside the target and projectile nuclei [10]. Up to and including central S-U collisions, this ratio decreases exponentially with L, consistent with a nuclear absorption cross-section of 6 mb. It took the better part of the last ten years, a variety of data for J/Ψ, Ψ' and Υ — from low energy pp and pA reactions to photoproduction and high p_T production at the Tevatron — and a good measure of other ingredients (nuclear structure functions, initial and final state scattering, formation time) to come to a consistent and theoretically substantiated interpretation [11,12]. The exponential attenuation is today seen as resulting from the interaction between the nuclear medium and a pre-resonance state, a coloured $c\bar{c}$-gluon configuration which evolves only later (and outside the nucleus) after some finite formation time into the physical, colour neutral J/Ψ or

Ψ' hadron. So J/Ψ suppression has provided a lot of insight into the dynamics of charmonium production, hadron formation and using the nucleus as a tool to measure short time scales, but leaving no room for QGP effects.

The extrapolation of this model to central Pb-Pb reactions was straightforward, essentially parameter free, and completely wrong (see Fig. 6)! The Pb-Pb data, whether plotted as a function of L or any other variable, shows significantly less J/Ψ's than hadronic absorption models would predict by extrapolating from light ion and pA results. While some debate still persists if the additional suppression is really 'anomalous' or not, new precision data which is currently being analysed should settle this question in the near future. Most likely, some additional physics will have to be included in order to describe the Pb data. However, whether this 'new physics' will require deconfinement, dynamical pre-cursor phenomena of the QGP transition, or just some overlooked hadronic effect, remains to be seen.

C Future fixed target program

Given the recent exciting developments, the future directions are perfectly clear. The current SPS fixed target program is unique in the world, it addresses a well focused set of fundamental questions, it has entered an extremely productive phase and it is now being brought to its full potential. With the exception of the Hyperon and the low mass lepton pair measurements, statistics is, in general, not a problem. A run at the lowest possible SPS energy, around 40 GeV/nucleon, in 1999 will increase the baryon density, possibly close to its maximum value. Signals related to chiral symmetry restoration, in particular the low mass lepton pairs, will in general be rather sensitive to baryon density. The low energy run can also make contact with the AGS regime and will allow the CERN experiments, which are quite distinct in their capabilities from the AGS detectors, to compare with and complement the program at lower energies. The SPS program is currently foreseen to terminate in the year 2000, however, extensions to address in particular charm production and the intermediate mass lepton pair excess are under discussion. Later, around the turn of this century, the new generation of heavy ion colliders will come into operation.

III HEAVY ION PHYSICS OF THE NEXT CENTURY

The study of ultra-relativistic heavy-ion collisions is a rather new, but rapidly evolving field. After the pioneering experiments at the BEVALAC and in DUBNA with relativistic heavy ions ($E/m \approx 1$), the first experiments started in 1986 with light ions almost simultaneously in Brookhaven (AGS) and at CERN (SPS). Really heavy ions (A $\approx$ 200) have been available in the AGS since the end of 1992 and at the SPS since the end of 1994. When the colliders RHIC ($\sqrt{s} = 200$ GeV/n) and LHC ($\sqrt{s} = 5.5$ TeV/n) come into operation in 1999 and 2005, respectively, the available energy in the centre-of-mass will have increased by almost five orders of magnitude within 20 years. This unprecedented pace was made possible only by (re)using accelerators, and to some extent even detectors, built over a much longer time scale for use in high-energy physics. The following sections will summarise the physics and experiments to come in these latest (and possible last) heavy-ion machines.

Heavy ion collisions at the colliders are expected to provide a qualitatively different environment from existing accelerators by creating a very hot, and therefore more clearly detectable QGP, via hard initial parton scatterings that can be calculated rather precisely. Extrapolating from present results to LHC, all parameters relevant to the formation of the QGP will be more favourable: the energy density, the size and lifetime of the system and relaxation times should all improve by a large factor, typically by an order of magnitude compared to Pb+Pb collisions at the SPS. We expect particle densities of several thousand per unit of rapidity, a freeze-out volume approaching 100,000 fm^3 and an initial energy density orders of magnitude larger than the one of normal nuclear matter. The initial temperature might be close to 1000 MeV, as compared to a value of 400 – 500 MeV at RHIC and about 200 MeV at the SPS. The energy densities and temperatures at LHC should be far above the deconfinement threshold, allowing us to probe the QGP in its asymptotically free 'ideal gas' form.

The analysis of extended strongly interacting matter at both colliders will move from the fixed target regime dominated by soft phenomena into a domain where hard interaction between the primary partons will lead to a rapid production of further partons, and the interactions within this dense partonic medium, with the resulting strong increase in entropy, are expected to produce the QGP. The abundant formation of 'minijets' with transverse momenta of a few GeV plays an essential role in this process. Perturbative QCD calculations can be used to construct and evaluate such parton interaction cascades; they indeed show the expected rapid rise towards thermalisation, on time scales considerably below 1 fm/c, to the extremely high initial temperatures mentioned.

In order to verify that a QGP was produced and to study its properties, we need probes sensitive to the earliest and hottest stages of the medium. Three such probes are currently known: Bound heavy quark resonances (quarkonia), hard jets, and thermal dileptons/photons. Only charmonia as deconfinement probes have been studied successfully at the SPS as discussed above; for all others, higher incident energies appear necessary to find a positive signal.

The deconfinement analysis of hot and dense matter must be extended to bottonium states. The Υ, with its very small radius, can be dissociated only at the highest energy density attainable at LHC (of order 30 GeV/fm^3), while the excited states Υ' and Υ'' are comparable to the charmonium resonances and will serve as important consistency checks.

Hard jets probe the produced medium through the energy loss of partons passing through dense matter. The theoretical aspects of this problem were recently studied in considerable detail. In particular, the rate of energy loss was found to depend quite sensitively on the size of the medium and there are now indications

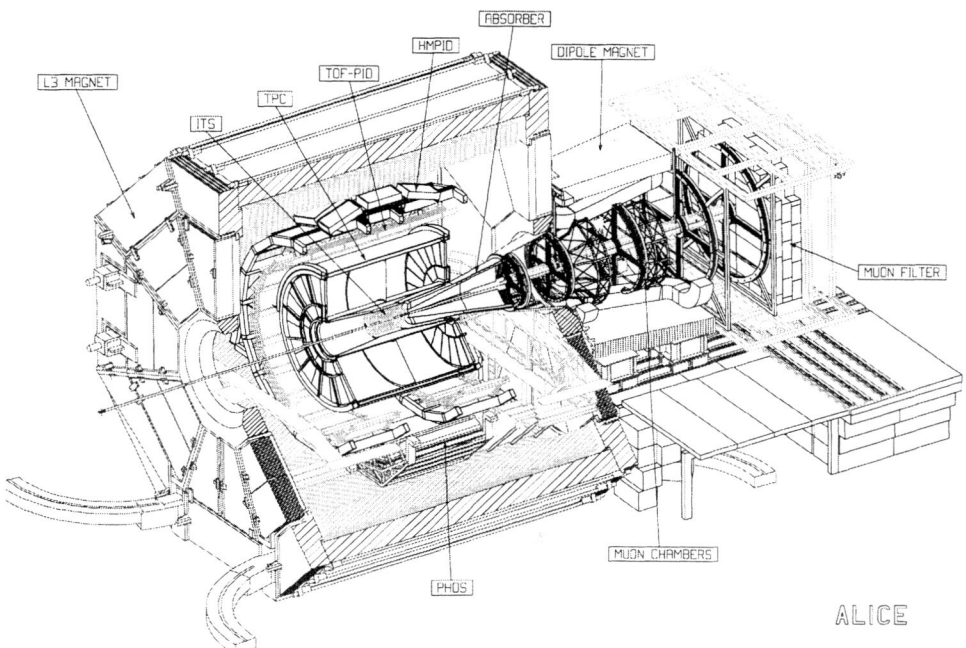

FIGURE 7. Schematic view of the ALICE detector which combines a central barrel for electron pair, photon, and multihadron studies, with a forward dimuon spectrometer.

that the energy loss in cold nuclear matter is much smaller than that in a hot QGP. The production and subsequent attenuation of fast partons will add a crucial new penetrating probe to diagnose the nature of the strongly interacting matter produced in heavy ion collisions.

The temperature of the primordial medium could be best determined through measurement the of spectra of real or virtual photons. The thermal photon spectrum will be an integral over the temperature history of the system. Superimposed are the soft photons from the late hadronic stage as well as the primary Drell-Yan or hard QCD photons. Whether there is a window in transverse momentum (around one to a few GeV) to actually measure such thermal dileptons or photons depends crucially on the density of the produced system; fortunately, conditions could be quite favourable at the high energy densities predicted for RHIC and LHC.

Another unique feature of heavy-ion collisions at the colliders is the possibility to measure a large number of observables with very good accuracy on an event-by-event basis: impact parameter, multiplicity, particle ratios and spectra and, of particular importance, size and lifetime from interferometry. Single event analysis, currently pioneered by NA49 at the SPS, will become a precision tool at very high multiplicity. One of the important design considerations for both the STAR detector at RHIC and the ALICE detector at LHC (Fig. 7) is to make full use of this opportunity. It will allow the study of correlations and non-statistical fluctuations which would otherwise be washed out when averaging over many events. Such fluctuations are, in general, associated with critical phenomena in the vicinity of a phase transition.

IV SUMMARY

The still very young field of ultra-relativistic heavy ion physics has proceeded since its inception in 1986 through three essential phases:

The initial round of 'exploratory' experiments has shown that appropriate detectors and analysis procedures can cope with the extreme particle densities produced in heavy ion collisions. They have qualitatively shown that an extended, interacting and very dense system has been formed that differs in many observables from the more elementary hadron-hadron reactions investigated in the past. Falling short of striking discoveries, this phase has nevertheless provided a *'principle proof of feasibility'* and has substantiated the expectation that heavy ion collisions are an appropriate tool to create equilibrium hadron matter and eventually the quark-gluon plasma.

The next phase was characterized by efforts to get a comprehensive and precise set of data and to come to a quantitative understanding of the experimental results. A close and very effective interaction between theory and experiment, models and data, has led to significant progress in identifying relevant ingredients and important microscopic processes.

The field is currently in its third, and most dramatic phase. Results from both AGS and SPS with really 'heavy' ion beams have produced puzzling results which strongly hint at a picture of high energy nuclear reactions almost too good to be true: i) a premordial phase of deconfined partons – the QGP ? – responsible for quarkonium suppression, followed by ii) a transition regime with gradual onset of chiral symmetry breaking, leading to changes in the properties of light hadrons, concluded by iii) a gas of hadronic matter governed by the simple laws of thermodynamics. On the short term, the ongoing experiments should provide us with additional and more complete data in order to substantiate (or refute) this scenario.

In the longer term, making use of RHIC and LHC for heavy ion collisions provides a unique opportunity for exploring the physics of QCD matter in a qualitatively very different region of extremely high energy density. RHIC and its four major experiments (STAR, PHENIX, BRAHMS and PHOBOS) will make the first step, starting in 2000 with Au+Au collisions at an energy an order of magnitude above what is currently available at the SPS. The LHC, with a centre-of-mass energy almost thirty times above RHIC, will lead by 2005 into a region comparable only to the highest energy cosmic ray events. Its single dedicated detector, the ALICE experiment, represents *the* long term future of the ultra-relativistic heavy ion program. Building a detector of the size and complexity required for LHC will be an unprecedented challenge, and its successful completion will need the continued, strong and emphatic support and participation of the heavy ion community world wide.

REFERENCES

1. P. Braun-Munzinger, J. Stachel, H. Wessels, N. Xu, Phys Lett. B 366 (1996) 1, J. Stachel, Nucl. Phys A610 (1996) 509c and P. Braun-Munzinger, I. Heppe, J. Stachel Phys. Lett. B 465 (1999) 15.
2. N. Xu et al, Nucl. Phys A610 (1996) 175c.
3. H. Appelshauser et al, Eur. Phys.J. C2 (1998) 661.
4. U. Heinz, J. Phys.G 25 (1999) 263.
5. F. Becattini, U. Heinz, Z. Phys. C 76 (1997) 269.
6. R. Stock, Phys. Lett. B 456 (1999) 277.
7. A. Drees, Nucl. Phys A610 (1996) 536c.
8. E. Scomparin et al, Nucl. Phys A610 (1996) 331c.
9. T. Matsui, H. Satz, Phys. Lett. B 178 (1986) 416.
10. M.C. Abreu et al, Phys. Lett. B 450 (1999) 456.
11. C. Lourenco, Nucl. Phys A610 (1996) 552c.
12. D. Kharzeev, Nucl. Phys A610 (1996) 418c.

Spontaneous Symmetry Breaking and Chiral Symmetry

Ling-Fong Li,
Physics Department, Carnegie Mellon University,
Pittsburgh, PA 15213, USA

Abstract

In this introductory lecture, some basic features of the spontaneous symmetry breaking are discussed. More specifically, σ-model, non-linear realization, and some examples of spontaneous symmetry breaking in the non-relativistic system are discussed in details. The approach here is more pedagogical than rigorous and the purpose is to get some simple explanation of some useful topics in this rather wide area.

I. Introduction

The symmetry principle is perhaps the most important ingredient in the development of high energy physics. Roughly speaking, the symmetries of the physical system lead to conservation laws, which give many important relations among physical processes. Many of the symmetries in Nature are however approximate symmetries rather than exact symmetries and are also very useful in the understanding of various phenomena in high energy physics. Among the broken approximate symmetries, the most interesting one is the Spontaneous Symmetry Breaking (SSB) which seems to have played a special role in high energy physics. Many important progress has come from the understanding of the SSB. The SSB is characterized by the fact that symmetry breaking shows up in the ground state rather than in the basic interaction. This makes it difficult to uncover this kind of approximate symmetries. Historically, SSB was first discovered around 1960 in the study of superconductivity in the solid state physics by Nambu and Goldstone.[1,2] One of consequences of SSB is the presence of the massless excitation,[3] called the Nambu-Goldstone boson, or just Goldstone boson for short. Later, Nambu[1] applied the idea to the particle physics. In combination with $SU(3) \times SU(3)$ current algebra, SSB has been quite successful in the understanding of the chiral symmetry in the low energy phenomenology of strong interaction. More importantly, in 1964 it was discovered by Higgs[5] and others[6,7] that in the context of gauge theory, SSB has the remarkable property that it can convert the long range force in the gauge theory into a short range force. Thus it avoids both the massless Goldstone bosons and the massless gauge bosons. Weinberg,[8] and Salam,[9] then applied this ideas to construct a model of electromagnetic and weak interactions. The significance of this model was not realized until t'Hooft[10] show in 1971 that it was renomalizable. Since then this model has enjoyed remarkable experimental success and now called the "Standard Model of Electroweak Interactions".[11] Undoubtedly, this will serve as benchmark for any new physics for years to come.

In this article I will give a simple introduction to the spontaneous symmetry breaking and its application to chiral symmetries in the hadronic interaction. The emphasis is on the qualitative understanding rather than completeness and mathematical rigor. Eventhough SSB has been quite successful in explaining many interesting phenomena, its implementation in the theoretical framework is more or less put in by hand and it is not at all clear what is the origin of SSB. Here I will also discuss some non-relativistic example where the physics is more tractable in the hope that they might give some hints about the true nature of SSB. Maybe good understanding of SSB might extend its applicability to some new frontier.

II. $SU(2) \times SU(2)$ σ-Model

The σ-model has a long and interesting history. It was originally constructed in 1960's as a tool to study the chiral symmetry in the system with pions and nucleons.[4] Later the spontaneous symmetry breaking and PCAC (partially conserved axial current) were incorporated.. Eventhough this model is not quite phenomenologically correct it remains the simplest example which realizes many important aspects of broken symmetries. Even though the strong interaction is now described by QCD, the σ-model of pions and nucleons is still useful as an effective interaction in the low energies where it is difficult to calculate directly from QCD. In addition, the σ-model has also been used quite often as a framework to test many interesting ideas in field theory and string theory. Here we will discuss the most basic features of the σ-model.

The Lagrangian for $SU(2) \times SU(2)$ σ-Model is given by

$$\mathcal{L} = \frac{1}{2}\left[(\partial_\mu \sigma)^2 + \left(\partial_\mu \vec{\pi}\right)^2\right] + \frac{\mu^2}{2}\left(\sigma^2 + \vec{\pi}^2\right) - \frac{\lambda}{4}\left(\sigma^2 + \vec{\pi}^2\right)^2 \qquad (1)$$
$$+ \overline{N} i \gamma^\mu \partial_\mu N + g\overline{N}\left(\sigma + i\gamma_5 \vec{\tau}\cdot\vec{\pi}\right) N$$

where $\vec{\pi} = (\pi_1, \pi_2, \pi_3)$ is the isotriplet pion fields, σ is the isosinglet field, and N is the isodoublet nucleon field. To discuss the symmetry property, it is more useful to write this Lagrangian as :

$$\mathcal{L} = \frac{1}{2} tr\left(\partial_\mu \Sigma \partial^\mu \Sigma^\dagger\right) + \frac{\mu^2}{4} tr\left(\Sigma \Sigma^\dagger\right) - \frac{\lambda}{8}\left[(\Sigma \Sigma^\dagger)\right]^2 \qquad (2)$$
$$+ \overline{N}_L i\gamma^\mu \partial_\mu N_L + \overline{N}_R i\gamma^\mu \partial_\mu N_R + g(\overline{N}_L \Sigma N_R + \overline{N}_R \Sigma^\dagger N_L)$$

where

$$\Sigma = \sigma + i\vec{\tau}\cdot\vec{\pi}, \qquad N_L = \frac{1}{2}(1-\gamma_5) N, \qquad N_R = \frac{1}{2}(1+\gamma_5) N \qquad (3)$$

This Lagrangian is now clearly invariant under transformation,

$$\Sigma \to \Sigma' = L\Sigma R^\dagger, \qquad N_L \to N_L' = LN_L, \qquad N_R \to N_R' = RN_R \qquad (4)$$

where

$$L = \exp\left(-i\vec{\tau}\cdot\vec{\theta}_L\right), \qquad R = \exp\left(-i\vec{\tau}\cdot\vec{\theta}_R\right) \qquad (5)$$

are two arbitrary 2×2 unitary matrices. Thus the symmetry group is $SU(2)_L \times SU(2)_R$ and representation contents under this group are

$$\Sigma \sim \left(\frac{1}{2},\frac{1}{2}\right), \qquad N_L \sim \left(\frac{1}{2},0\right), \qquad N_R \sim \left(0,\frac{1}{2}\right)$$

<u>Remark:</u>The nucleon mass term $\overline{N}_L N_R + h.c.$ transforms as $\left(\frac{1}{2},\frac{1}{2}\right)$ representation and is not invariant. One way to construct invariant nucleon mass term is to introduce another doublet of fermions with opposite parity,

$$N_L' \sim \left(0,\frac{1}{2}\right), \qquad N_R' \sim \left(\frac{1}{2},0\right)$$

so that the term $\left(\overline{N}'_L N_R + \overline{N}'_R N_L + h.c.\right)$ is invariant. This will give same mass to both doublets and is usually called parity doubling. As we shall see later, another way to give mass to nucleon is by spontaneous symmetry breaking which does not require another doublet.

The general form of Noether current is of the form

$$J_\mu \sim \sum_i \frac{\partial \mathcal{L}}{\partial (\partial_\mu \phi_i)} \delta \phi_i$$

where $\delta\phi_i$ is the infinitesimal change of the fields under the symmetry transformations. We have for the left-handed transformation,

$$\delta_L \sigma = \vec{\theta}_L \cdot \vec{\pi}, \qquad \delta_L \vec{\pi} = -\vec{\theta}_L \sigma + \vec{\theta}_L \times \vec{\pi}, \tag{6}$$

$$\delta_L N_L = -i\frac{\vec{\theta}_L \cdot \vec{\tau}}{2} N_L, \qquad \delta_L N_R = 0$$

and

$$J^a_{L\mu} = \varepsilon^{abc} \pi^b \partial_\mu \pi^c + [\sigma \partial_\mu \pi^a - \pi^a \partial_\mu \sigma] + \overline{N}_L \gamma_\mu \frac{\tau^a}{2} N_L \tag{7}$$

Similarly,

$$\delta_R \sigma = -\vec{\theta}_R \cdot \vec{\pi}, \qquad \delta_R \vec{\pi} = \vec{\theta}_R \sigma + \vec{\theta}_R \times \vec{\pi}, \tag{8}$$

$$\delta_R N_L = 0, \qquad \delta_R N_R = -i\frac{\vec{\theta}_R \cdot \vec{\tau}}{2} N_R$$

and

$$J^a_{R\mu} = \varepsilon^{abc} \pi^b \partial_\mu \pi^c - [\sigma \partial_\mu \pi^a - \pi^a \partial_\mu \sigma] + \overline{N}_R \gamma_\mu \frac{\tau^a}{2} N_R \tag{9}$$

The corresponding charges are given by

$$Q^a_L = \int d^3x J^a_{L0}, \qquad Q^a_R = \int d^3x J^a_{R0}. \tag{10}$$

Using the canonical commutation relations, we can derive

$$[Q^a_L, Q^b_L] = i\varepsilon_{ijk} Q^c_L, \qquad [Q^a_R, Q^b_R] = i\varepsilon_{ijk} Q^c_R, \qquad [Q^a_L, Q^b_L] = 0 \tag{11}$$

which is the $SU_L(2) \times SU_R(2)$ algebra.

The vector and axial charges are given by

$$Q^a = Q^a_R + Q^a_L, \qquad Q^a_5 = Q^a_R - Q^a_L \tag{12}$$

In particular, the axial charges are

$$Q^5_i = \int d^3x A^0_i(x) = \int d^3x \left[i(\sigma \partial_0 \pi_i - \pi_i \partial_0 \sigma) + N^\dagger \frac{\sigma_i}{2} \gamma_5 N\right] \tag{13}$$

Remark: Another way to describe the symmetry of the σ-model is the $O(4)$ symmetry, which is isomorphic to $SU(2) \times SU(2)$ locally and is characterized by 4×4 orthogonal matrix,

$$RR^T = R^T R = 1$$

The infinitesimal transformation is

$$R_{ij} = \delta_{ij} + \varepsilon_{ij}, \quad \text{with} \quad \varepsilon_{ij} = -\varepsilon_{ji}$$

The scalar field $\phi_i = (\pi_1, \pi_2, \pi_3, \sigma)$ transform as 4-dimensional vector,

$$\phi_i \to \phi'_i = R_{ij} \phi_j \simeq \phi_i + \varepsilon_{ij} \phi_j$$

The combination $\phi_i \phi_i = \sigma^2 + \vec{\pi}^2$ is just the length of the vector ϕ_i and is clearly invariant under the rotations in 4-dimension. If we take

$$\varepsilon_{ij} = \varepsilon_{ijk} \alpha_k, \quad \varepsilon_{4i} = \beta_i, \quad i,j,k = 1,2,3$$

we get

$$\vec{\pi}' = \vec{\pi} + \vec{\alpha} \times \vec{\pi} + \vec{\beta} \sigma, \quad \sigma' = \sigma + \vec{\beta} \cdot \vec{\pi}$$

Thus we see from Eqs(6,8) that the parameters $\vec{\alpha}$ correspond to vector transformations and $\vec{\beta}$ the axial transformation.

A. Spontaneous Symmetry Breaking

The classical ground state is determined by minimum of the self interaction of scalars,

$$V(\sigma, \vec{\pi}) = -\frac{\mu^2}{2}\left(\sigma^2 + \vec{\pi}^2\right) + \frac{\lambda}{4}\left(\sigma^2 + \vec{\pi}^2\right)^2 \tag{14}$$

The minimum of the potential is located at

$$\sigma^2 + \vec{\pi}^2 = \frac{\mu^2}{\lambda} \equiv v^2 \tag{15}$$

which is a 3-sphere, S^3 in the 4-dimensional space formed by the scalar fields. Each point on S^3 is invariant under $O(3)$ rotations. For example, the point $(0,0,0,v)$ is invariant under the rotations of the first 3 components of the vector. Then, after a point on S^3 is chosen to be the classical ground state, the symmetry is broken spontaneously from $O(4)$ to $O(3)$. Note that different points on S^3 are related to each other by the action of those rotations which are in $O(4)$ but not in $O(3)$. These rotations are usually denoted by $O(4)/O(3)$.(This is called the coset space.). Thus we can identify 3-sphere with $O(4)/O(3)$.

For the quantum theory, we need to expand the fields around the classical values

$$\sigma = v + \sigma', \quad \vec{\pi}' = \vec{\pi}, \quad \text{where} \quad <\sigma> = v \tag{16}$$

Here v is usually called the vacuum expectation value (VEV). Then we see that

$$V\left(\sigma, \vec{\pi}\right) = \mu^2 \sigma'^2 + \lambda v \sigma' \left(\sigma'^2 + \vec{\pi}'^2\right) + \frac{\lambda}{4}\left(\sigma'^2 + \vec{\pi}'^2\right)^2 \qquad (17)$$

$$g\overline{N}\left(\sigma + i\gamma_5 \vec{\tau}\cdot\vec{\pi}\right)N = gv\overline{N}N + g\overline{N}\left(\sigma' + i\gamma_5 \vec{\tau}\cdot\vec{\pi}'\right)N$$

Thus π's are massless and N's are massive.
Remark:If we had made another choice for VEV,e.g.

$$<\pi_3> = v, \qquad <\pi_1> = <\pi_2> = <\sigma> = 0$$

The physics is still the same as we will now illustrate. In this case, we write $\pi'_3 = \pi_3 + v$ to get

$$V\left(\sigma, \vec{\pi}\right) = \mu^2 \pi'^2 + (\text{cubic terms and higher})$$

$$g\overline{N}\left(\sigma + i\gamma_5 \vec{\tau}\cdot\vec{\pi}\right)N = gv\overline{N}\tau_3\gamma_5 N + \cdots$$

Thus we still have 3 massless scalar fields. For the nucleon, if we define

$$N_L = \exp\left(-i\pi\frac{\tau_3}{2}\right)N'_L, \qquad N_R = N'_R$$

we get

$$gv\overline{N}\tau_3\gamma_5 N = gv\overline{N}'N'$$

which the mass term for the new field. It is easy to see that Q_5^1, Q_5^2, Q^3 form an unbroken $SU(2)$ algebra.

There are several interesting features worth noting:

(1) $\vec{\pi}$'s are massless. This is a consequence of the Goldstone theorem which states that spontaneous symmetry breaking (SSB)of a continuous symmetry will give massless particle or zero energy excitation. This theorem will be discussed in more detail in next subsection.

(2)After SSB, the original multiplet $\left(\sigma, \vec{\pi}\right)$ splits into massless $\vec{\pi}'$ s and massive σ. Also the nucleons become massive. Thus eventhough the interaction is $SU(2)_L \times SU(2)_R$ symmetric the spectrum is only $SU(2)$ symmetric. This is the typical consequence of SSB. In some sense, the original symmetry is realized by combining the $SU(2)$ multiplet, e.g. N ,with the massless Goldstone bosons to form the multiplets of $SU(2)_L \times SU(2)_R$. This, as we will discuss later, is the basis of the low energy theorem.

(3)The axial current in Eq(13) after the SSB will have a term linear in π field,

$$A_i^\mu = iv\partial_\mu \pi_i + \cdots \qquad (18)$$

which is responsible for the matrix element,

$$<0|A_i^\mu|\pi_j(p)> = ip^\mu v \qquad (19)$$

Using this matrix element in π decay, we can identify the VEV v with the pion decay constant f_π. This coupling between axial current A_i^μ and π_i will give rise to a massless pole.

(4) The appearance of the cubic term $\sigma'\left(\sigma'^2 + \vec{\pi'}^2\right)$ and the mass term $gv\overline{N}N$ is the result of the spontaneous symmetry breaking. Since these terms have dimension 3, they are usually called soft breaking, in contrast to the dimension 4 hard breaking terms.

(5) In the scalar self interaction, quartic, cubic, and quadratic terms have only 2 parameters, λ, μ. This means that these 3 terms are not independent, and there is a relation among them. This is an example of low energy theorem for theory with spontaneous symmetry breaking.

B. Low energy theorem

The SSB leads to many relations which are quite different from the usual symmetry breaking. The most distinct ones are relations among amplitudes involving Goldstone bosons in low energies. As we have mentioned before, these relations are consequence of the fact that Goldstone bosons are massless and can be tagged on to other particles to form a larger multiplet. Since Goldstone bosons do carry energies, this is possible only in limit that Goldstone bosons have zero energies.

Consider the following processes involving the Goldstone bosons in the external states.

$(i)\pi^0(p_1) + \sigma(p_2) \to \pi^0(p_3) + \sigma(p_4)$

The tree-level contributions are coming from diagrams in Fig1,

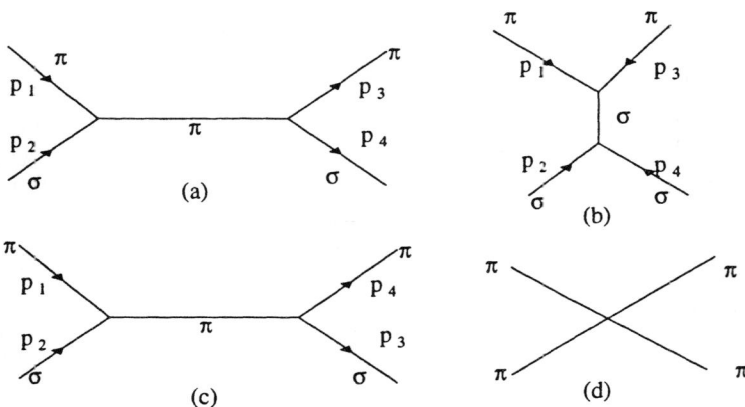

Fig 1 Tree graphs for $\pi\pi$ scattering

The amplitudes for these diagrams are given by,

$$M_a = (-2i\lambda v)^2 \frac{i}{s}, \quad M_b = 3(-2i\lambda v)^2 \frac{i}{t - m_\sigma^2}, \quad M_c = (-2i\lambda v)^2 \frac{i}{u}, \quad M_d = -2i\lambda \tag{20}$$

$$M = M_a + M_b + M_c + M_d = 4i\lambda^2 v^2 \left[\frac{1}{s} + \frac{3}{t - m_\sigma^2} + \frac{1}{u} + \frac{1}{2\lambda v^2}\right] \tag{21}$$

Here s, t, and u are the usual Mandelstam variables, $s = (p_1 + p_2)^2$, $t = (p_1 - p_3)^2$, $u = (p_1 - p_4)^2$. In the limit where pions have zero momenta, $p_1 = p_3 = 0$, we get $s = u = m_\sigma^2$, $t = 0$ and

$$M = 4i\lambda^2 v^2 \left[\frac{1}{m_\sigma^2} + \frac{1}{m_\sigma^2} - \frac{3}{m_\sigma^2} + \frac{1}{m_\sigma^2}\right] = 0 \tag{22}$$

where we have used $m_\sigma^2 = 2\lambda v^2$. Thus the amplitude vanishes in the soft pion limit, i.e. $p_\pi \to 0$.

$(ii) \pi^0 \pi^0 \to \pi^0 \pi^0$
Similar calculation gives

$$M = M_a + M_b + M_c + M_4 = -2i\lambda \left[\frac{s}{s - m_\sigma^2} + \frac{t}{t - m_\sigma^2} + \frac{u}{u - m_\sigma^2}\right]$$

In the soft pion limit, $p_i \to 0$, we get

$$M \simeq \frac{2i\lambda}{m_\sigma^2}(s + t + u) = \frac{i}{v^2}(s + t + u) \to 0. \tag{23}$$

This is the same as the limit, $m_\sigma^2 \to \infty$, because soft pion means pion momentum much smaller than m_σ^2. These are simple examples of the low energy theorem which says that physical amplitudes vanish in the limit where momenta of Goldstone bosons go to zero.

In examples above, the vanishing of these amplitudes results from some cancellation among different contributions. Since this is a general property of the Goldstone boson, there should be a better way of getting this. It turns that one can change the variables representing the scalar fields such that Goldstone bosons always enter with derivative coupling. Then the vanishing of the amplitudes involving Goldstone boson is manifest. This can be accomplished by the field redefinition which we will now describe briefly[13]. Suppose we start from a Lagrangian with field ϕ and make a transformation to a new field η, with the relation,

$$\phi = \eta F(\eta)$$

where $F(\eta)$ is some power series in η. If we impose the condition that $F(0) = 1$, the free Lagrangian for η will be the same as that for ϕ. Then according to a general theorem valid with rather weak restrictions on the Lagrangian and $F(\eta)$, the on-shell matrix elements calculated with η fields and with ϕ fields are the same. We will use this field redefinition to write the Goldstone boson interaction in terms of derivative coupling. Consider a simplified Lagrangian given by

$$\mathcal{L} = \frac{1}{2}\left[(\partial_\mu \sigma)^2 + (\partial_\mu \pi)^2\right] + \frac{\mu^2}{2}\left(\sigma^2 + \pi^2\right) - \frac{\lambda}{4}\left(\sigma^2 + \pi^2\right)^2 \qquad (24)$$

which is just the $O(2)$ version of the σ-model without the nucleon. As before, the SSB will require the shift of the σ-field, as in Eq(16) and π field is massless (Goldstone boson). Equivalently, we can use a complex field defined by $\phi = \frac{1}{\sqrt{2}}(\sigma + i\pi)$ so that the Lagrangian is of the form

$$\mathcal{L} = \partial_\mu \phi^\dagger \partial^\mu \phi + \mu^2 \phi^\dagger \phi - \lambda \left(\phi^\dagger \phi\right)^2 \qquad (25)$$

The symmetry transformation is then $\phi \to \phi' = e^{i\alpha}\phi$, α is some constant. Now we use the polar coordinates for the complex field

$$\phi(x) = \frac{1}{\sqrt{2}}(\rho(x) + v)\exp\left(\frac{i\theta(x)}{v}\right) \qquad (26)$$

to write the Lagrangian in the form,

$$\mathcal{L} = \frac{1}{2}(\partial_\mu \rho)^2 + \frac{(\rho+v)^2}{2v^2}(\partial_\mu \theta)^2 + \frac{\mu^2}{2}(\rho+v)^2 - \frac{\lambda}{4}(\rho+v)^4 \qquad (27)$$

This clearly shows that $\theta(x)$ is massless and has only derivative couplings. This follows from the fact that the $U(1)$ (or $SO(2)$) symmetry $\phi \to e^{i\alpha}\phi$, corresponds to $\theta \to \theta + v\alpha$, which is inhomogeneous. So $\theta(x)$ needs to have a derivative in order to be invariant under such inhomogeneous transformation. Note that this Lagrangian, due to the presence of terms like $(\partial_\mu \theta)^2 \rho^2$ is not renormalizable. But this Lagrangian will be used only as an effective theory to study the low energy phenomenology while the renomalizability deals with high energy behavior.

C. Goldstone Theorem

From Eq(17) we see that the pions $\vec{\pi}$ are massless. This is a consequence of the Goldstone theorem which states that spontaneous breaking of a continuous symmetry will give a massless particle or zero energy excitation. We will first illustrate this by showing that quadratic terms in $\vec{\pi}$ are absent in the tree level as a consequence of spontaneous symmetry breaking of the original chiral symmetry.. We will use the $O(4)$ notation, $\phi^i = (\pi_1, \pi_2, \pi_3, \sigma)$ The invariance under the chiral transformation implies that

$$\delta V = \frac{\partial V}{\partial \phi_i}\delta \phi_i = \frac{\partial V}{\partial \phi_i}\varepsilon_{ij}\phi_j = 0 \qquad (28)$$

Differentiating Eq(28) with respect to ϕ_k and then evaluating this at the minimum, we see that
$$\frac{\partial^2 V}{\partial \phi_i \partial \phi_k}\Big|_{\min}\varepsilon_{ij}\langle\phi_j\rangle = 0, \tag{29}$$
For the case $<\phi_j> = \delta_{i4}v$, we see that in the expansion of V around the minimum, $\sigma = v, \pi_i = 0$, there are no terms of the form, $\pi_i\pi_j, \sigma'\pi_i$, with $\sigma' = \sigma - v$. Therefore $\pi_i's$ are massless in the tree level. We can extend this to more general case where the effective potential is written as $V(\phi_i)$. This potential is invariant under some symmetry group G, which transforms ϕ_i as
$$\phi_i \to \phi_i' = \phi_i + \alpha^a t_{ij}^a \phi_j \quad \text{or} \quad \delta\phi_i = \alpha^a t_{ij}^a \phi_j \tag{30}$$
where $|\alpha_i| \ll 1$ are the parameters for the infinitesimal transformations and t^a matrix for the representation where ϕ_i belongs. The invariance under these transformation implies that
$$\frac{\partial V}{\partial \phi_i}\alpha^a t_{ij}^a \phi_j = 0 \tag{31}$$
The minimum of the potential is located at $\phi_i = v_i$, which satisfies the equation,
$$\left(\frac{\partial V}{\partial \phi_i}\right)_{\phi=v} = 0 \tag{32}$$
Differentiating Eq (31) with respect to ϕ_k and evaluating this at the minimum, $\phi_i = v_i$, we get
$$\left(\frac{\partial^2 V}{\partial \phi_k \partial \phi_i}\right)_{\phi=v}(t_{ij}^a v_j) = 0 \tag{33}$$
This means that the vector $u_i^a = t_{ij}^a v_j$, if non-zero, is an eigenvector of the mass matrix
$$m_{ij}^2 = \left(\frac{\partial^2 V}{\partial \phi_k \partial \phi_i}\right)_{\phi=v} \tag{34}$$
with zero eigenvalue(massless). Thus the number of massless Goldstone bosons is just the number of independent vectors of the form, $u_i^a = t_{ij}^a v_j$. In other words, if $u_i^a \neq 0$, the combination
$$\chi^a = \sum_{ij}\phi_i t_{ij}^a v_j \tag{35}$$
is the Goldstone boson, up to a normalization constant.

These arguments only show that $\vec{\pi}$'s are massless in the tree level. It turns out that this property is true independent of perturbation theory and can be illustrated in case of σ-model as follows. The axial charge is of the form,
$$Q_i^5 = i\int d^3x \left[\pi_i \partial_0 \sigma - \sigma \partial_0 \pi_i + \cdots\right] \tag{36}$$

Since $\partial_0 \pi_i$ and $\partial_0 \sigma$ are just the momenta conjugate to π_i and σ, we can derive,

$$[Q_i^5, \pi_j(0)] = \delta_{ij}\sigma(0). \tag{37}$$

Between vacuum states, this yields

$$<0|\,[Q_i^5, \pi_j(0)]\,|0> = \delta_{ij} <0|\sigma(0)|0>. \tag{38}$$

For the case of SSB, we have

$$<0|\sigma(0)|0> \neq 0 \tag{39}$$

Note that this condition implies that the axial charges Q_i^5's do not annihilate the vacuum, $Q_i^5|0> \neq 0$. Using $Q_i^5 = \int A_i^0(x)\,d^3x$ we can write the LHS of Eq (38) as

$$\langle 0|[Q_i^5, \pi_j(0)]|0\rangle = i\int d^3x \langle 0|\,[A_i^0(x), \pi_j]\,|0\rangle = i\sum_n \delta^3(\vec{p_n})\,[\langle 0|A_i^0(0)|n\rangle$$
$$\langle n|\pi_j|0\rangle e^{-iE_n t} - \langle n|\pi_j|0\rangle\langle 0|A_i^0(x)|n\rangle e^{iE_n t}] \tag{40}$$

This has explicit dependence on t, while the right hand side, $<0|\sigma|0>$, is independent of time. The only way these two features can be consistent is to have a state with the property that

$$E_n \to 0 \quad \text{as} \quad \vec{p_n} \to 0. \tag{41}$$

This is the content of the Goldstone theorem. For the relativistic system, the energy and momentum is related by $E_n = \sqrt{\vec{p_n}^2 + m_n^2}$. Then Eq (41) implies the existence of massless particle in the system. More specifically, there are physical states $|\pi_l\rangle$ with the property that

$$\langle 0|A_i^0(0)|\pi_l\rangle\langle \pi_l|\pi_j|0\rangle \neq 0$$

and are massless from Goldstone theorem. It is convenient to choose the normalization such that $<\pi_l|\pi_j|0> = \delta_{ij}$ and write

$$\langle 0|A_i^\mu(0)|\pi_l(p)\rangle = if_\pi p^\mu \delta_{il} \quad \text{with } f_\pi \text{ a constnat} \tag{42}$$

It is easy to see that
$$f_\pi = \langle 0|\sigma|0\rangle \tag{43}$$

For the non-relativistic system Eq(41) simply says that the dispersion relation $E(p)$ has zero energy excitation.

D. Non-linear σ-model

In the σ-model without the nucleons, we have 3 massless π's and a massive σ field. For the energies much smaller than m_σ, the massless Goldstone bosons are the important physical degrees of freedoms and it is desirable to write down an effective theory with π's only. As we have seen, the theory with SSB has many physical consequences, e.g. low energy theorem for the Goldstone bosons. The removal of σ-field should preserve symmetry so that these results are maintained. Also, phenomenologically there are no good evidence for the existence of the σ meson which is the partner of π's in the chiral symmetry. We now discuss the explicit steps for carrying out this process. Write the scalar fields as a vector in 4-dimensional space,

$$\phi_i = (\phi_1, \phi_2, \phi_3, \phi_4,) = \left(\vec{\pi}, \sigma\right) \qquad (44)$$

We want to parametrize the ϕ fields in such a way that the non-Goldstone field to be eliminated later is $O(4)$ invariant. One simple parametrization for this purpose is

$$\phi_i = R_{i4}(x) s(x), \qquad i = 1, \cdots 4 \qquad (45)$$

where R_{ab} a 4×4 orthogonal matrix, $RR^T = R^T R = 1$, which gives $R_{i4} R_{i4} = 1$ and

$$\phi_i \phi_i = s^2. \qquad (46)$$

So $s(x)$ is the magnitude of the vector ϕ_i and is clearly $O(4)$ invariant. Thus it can be eliminated without effecting the symmetry. One simple choice for R_{i4} is, (?)

$$R_{a4}(x) = \frac{2\eta_a(x)}{1+\vec{\eta}^2}, \qquad a=1,2,3 \qquad R_{44}(x) = \frac{1-\vec{\eta}^2}{1+\vec{\eta}^2} \qquad (47)$$

Note that we can invert these relations to get

$$\vec{\eta} = \frac{\vec{\pi}}{\sigma + s} \qquad (48)$$

The Lagrangian is of the form

$$\mathcal{L} = \frac{1}{2}\left[(\partial_\mu s)^2 + 4s^2 \frac{\left(\partial_\mu \vec{\eta}\right)^2}{\left(1+\vec{\eta}^2\right)^2}\right] + \frac{1}{2}\mu^2 s^2 - \frac{\lambda}{4}s^4 \qquad (49)$$

So $\eta'_i s$ are the massless Goldstone bosons. To study the physics of Goldstone bosons at low energies, $E \ll m_\sigma$, we can replace the s field by a constant, $s(x) = v$ to get

$$\mathcal{L} = 2v^2 \frac{\left(\partial_\mu \vec{\eta}\right)^2}{\left(1+\vec{\eta}^2\right)^2} \qquad (50)$$

In order to get the correct nomoralization we rescale, $\vec{\eta} = \frac{\vec{\pi}'}{2v}$ so that

$$\mathcal{L} = \frac{1}{2}\frac{\left(\partial_\mu \vec{\pi}'\right)^2}{\left(1+\frac{\vec{\pi}'^2}{4v^2}\right)^2} = \frac{1}{2}\left(\partial_\mu \vec{\pi}'\right)^2 - \frac{1}{4v^2}\left(\vec{\pi}'\right)^2\left(\partial_\mu \vec{\pi}'\right)^2 + \cdots \cdots \qquad (51)$$

Here the interaction terms will always contain derivatives and amplitudes involving $\vec{\pi}'$ s will vanish in the limit of zero momenta (low energy theorem). According to the theorem of field redefinition, this describes the same physics as the usual σ-model Lagrangian in the Goldstone sector. For example, we can check that in the simple case of $\pi^0 \pi^0$ scattering in the tree level the amplitude from this Lagrangian in Eq (51) is

$$M = \frac{2i}{v^2}[p_1 \cdot p_2 - p_1 \cdot p_3 - p_1 \cdot p_4] = \frac{i}{v^2}(s + t + u). \tag{52}$$

This is the same result as in Eq(23), obtained in the σ-model with $m_\sigma \to \infty$.

The Lagrangian in Eq (51) which contains on the Goldstone boson fields, is one example of non-linear realization of chiral symmetry, which will be discussed in detail in the next section. Here we want to mention a useful geometric interpretation of Lagrangian in Eq (51). When we eliminate the $O(4)$ invariant field by setting $s(x) = v$, ϕ_i's satisfy the relation,

$$\phi_1^2 + \phi_2^2 + \phi_3^2 + \phi_4^2 = v^2$$

which is just the sphere with radius v in 4-dimensional Euclidean space, S^3. The variables η_1, η_2, and η_3 are just one particular choice of the coordinates of the space S^3. The transformation of Lagrangian from ϕ fields to η fields can be understood in terms of metric tensor in S^3. For simplicity, consider just the kinetic terms in $\mathcal{L}$,

$$\mathcal{L} = \frac{1}{2}(\partial_\mu \phi^i)(\partial_\mu \phi^j) g_{ij}$$

where $g_{ij} = \delta_{ij}$ is the trivial metric in the 4-dimensional Euclidean space. Then the transformation $\phi^i \to \phi^i(\eta)$ gives

$$\partial_\mu \phi^i = \frac{\partial \phi^i}{\partial \eta^a} \partial_\mu \eta^a \tag{53}$$

and

$$\mathcal{L} = \frac{1}{2}\delta_{ij}\frac{\partial \phi^i}{\partial \eta^a}\frac{\partial \phi^j}{\partial \eta^b}(\partial^\mu \eta^a)(\partial^\mu \eta^b) = \frac{1}{2}g_{ab}(\eta)(\partial^\mu \eta^a)(\partial^\mu \eta^b) \tag{54}$$

where

$$g_{ab}(\eta) = g_{ij}\frac{\partial \phi^i}{\partial \eta^a}\frac{\partial \phi^j}{\partial \eta^b} = \frac{4v^2 \delta_{ab}}{(1+\eta^2)^2} \tag{55}$$

is the induced metric on S^3. Thus in the Lagrangian the coefficient of $(\partial_\mu \eta^a)(\partial^\mu \eta^b)$ is the metric of the space S^3, which is just the coset space $O(4)/O(3)$. In the general case where the symmetry breaking is of the form, $G \to H$, the non-linear Lagrangian can be written down with the metric on the manifold G/H as in Eq (54).

It is interesting to see how the transformations of $SU(2) \times SU(2)$ are realized on this manifold S^3. For the infinitesimal isospin rotation, we have

$$\delta \vec{\pi} = \vec{\alpha} \times \vec{\pi}, \quad \delta\sigma = 0 \quad \alpha : \text{group parameters} \tag{56}$$

which implies from Eq(48) that

$$\delta \vec{\eta} = \vec{\alpha} \times \vec{\eta} \tag{57}$$

This is just a rotation on the vector $\vec{\eta}$ and the Lagrangian in Eq(??) with metric given in Eq(55) is clearly invariant under such transformation. The axial transformation on $\vec{\pi}$ and σ is of the form

$$\delta \vec{\pi} = \vec{\beta} \sigma, \qquad \delta \sigma = -\vec{\beta} \cdot \vec{\pi}, \qquad \vec{\beta} : \text{group parameters} \tag{58}$$

which gives

$$\delta \vec{\eta} = \frac{\vec{\beta}}{2}(1 - \eta^2) + \vec{\eta}\left(\vec{\beta} \cdot \vec{\eta}\right). \tag{59}$$

This transformation is non-linear and inhomogeneous.. But we can get simple transformation for the combination,

$$\delta \left(\frac{\partial_\mu \vec{\eta}}{1 + \eta^2}\right) = \left(\vec{\eta} \times \vec{\beta}\right) \times \left(\frac{\partial_\mu \vec{\eta}}{1 + \eta^2}\right) \tag{60}$$

This looks very much like an isospin rotation except that the parameters for the rotation now depend on the fields η and it is easy now to see that the Lagrangian in Eq(54) is invariant under the axial transformations.

Remark: We can transform the metric in Eq(55) into the more familiar Robertson-Walker metric used in cosmology as follows. First we use the spherical coordinates for $\vec{\eta}$ to write the metric in the line element as

$$(dl)^2 = g_{ab}d\eta^a d\eta^b = \frac{4v^2}{(1 + \eta^2)^2}\left[(d\eta)^2 + \eta^2(d\theta)^2 + \eta^2 \sin^2\theta (d\phi)^2\right]. \tag{61}$$

Define the new variable r by

$$r = \frac{2\eta}{1 + \eta^2}. \tag{62}$$

In terms of new variable the line element is of form,

$$(dl)^2 = \frac{(dr)^2}{1 - r^2} + r^2(d\theta)^2 + r^2 \sin^2\theta (d\phi)^2 \tag{63}$$

which is just the usual Robertson-Walker metric for the case of positive curvature.

III. Non-linear Realization

As we have seen in the last section, it is useful to write down a Lagrangian with only Goldstone bosons as an effective theory to describe physics at low energies. In this section we will discuss the general description of this procedure, which is usually called the non-linear realization. The usual discussion of this subject is generally rather formal and abstract. The discussion here will emphasize the intuitive understanding rather than the mathematical rigor.

In order to make the discussion here somewhat self contained, we will first discuss some simple results from group theory ([17]) which are useful for the understanding of non-linear realization. Then we discuss the general features of non-linear realization.

A. Useful results from group theory

Here we will recall the rearrangement theorem which is central to most of the group theoretical result and then discuss the concept of coset space which forms the basis of the non-linear realization.

(i)**Rearrangement Theorem**
Let $G = \{g_1, \cdots g_n\}$ be a finite group. If we multiply the whole group by an arbitrary group element g_i, i.e. $\{g_i g_1, \cdots g_i g_n\}$, the resulting set is just the group G itself.

(ii)**Coset space**
Coset space decomposes a group into non-overlapping sets with respect to a subgroup. Let $H = \{h_1, \ldots h_l\}$ be a non-trivial subgroup of G. For any element g_i in G but not in H, the left coset $g_i H$, or coset for short, is just $\{g_i h_1, \ldots g_i h_l\}$. The coset $g_i H$ will not have any element in common with the subgroup H, and any two such cosets are either identical or have no elements in common. This can be seen as follows. Consider cosets $g_1 H$, and $g_2 H$. Suppose that there is one element in common,

$$g_1 h_i = g_2 h_j \qquad \text{for some } i, j$$

Then we can write

$$g_1^{-1} g_2 = h_i h_j^{-1}$$

which means $g_1^{-1} g_2$ is one of the element of subgroup H. Then by the rearrangement theorem applied to the subgroup H, we get

$$g_1^{-1} g_2 H = H, \qquad \Rightarrow \qquad g_1 H = g_2 H$$

i.e. these two cosets have the same group elements. The group G is now decomposed into these non-overlapping cosets and the collection of all the distinct cosets $g_1 H, \cdots g_k H$, together with H, will contain all the group elements of G. This is denoted by G/H. This can be generalized to Lie group where rearrangement theorem is valid.

Example of coset space: Consider points on 2-dimensional plane which form a group under the addition,

$$(x_1, y_1) \cdot (x_2, y_2) = (x_1 + x_2, y_1 + y_2)$$

Clearly, points on the y-axis, $(0, y)$, form a subgroup, denoted by H. Then the vertical line of the form, (a, y) with a fixed is a coset with respect to the subgroup H. It is clear that the whole 2-dimensional plane can be decomposed into collection of such vertical lines. We can label these cosets, vertical lines, by choosing one element from each coset. Clearly there are many ways to choose such representatives. One convenient parametrization is to choose those points on the x-axis, so that the cosets are of the form $x_i H$. Each group element (x, y) can be written as the product,

$$(x, y) = (x, 0) \cdot (0, y)$$

where $(0, y) \in H$. Under the action of an arbitrary group element $g = (a, b)$ this will give

$$\begin{aligned} g(x, y) &= (a, b) \cdot (x, y) = (a, b) \cdot (x, 0) \cdot (0, y) = (a+x, 0) \cdot (0, b) \cdot (0, y) \\ &= (a+x, 0) \cdot (0, b+y) \end{aligned}$$

(the computation here is organized in such a way that it parallel to the more complicate case in the non-linear realization.) Thus the group element $g = (a, b)$ will move the points in the coset xH to points in the coset $(a+x)H$. In terms of coset parameters, we have

$$g : x \rightarrow x + a.$$

B. Non-linear Realization of Symmetries

We will first discuss the general machinery of non-linear realization ($^{14\,15}$) and then take up the special case of chiral symmetry where the parity symmetry will make the realization much simpler..

1. General case

Suppose the symmetry group G is spontaneously broken to a subgroup H. where both G and H are Lie groups. Choose the generators of G to be of the form $\{V_1,, \ldots V_l, A_1, \ldots A_k\}$ such that $\{V_1, V_2, \ldots V_l,\}$ are the generators of the subgroup H. As usual, the group elements in H can be written in the form,

$$\exp\left(i\vec{\alpha} \cdot \vec{V}\right)$$

where $\alpha_1, \alpha_2, \ldots, \alpha_l$, are the group parameters for H and are taken to be real. From the coset decomposition, we can write an arbitrary group element in G as

$$g = e^{i\vec{\xi} \cdot \vec{A}} e^{i\vec{\alpha} \cdot \vec{V}} \qquad (64)$$

Here $\xi_1, \cdots \xi_l$, are the parameters which label the coset $e^{i\vec{\xi} \cdot \vec{A}} H$. Note that since the vacuum is invariant under H, we have

$$g|0> = e^{i\vec{\xi} \cdot \vec{A}} e^{i\vec{\alpha} \cdot \vec{V}}|0> = e^{i\vec{\xi} \cdot \vec{A}}|0> \qquad (65)$$

i.e. $\vec{\xi}$ also labels the different vacua, which are degenerate. Recall that the different vacua form the manifold G/H. Thus the coset parameters are also the parameters for the manifold G/H. Under the action of an arbitrary group element $g_1 \in G$, we have the combination

$$g_1 g = g_1 e^{i\vec{\xi}\cdot\vec{A}} e^{i\vec{\alpha}\cdot\vec{V}}. \tag{66}$$

Since $g_1 e^{i\vec{\xi}\cdot\vec{A}}$ is also a group element in G, we can write a coset decomposition,

$$g_1 e^{i\vec{\xi}\cdot\vec{A}} = e^{i\vec{\xi}'\cdot\vec{A}} e^{i\vec{\alpha}'\cdot\vec{V}} \tag{67}$$

and then

$$g_1 \left(e^{i\vec{\xi}\cdot\vec{A}} e^{i\vec{\alpha}\cdot\vec{V}} \right) = e^{i\vec{\xi}'\cdot\vec{A}} e^{i\vec{\alpha}'\cdot\vec{V}} e^{i\vec{\alpha}\cdot\vec{V}} = e^{i\vec{\xi}'\cdot\vec{A}} e^{i\vec{\alpha}''\cdot\vec{V}} \tag{68}$$

where we have used the group property of H to write

$$e^{i\vec{\alpha}'\cdot\vec{V}} e^{i\vec{\alpha}\cdot\vec{V}} = e^{i\vec{\alpha}''\cdot\vec{V}} \tag{69}$$

Note that the new coset parameters $\vec{\xi}'$ and parameters $\vec{\alpha}'$ for the subgroup H, all depend on the original coset parameters $\vec{\xi}$,

$$\vec{\xi}' = \vec{\xi}'\left(\vec{\xi}, g_1\right), \qquad \vec{\alpha}' = \vec{\alpha}'\left(\vec{\xi}, g_1\right) \tag{70}$$

In this way, the group element g_1 transform the coset parameters from $\vec{\xi} \to \vec{\xi}'$ in the coset space G/H. As we will see later, these coset parameters will be identified with the Goldstone bosons. These transformations on $\vec{\xi}$, and $\vec{\alpha}$ induced by the group elements will have the same group properties as the group elements and are called the *non-linear realization* of the group. This is in contrast to the usual representation of the group where group elements are represented by matrices. In the transformation in Eq(70) $\vec{\xi}'$ is generally not a linear function of $\vec{\xi}$. But for the special case where $g = h$ is a group element from the unbroken subgroup H, we get, from Eq(64),

$$hg = he^{i\vec{\xi}\cdot\vec{A}} e^{i\vec{\alpha}\cdot\vec{V}} = \left(he^{i\vec{\xi}\cdot\vec{A}} h^{-1}\right)\left(he^{i\vec{\alpha}\cdot\vec{V}}\right) = \left(he^{i\vec{\xi}\cdot\vec{A}} h^{-1}\right) e^{i\vec{\alpha}\cdot\vec{V}'} \tag{71}$$

where

$$he^{i\vec{\alpha}\cdot\vec{V}} = e^{i\vec{\alpha}\cdot\vec{V}'} \tag{72}$$

In general, the broken generators $\vec{A}$ transform as some representation D with respect to the subgroup H,

$$hA_i h^{-1} = A_j D_{ji}(h) \tag{73}$$

For example, in the case of $G = SU(2) \times SU(2)$ model, the broken generators, A_1, A_2, A_3 transform as triplet under the unbroken subgroup $H = SU(2)$. We can then write

$$he^{i\vec{\xi}\cdot\vec{A}} h^{-1} = \exp\left(i\xi_i h A_i h^{-1}\right) = \exp\left(i\xi_i A_j D_{ji}(h)\right) = \exp\left(i\xi_j' A_j\right) \tag{74}$$

where
$$\xi'_j = D_{ji}(h)\xi_i \tag{75}$$

This means that $\xi'_i s$ transform linearly under the subgroup H. Also it is easy to see that the parameters $\vec{\alpha}$ are independent of the coset parameters ξ_i. But if the group element is of the form $e^{i\vec{\xi'}\cdot\vec{A}}$ the transformation law for the coset parameters ξ is non-linear and quite complicate.

2. Chiral symmetry

For the case of chiral symmetry, there is significant simplification due to parity operation. We will illustrate this in the simple case of $SU(2)_L \times SU(2)_R$ symmetry. The parity operation is of the form,

$$P: \vec{V} \to \vec{V}, \quad \vec{A} \to -\vec{A}$$

Consider the case where group element g consists of left-handed transformation,

$$g = \exp\left(i\vec{\theta}\cdot(\vec{V}-\vec{A})\right) \equiv L \tag{76}$$

and write the transformation of coset parameters as

$$ge^{i\vec{\xi}\cdot\vec{A}} = Le^{i\vec{\xi}\cdot\vec{A}} = e^{i\vec{\xi'}\cdot\vec{A}}e^{i\vec{\alpha'}\cdot\vec{V}} \tag{77}$$

Under the parity transformation, the left-handed transformation is changed into right-handed one,

$$P(L) = P\left(e^{i\vec{\theta}\cdot(\vec{V}-\vec{A})}\right) = e^{i\vec{\theta}\cdot(\vec{V}+\vec{A})} \equiv R \tag{78}$$

Then applying the parity transformation to Eq(77), we get

$$Re^{-i\vec{\xi}\cdot\vec{A}} = e^{-i\vec{\xi'}\cdot\vec{A}}e^{i\vec{\alpha'}\cdot\vec{V}} \tag{79}$$

Using the notation

$$\Sigma \equiv e^{i\vec{\xi}\cdot\vec{A}}, \quad h = e^{i\vec{\alpha'}\cdot\vec{V}} \tag{80}$$

we can combine the Eqs(77,79) into

$$\Sigma' = L\Sigma h^\dagger = h\Sigma R^\dagger \tag{81}$$

Note that as we have mentioned before, the parameters $\vec{\alpha'}$ depend on the coset parameters $\vec{\xi}$, the transformation law for Σ here is non-linear because the factor h depends on $\vec{\alpha'}$. However, the combination $U = \Sigma^2$ will have a simple transformation law,

$$U' = LUR^\dagger. \tag{82}$$

33

Since L, R are independent of the coset parameters $\vec{\xi}$, this transformation law is linear and will be useful for constructing Lagrangian.

We are interested in the cases where spontaneous symmetry breaking is generated by the scalar fields. Some of these scalar fields become the massless Goldstone bosons, like the pions in the σ-model and others remain massive, like σ-field. Consider the scalar fields in the σ-model, where we will use the notation,

$$\vec{\phi} = (\pi_1, \pi_2, \pi_3, \sigma) = (\phi_1, \phi_2, \phi_3, \phi_4) \tag{83}$$

The vacuum expectation value which gives the classical ground state is of the form,

$$<\phi>_0 = \begin{pmatrix} 0 \\ 0 \\ 0 \\ v \end{pmatrix} \tag{84}$$

Suppose we make a very general assumption that from a given point in ϕ-space we can reach any other point by some group transformation.(Space is transitive). Then the general field configuration can be written as

$$\phi(x) = \begin{pmatrix} \phi_1(x) \\ \phi_2(x) \\ \phi_3(x) \\ \phi_4(x) \end{pmatrix} = g \begin{pmatrix} 0 \\ 0 \\ 0 \\ \sigma \end{pmatrix} \quad \text{for some } g \in G \tag{85}$$

<u>Remark</u>:Strictly speaking we should use the representation matrices $D(g)$ of the group element g rather than g itself. However for simplicity of notation, g here is a shorthand for $D(g)$. From the coset decomposition in Eq(64), we can write

$$g = e^{i\vec{\pi} \cdot \vec{A}} e^{i\vec{\alpha} \cdot \vec{V}} \tag{86}$$

Here we have chosen coset parameters to be the pion fields. Then we can write the scalar fields as

$$\phi(x) = g \begin{pmatrix} 0 \\ 0 \\ 0 \\ \sigma \end{pmatrix} = e^{i\vec{\pi} \cdot \vec{A}} e^{i\vec{\alpha} \cdot \vec{V}} \begin{pmatrix} 0 \\ 0 \\ 0 \\ \sigma \end{pmatrix} = e^{i\vec{\pi} \cdot \vec{A}} \begin{pmatrix} 0 \\ 0 \\ 0 \\ \sigma \end{pmatrix} \tag{87}$$

where we have used the fact that the vector $(0, 0, 0, \sigma)$ is proportional to the vacuum configuration $(0, 0, 0, v)$ and is invariant under the subgroup $H = \{e^{i\vec{\alpha} \cdot \vec{V}}\}$. Since the Goldstone bosons are identified as the coset parameters, they transform the same way as $\vec{\xi}$ in Eq(70), i.e.

$$g_1 e^{i\vec{\pi} \cdot \vec{A}} = e^{i\vec{\pi}' \cdot \vec{A}} e^{i\vec{\alpha}' \cdot \vec{V}} \tag{88}$$

where

$$\vec{\pi}' = \vec{\pi}'(\pi, g_1), \quad \vec{\alpha}' = \vec{\alpha}'(\pi, g_1) \tag{89}$$

For the case $g = h \in H$, we have from Eq(75),
$$\pi'_j = D_{ji}(h)\pi_i \tag{90}$$

Remark: The scalar fields here have property that after separating out the Goldstone bosons, the remainders are proportional to the vacuum expectation value and is then invariant under the subgroup H. This is true only for scalar fields in the vector representation in $O(n)$ or $SU(n)$ groups and is not true for scalars in more general representation. In more general cases, we can separate out the Goldstone bosons by writing $\phi(x)$ as

$$\phi(x) = e^{i\vec{\pi}\cdot\vec{A}}\chi(x) \tag{91}$$

where $\chi(x)$ contains all the massive fields. Under the action of group element g_1, we have

$$g_1\phi = g_1 e^{i\vec{\pi}\cdot\vec{A}}\chi(x) = e^{i\vec{\pi}'\cdot\vec{A}} e^{i\vec{\alpha}'\cdot\vec{V}}\chi(x) = e^{i\vec{\pi}'\cdot\vec{A}}\chi'(x) \tag{92}$$

where

$$\chi'(x) = e^{i\vec{\alpha}'\cdot\vec{V}}\chi(x) \tag{93}$$

To see how the Goldstone bosons transform under the axial transformation, we set $L = R^\dagger = e^{i\vec{\theta}\cdot\vec{A}}$, in Eq(81) to get

$$e^{i2\vec{\pi}'\cdot\vec{A}} = e^{i\vec{\theta}\cdot\vec{A}} e^{i2\vec{\pi}\cdot\vec{A}} e^{i\vec{\theta}\cdot\vec{A}} \tag{94}$$

To understand this equation better, we first take the infinitesimal transformation, $|\vec{\theta}| \ll 1$,

$$e^{i2\vec{\pi}'\cdot\vec{A}} = e^{i2\vec{\pi}\cdot\vec{A}} + \left(i\vec{\theta}\cdot\vec{A}\right) e^{i2\vec{\pi}\cdot\vec{A}} + e^{i2\vec{\pi}\cdot\vec{A}}\left(i\vec{\theta}\cdot\vec{A}\right) + \cdots \tag{95}$$

and then expand this in powers of $\vec{\pi}'$s,

$$1 + 2i\vec{\pi}'\cdot\vec{A} + \cdots = 1 + 2i\vec{\pi}\cdot\vec{A} + 2i\vec{\theta}\cdot\vec{A} + \cdots$$

Comparing both sides we see that

$$\vec{\pi}' = \vec{\pi} + \vec{\theta} + \cdots \tag{96}$$

Thus there is a inhomogeneous term in the transformation law for the Goldstone bosons $\vec{\pi}$ and this is why Goldstone bosons have derivative coupling.

Since the transformation law for $U = e^{i2\vec{\pi}\cdot\vec{A}}$ is linear and simple, it is easier to construct the chirally invariant interaction in terms of U rather than Σ. It is easy to see that the only invariants without derivatives will involve trace of some powers of $UU^\dagger$, which is just an identity matrix. (This also implies that the Goldstone boson coupling will involve derivatives). Thus the interaction with lowest numbers of derivative is of the form

$$\mathcal{L} = tr\left(\partial_\mu U \partial^\mu U\right) \tag{97}$$

Covariant derivative: The parity symmetry in the σ-model is responsible for getting the simple combination $U(x)$ which transforms linearly. For the more general case where there is no such simplification, to construct invariant terms involving derivatives is quite complicate because the non-linear transformation law will involve Golstone boson fields which are space-time dependent. This means that we need to construct the covariant derivatives. Futhermore, to exhibit the low energy explicitly we need to couple Golstone field with derivative to other matter fields. We will now discuss briefly in the simple case of σ-model. Write the scalar fields $\phi(x)$ in the form,

$$\phi(x) = e^{i\vec{\pi}(x)\cdot\vec{A}} \begin{pmatrix} 0 \\ 0 \\ 0 \\ \sigma(x) \end{pmatrix} = \Sigma(x)\chi(x) \tag{98}$$

with $\Sigma(x) = e^{i\vec{\pi}(x)\cdot\vec{A}}$. As before under the action of group element g, we have

$$g\Sigma(x) = \Sigma'(x)h(x) \quad \text{with} \quad h(x) = e^{i\vec{\alpha}\cdot\vec{V}} \tag{99}$$

Since ϕ transforms linearly, we have

$$\phi' = g\phi$$

which implies

$$\Sigma'\chi' = g\Sigma\chi = \Sigma'h\chi \quad \text{or} \quad \chi' = h\chi \tag{100}$$

This means that χ transforms non-linearly because $\vec{\alpha}$ in h depend on π fields. Making use of the simplification for the case of chiral symmetry we have, from transformation law,(81)

$$\partial_\mu \Sigma' = L(\partial_\mu \Sigma h^{-1} + \Sigma \partial_\mu h^{-1}) = (\partial_\mu \Sigma h + \Sigma \partial_\mu h)R^\dagger \tag{101}$$

and

$$\Sigma'^{-1}\partial_\mu \Sigma' = h\left(\Sigma^{-1}\partial_\mu \Sigma\right)h^{-1} + h\partial_\mu h^{-1} \tag{102}$$

$$\partial_\mu \Sigma' \Sigma'^{-1} = h\left(\partial_\mu \Sigma \Sigma^{-1}\right)h^{-1} - h\partial_\mu h^{-1} \tag{103}$$

If we define

$$v_\mu = \frac{1}{2}\left[\Sigma^{-1}\partial_\mu \Sigma - \partial_\mu \Sigma \Sigma^{-1}\right] \tag{104}$$

$$a_\mu = \frac{1}{2}\left[\Sigma^{-1}\partial_\mu \Sigma + \partial_\mu \Sigma \Sigma^{-1}\right] \tag{105}$$

we get

$$v'_\mu = hv_\mu h^{-1} + h\partial_\mu h^{-1}$$

$$a'_\mu = ha_\mu h^{-1}$$

This means that v_μ transforms like "gauge field", while a_μ transforms as global adjoint field. Therefore v_μ can be used to construct the covariant derivative and a_μ is like a global axial vector field.

Nucleon Field The nucleon field in the linear σ-model has the transformation properties,

$$N_L \rightarrow N'_L = L N_L, \quad \text{and} \quad N_R \rightarrow N'_R = R N_R$$

Thus to couple nucleon fields to the Goldstone bosons, we could write down the following $SU(2) \times SU(2)$ invariant coupling,

$$\mathcal{L}_{int} = g \left(\overline{N}_L \Sigma N_R + \overline{N}_R \Sigma^\dagger N_L \right).$$

However, this is not of the form of derivative coupling which exhibits the low energy theorem explicitly. For this purpose and general non-linear realization, we define a new nucleon field by,

$$N = e^{i\vec{\pi}\cdot\vec{A}} \widetilde{N}$$

Under the action of the group element g, we get

$$gN = ge^{i\vec{\pi}\cdot\vec{A}} \widetilde{N} = e^{i\vec{\pi}'\cdot\vec{A}} e^{i\vec{\alpha}\cdot\vec{V}} \widetilde{N} = e^{i\vec{\pi}'\cdot\vec{A}} \widetilde{N}'$$

where

$$\widetilde{N}' = e^{i\vec{\alpha}\cdot\vec{V}} \widetilde{N} = h\widetilde{N}$$

Thus $\widetilde{N}$ transforms according to the representation of the subgroup H but with the group parameters depend on π fields, $\vec{\alpha}\left(\vec{\pi}, g\right)$. Then from the transformation properties given in Eqs(??), we can write down the derivative coupling as

$$\mathcal{L}_N = \overline{\widetilde{N}}\gamma^\mu \left(i\partial_\mu - v_\mu\right) \widetilde{N} + g\overline{\widetilde{N}}\gamma^\mu a_\mu \widetilde{N}$$

which will yield the low energy theorem explicitly.

IV. Examples in the Non-relativistic System

In the frame work of relativistic field theory, e.g. in $SU(2) \times SU(2)$ σ-model, spontaneous symmetry breaking seems to be put in by hand, i.e. setting the quadratic terms to have negative sign in the scalar potential in order to develope vacuum expectation value. This is rather ad hoc and no physical reason is given for why this is the case. We will now discuss some simple non-relativistic examples of spontaneous symmetry breaking in order to shed some light on this ([18]).

A. Infinite range Ising model

Consider a system of N spins on an one dimensional lattice with Hamiltonian,

$$H = -\frac{J}{N} \sum_{i<j}^{N} s_i s_j - B \sum_{i}^{N} s_i \tag{106}$$

where $s_i = \pm 1$, J is the coupling constant for spin-spin interaction and B is the external magnetic field. In this Hamiltonian, for calculational simplicity we allow every spin to interact with every other spin, while more realistic situation will be the short range nearest neighbor interaction. But the interest here is to see how the spontaneous symmetry breaking come about and we will ignore this. The partition function is given by

$$Z = Tr\left(e^{-\beta H}\right) = \sum_{s_i = \pm 1} \exp\left(\frac{\beta J}{2N}\left(\sum_i s_i\right)^2 + \beta B \sum_i s_i\right) \qquad (107)$$

Using the identity for the Gaussian integral,

$$\int_{-\infty}^{+\infty} dx e^{-ax^2 + bx} = \sqrt{\frac{\pi}{a}} e^{-\frac{b^2}{4a}}$$

we can write the partition function as

$$Z = \sum_{s_i = \pm 1} \sqrt{\frac{N\beta J}{2\pi}} \int_{-\infty}^{+\infty} dx \exp\left[-\frac{N\beta J}{2}x^2 + \beta(Jx + B)\left(\sum_i s_i\right)\right] \qquad (108)$$

Now we can sum over each s_i independently,

$$\sum_{s_i = \pm 1} \exp\left(\beta(Jx + B)\left(\sum_i s_i\right)\right) = \exp\{N \log[2 \cosh \beta(Jx + B)]\}$$

and

$$Z = \sqrt{\frac{N\beta J}{2\pi}} \int_{-\infty}^{+\infty} dx \exp\left(-\frac{N\beta J}{2}x^2 + N \log[2 \cosh \beta(Jx + B)]\right) \qquad (109)$$

From the partition function Z, we can compute the average spin,

$$S = \frac{1}{N} <\sum_i s_i> = -\frac{1}{\beta N}\frac{\partial}{\partial B} \ln Z. \qquad (110)$$

If $S \neq 0$ in the limit the external field vanishes, $B \to 0$, then we have spontaneous symmetry breaking. Since we are interested in the case where N is very large, we can use saddle point method to compute Z. Write

$$Z = \sqrt{\frac{N\beta J}{2\pi}} \int_{-\infty}^{+\infty} dx \exp\{-N\beta f(x)\} \qquad (111)$$

where

$$f(x) = \frac{Jx^2}{2} - \frac{1}{\beta}\log[2 \cosh \beta(Jx + B)] \qquad (112)$$

The minimum of $f(x)$ is given by

$$f'(x) = 0, \quad \Rightarrow \quad x = \tanh \beta(Jx + B) \qquad (113)$$

Let $x_i, i = 0, 1, 2, \cdots$ be the solutions of this transcendental equation, then

$$Z = \sqrt{\frac{N\beta J}{2\pi}} \sum_i \exp\{-N\beta f(x_i)\} \sqrt{\frac{2\pi}{N\beta f''(x_i)}} \tag{114}$$

Suppose x_0 is the smallest of these solutions, it will dominate the partition function for large N. Then the average magnetization is then

$$S = -\frac{1}{\beta N} \frac{\partial}{\partial B} \ln Z = \tanh \beta (Jx_0 + B) = x_0 \tag{115}$$

where we have used the equation satisfied by x_0. Thus the minimum of $f(x)$ will correspond to the average magnetization.

To study spontaneous symmetry breaking, we set $B = 0$, in Eq(113), and get

$$x = \tanh \beta J x \tag{116}$$

It turns out that this equation has only the trivial solution, $x = 0$ if $\beta J < 1$ and non-trivial solution exists only for $\beta J > 1$. To understand this feature, we expand $f(x)$ in powers of x for the case $B = 0$,

$$\begin{aligned} f(x) &= \frac{Jx^2}{2} - \frac{1}{\beta} \log\left[1 + \frac{1}{2}(\beta J x)^2 + \frac{1}{4!}(\beta J x)^4 + \cdots\right] \\ &= \frac{J}{2}(1 - \beta J)x^2 + \frac{1}{12}\beta^3 J^4 x^4 + \cdots \end{aligned} \tag{117}$$

Thus $\beta J > 1$ corresponds to negative quadratic term, which is the familiar situation in the scalar potential in the σ-model and the like. In terms of temperature this condition, we have

$$J > kT. \tag{118}$$

Here J is the coupling which wants to align the spins in the same direction while the effect of temperature is to randomize spins. Thus the condition in Eq(118) simply means that the interaction of spins has to overcome the thermalization in order to produce significant spin alignment. The temperature $T_c = \frac{J}{k}$ is usually called the critical temperature and spontaneous symmetry breaking is possible only for $T < T_c$. In this simple example, the non-zero magnetization S breaks the symmetry, $s_i \rightarrow -s_i$. and originates from the competition between spin-spin interaction which aligns the spin and the thermalization which tends to destroy the alignment.

Remarks:(1)From the partition function in Eq(114) we see that the probability to find the system to have x_i is given by the Boltzmann factor

$$P(x_i) = \frac{\exp(-\beta N f(x_i))}{\sum_i \exp(-\beta N f(x_i))} \tag{119}$$

If x_0 is the absolute minimum for $f(x)$, then in the thermodynamic limit $N \rightarrow \infty$, we have $P(x_0) \rightarrow 1$ and the probability for all the other x_i will be zero.

(2) For the case T is near T_c, the minimum of $f(x)$ is located at small values of x. Thus we can expand $f(x)$ in power series,

$$f(x) = \frac{1}{2\beta_c}\left(1 - \frac{\beta}{\beta_c}\right)x^2 + \frac{1}{12}\frac{\beta^3}{\beta_c^4}x^4 + \cdots \qquad (120)$$

The minimum is then

$$x_0 = \sqrt{\frac{3\beta_c^3}{\beta^3}\left(1 - \frac{\beta}{\beta_c}\right)} = \sqrt{\frac{3T^3}{T_c^3}\left(1 - \frac{T_c}{T}\right)} \qquad (121)$$

This means that near the critical temperature $T \to T_0$, the dependence of average magnetization on $(T - T_0)$ is non-analytic. This is a typical behavior of physical quantities near the critical point.

B. Superfluid

The superfluid He^4 provides a simple example of Goldstone excitation where the excitation energy, $\varepsilon(k)$ goes to zero when the wave number $k \to 0$. The helium atoms are tightly bounded and the long-distance attractive force between atoms are very weak while the short distance is strongly repulsive. Thus a system of helium atoms can be described as a gas of weakly interacting bosons with Hamiltonian,

$$H = -\frac{1}{2m}\int d^3x \psi^\dagger \nabla^2 \psi + \frac{1}{2}\int d^3x d^3y \psi^\dagger(x)\psi^\dagger(y) v(x-y)\psi(x)\psi(y) \qquad (122)$$

Here $v(x)$ is the potential describes the effective interaction between helium atoms and $\psi(x)$ is the field operator for the helium atom and satisfies the commutation relation,

$$\left[\psi(x), \psi^\dagger(y)\right] = \delta^3(x-y)$$

We will assume that the system is in a large box of volume Ω with periodical boundary condition. Clearly, this Hamiltonian is invariant under the transformation,

$$\psi(x) \to \psi'(x) = e^{i\alpha}\psi(x) \qquad (123)$$

This is just a $U(1)$ symmetry which says that the number of He atoms is conserved. The conserved charge is just the number operator,

$$Q = \int d^3x \psi^\dagger(x)\psi(x) \qquad (124)$$

with the commutation relations,

$$[Q, \psi(x)] = \psi(x), \qquad \left[Q, \psi^\dagger(x)\right] = -\psi^\dagger(x). \qquad (125)$$

We can expand $\psi(x)$ in plane waves,

$$\psi(x) = \frac{1}{\sqrt{\Omega}} \sum_{\vec{k}} a_{\vec{k}} e^{i\vec{k}\cdot\vec{x}} \tag{126}$$

where $a_{\vec{k}}$ and $a_{\vec{k}}^{\dagger}$ are the usual creation and annihilation operators satisfying the commutation relations,

$$\left[a_{\vec{k}}, a_{\vec{k}'}\right] = 0, \quad \left[a_{\vec{k}}, a_{\vec{k}'}^{\dagger}\right] = \delta_{\vec{k},\vec{k}'} \tag{127}$$

The Hamilton is then of the form,

$$H = \sum_k \frac{\hbar^2 k^2}{2m} a_k^{\dagger} a_k + \frac{1}{2\Omega} \sum_{k_i} \tilde{v}(k_1 - k_3) \delta_{k_1+k_2, k_3+k_4} a_{k_1}^{\dagger} a_{k_2}^{\dagger} a_{k_3} a_{k_4} \tag{128}$$

where

$$\tilde{v}(k) = \int d^3x e^{i\vec{k}\cdot\vec{x}} v(x) \tag{129}$$

In these two equations and there after we have, for notational simplicity, neglected the vector symbol for the wave vectors k_i' s. Since $v(x)$ is real we have $\tilde{v}(k) = \tilde{v}(-k)$. For the trivial case where there is no interaction, $v(x) = 0$, the ground state is just the one in which all particles are in the $\vec{k} = 0$ state,

$$|\Psi_0>_{v=0} = \frac{\left(a_0^{\dagger}\right)^N}{\sqrt{N!}}|0> \quad \text{where} \quad a_k|0> = 0 \quad \forall k \tag{130}$$

It is clear that if the interaction is small enough, in the ground state and low-lying excited states, most of the particles will be in the $\vec{k} = 0$ state, i.e.

$$<n_0> \gg <n_k> \quad \text{with} \quad k \neq 0. \tag{131}$$

where $n_k = a_k^{\dagger} a_k$. We are interested in the cases where N, the total number of particles, is very large. Thus $n_0 \sim N$ is very large. From the properties of the creation and annihilation operators,

$$a_0|n_0> = \sqrt{n_0}|n_0 - 1>, \quad a_0^{\dagger}|n_0> = \sqrt{n_0 + 1}|n_0 + 1> \tag{132}$$

we will make the assumption that the matrix elements of a_0 are of order $\sqrt{n_0}$ and $a_0^{\dagger}$ of order $\sqrt{n_0 + 1}$. Thus in the limit $n_0 \sim N \to \infty$, commutator of a_0 and $a_0^{\dagger}$, is of order unity while $a_0, a_0^{\dagger}$ are of order $\sqrt{N}$,

$$\left[a_0, a_0^{\dagger}\right] = 1 \ll a_0 \quad \text{or} \quad a_0^{\dagger} \sim \sqrt{n_0} \tag{133}$$

Thus we can neglect the commutator. Since a_0, and $a_0^{\dagger}$ commute with all the other operators,

$$[a_0, a_k] = \left[a_0^{\dagger}, a_k\right] = 0 \quad \text{for } k \neq 0.$$

We can then take a_0 and $a_0^\dagger$ to be c-numbers (Schur's lemma),

$$a_0 = a_0^\dagger = \sqrt{n_0} \tag{134}$$

Thus we will replace a_0 and $a_0^\dagger$ by $\sqrt{n_0}$. Then the coefficients of terms quadratic in a_k and $a_k^\dagger$, $k \neq 0$, in the interaction will be of order n_0 and those of quartic term is of order 1. Therefore we can make the approximation that neglect the quartic terms and get the Hamiltonian in the form,

$$H = \sum_{k \neq 0} \frac{\hbar^2 k^2}{2m} a_k^\dagger a_k + \frac{n_0}{2\Omega} \sum_{k \neq 0} [\bar{v}(k) (a_k^\dagger a_{-k}^\dagger + a_k a_{-k}) \tag{135}$$

$$+ 2 \bar{v}(0) a_k^\dagger a_k + 2 \bar{v}(k) a_k^\dagger a_k] + \frac{n_0^2}{2\Omega} \bar{v}(0) \tag{136}$$

or

$$H = \sum_{k \neq 0} \omega_k a_k^\dagger a_k + \frac{n_0}{2\Omega} \sum_{k \neq 0} \bar{v}(k) (a_k^\dagger a_{-k}^\dagger + a_k a_{-k}) + \frac{N^2}{2\Omega} \bar{v}(0) \tag{137}$$

where

$$\omega_k = \frac{\hbar^2 k^2}{2m} + \frac{n_0}{\Omega} \bar{v}(k) \tag{138}$$

and we have used

$$N^2 = \left(n_0 + \sum_{k \neq 0} a_k^\dagger a_k \right)^2 \approx n_0^2 + 2 n_0 \sum_{k \neq 0} a_k^\dagger a_k \tag{139}$$

Note that this Hamiltonian does not conserve the particle number but it conserves the momentum because the removal of $k = 0$ mode effects the particle number but not the momentum. Since this Hamiltonian contains only quadratic terms, we can solve this by Bogoliubove transformation as follows. Define the quasi-particle operators by

$$\alpha_k = \cosh \theta_k \, a_k + \sinh \theta_k \, a_{-k}^\dagger \tag{140}$$

then we have

$$\alpha_k^\dagger = \cosh \theta_k \, a_k^\dagger + \sinh \theta_k \, a_{-k} \tag{141}$$

where θ_k is an arbitrary parameter at our disposal. We now write the Hamiltonian in terms of the quasi particle operators by inverting the relations in Eq(??),

$$a_k = \cosh \theta_k \, \alpha_k - \sinh \theta_k \, \alpha_{-k}^\dagger, \quad a_{-k}^\dagger = -\sinh \theta_k \, \alpha_k + \cosh \theta_k \, \alpha_{-k}^\dagger \tag{142}$$

and choose the parameter θ_k so that the coefficient of the non-diagonal terms $\left(\alpha_k^\dagger \alpha_{-k}^\dagger + \alpha_k \alpha_{-k} \right)$ is zero. The computation is straightforward and the result is

$$\tanh 2\theta_k = \frac{\frac{n_0 \bar{v}}{\Omega}}{\omega_k} \tag{143}$$

and the Hamiltonian is

$$H = \sum_{k \neq 0} \varepsilon_k \alpha_k^\dagger \alpha_k + \frac{N^2 v(0)}{2\Omega} + \frac{1}{2} \sum_{k \neq 0} (\varepsilon_k - \omega_k) \quad (144)$$

where

$$\varepsilon_k = \sqrt{\omega_k^2 - \left(\frac{n_0 \bar{v}}{\Omega}\right)^2} = \sqrt{\left(\frac{\hbar^2 k^2}{2m}\right)^2 + 2\left(\frac{\hbar^2 k^2}{2m}\right)\left(\frac{n_0 \bar{v}(k)}{\Omega}\right)} \quad (145)$$

This is just the Hamiltonian for the uncoupled harmonic oscillators and the eigenvalues are

$$E = \sum_k n_k \varepsilon_k \quad (146)$$

The quai-particle energy excitation has the property that

$$\varepsilon_k \to 0, \quad \text{as} \quad k \to 0 \quad (147)$$

which is just the Goldstone excitation. Clearly, the ground state $|\Psi_0>$ is the one which is annihilated by all quasi particle operators α_k,

$$\alpha_k |\Psi_0> = 0 \quad \forall k \quad (148)$$

and the excited states are of the form,

$$\left(\alpha_{k_1}^\dagger\right)^{n_1} \left(\alpha_{k_2}^\dagger\right)^{n_2} \cdots |\Psi_0> \quad (149)$$

It is straightforward to show that the quasi particle ground state can be written in terms of the original creation operators $a_k^\dagger$ as

$$|\Psi_0> = \sqrt{Z} \exp\{-\sum_{k_i} \tanh \theta_{k_i} a_{k_i}^\dagger a_{-k_i}^\dagger\}|0> \quad (150)$$

where

$$Z = \prod_{k_i} \left(1 - \tanh^2 \theta_{k_i}\right) \quad (151)$$

This shows that the quasi particle ground state is a complicate combination of the vacuum, 2 particle states, 4 particle states,···etc. To elucidate the Goldstone theorem, we note that the vacuum expectation value of $\psi(0)$ in the ground state is non-zero,

$$<\Psi_0|\psi(0)|\Psi_0> = \frac{1}{\sqrt{\Omega}} <\Psi_0|a_0|\Psi_0> = \sqrt{\frac{n_0}{\Omega}} \neq 0. \quad (152)$$

where we have used the fact that $<\Psi_0|a_k|\Psi_0> = 0$, for $k \neq 0$, from the momentum conservation. From the commutation relation in Eq(125) we see that this is the symmetry breaking condition which implies that

$$Q|\Psi_0> \neq 0. \quad (153)$$

The quasi particle excitation which has the property that its energy ε_k goes to zero in the limit $k \to 0$, is the Goldstone excitation implies by the Goldstone's theorem.

References

[1] Y. Nambu, Phy. Rev. Lett., **4**, 140 (1960). Y. Nambu and G. Jona-Lasino, Phys. Rev.,**122**, 345 (1961)

[2] J. Goldstone, Nuovo Cimento **19**, 154 (1961)

[3] J. Goldstone, A. Salam and S. Weinberg, Phys. Rev. **127**, 965 (1962)

[4] M. Gell-Mann and M. Levy, Nouovo Cimento **16**, 705 (1960)

[5] P. W. Higgs, Phys. Lett., **12** 132 (1964), Phys. Rev. Lett., **13** 508 (1964)

[6] F. Englert and R. Barout, Phy. Rev. Lett. **13**, 321 (1964)

[7] G. S. Guralnik, C. R. Hagen, and T. W. Kibble, Phys. Rev. Lett. **13**, 585 (1964)

[8] S. Weinberg, Phys. Rev. Lett. **27**, 1264 (1967)

[9] A. Salam, In Elementary Particle Theory Proceedings of the 8th Nobel Symposium , edited by N. Svartholm(Almqvist and Wiksell, Stockholm)

[10] G. 't Hooft, Nucl. Phys.**B33**, 173 (1971), Nucl. Phys.**B35**, 167 (1971)

[11] See for example, Ta-Pei Cheng, and Ling-Fong Li, Gauge Theory of Elementary Particle Physics, Oxford University Press, (1984).

[12] See for example, S. Alder, and R. Dashen, Current Algebra, W. A. Benjamin (1968), H. Georgi, Weak Interaction and Modern Particle Physics, Bejamin/Cummings, (1984), J. Donoghue, E. Golowich, and B. Holstein , Dynamics of Standard Model, Cambridge University Press, (1992).

[13] :R. Haag, Phys. Rev. ,**112** 669, (1958)

[14] :S. Coleman, J. Wess, and B. Zumino, Phy Rev **177**, 2239, (1969).

[15] :C. Callan, S. Coleman, J. Wess, and B. Zumino, **177**, 2247, (1969)

[16] S. Weinberg, The Quantum Field Theory II, Cambridge University Press, (1996)

[17] see for example, M. Tinkham, Group Theory and Quantum Mechanics, McGraw-Hill, (1964).

[18] J. Negele and H. Orland, Quantum Many-Particle Systems, Addison-Wesley (1998).

CP VIOLATION

G. Valencia

*Department of Physics
Iowa State University
Ames, IA 50011*[1]

Abstract.
In these lectures I first review the basic ingredients of a CP violating theory and a CP violating observable. I then review the phenomenology of K^0 decays and describe the ingredients that go into a calculation of ϵ'/ϵ within the standard model. In the last lecture I discuss the basics of CP violation in hyperon decay and in non-leptonic neutral B meson decay.

GENERALITIES ON CP VIOLATION

I start by reviewing some of the basic ingredients that go into a CP violating theory and observable. More detailed discussions can be found, for example, in Refs. [1-4]. CP symmetry is the combination of charge conjugation and parity. Charge conjugation is the symmetry that interchanges particles and anti-particles up to an arbitrary phase convention. The effect of parity is to reverse the spatial coordinates leaving time unchanged, $x^\mu \leftrightarrow x_\mu$. Vector fields appear in a Lagrangian with the Lorentz index contracted with some other vector (as the Lagrangian is a Lorentz scalar), and spatial indices are integrated over. For this reason, the intrinsic parity of the field is the relevant quantity. For example, a combination of a vector field and a quark bilinear that may appear in a Lagrangian, transforms under parity as $W^\mu \bar{q}\gamma_\mu q \leftrightarrow W_\mu \bar{q}\gamma^\mu q$. The pion is a pseudoscalar, which means that under parity $\pi \leftrightarrow -\pi$.

The combined CP operation acts on a single pion (kaon) state as follows, $CP|\pi^0> = -|\pi^0>$, $CP|\pi^\pm> = -|\pi^\mp>$, and $CP|K^0> = -|\bar{K}^0>$. This last equation defines the phase convention that I will use. Similarly, for a two pion state, $CP|\pi^+\pi^-> = (-)^\ell|\pi^-\pi^+>$. For example, if the two pions originate from a K^0 decay they will be in a state with total angular momentum $J = 0$, hence $\ell = 0$ and this two pion state is CP even.

I want to illustrate in a simple way some of the statements one usually hears about CP violation.

---

[1] Supported in part by DOE under contract number DE-FG02-92ER40730.

To violate CP, a theory needs a phase in the Lagrangian.
This can be easily illustrated with an example. Suppose we have a theory with a charged current interaction of a vector field and a quark bilinear such as

$$\mathcal{L} = g W_\mu^+ \bar{u} \gamma^\mu d \tag{1}$$

For this theory to make sense, however, the Lagrangian has to be hermitian, so we need to add a second interaction

$$\mathcal{L} = g W_\mu^+ \bar{u} \gamma^\mu d + g^\star W_\mu^- \bar{d} \gamma^\mu u \tag{2}$$

These two terms are responsible for the vertices of Figure 1a, and 1b respectively. As

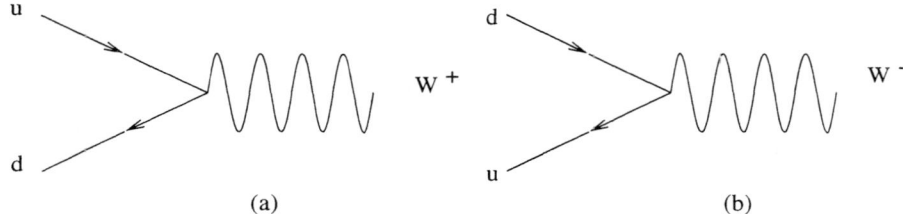

FIGURE 1. CP conjugate processes arising from the two terms in Eq. 2.

we saw before, however, these two terms are CP conjugate, under CP, $W_\mu^+ \bar{u} \gamma^\mu d \leftrightarrow W_\mu^- \bar{d} \gamma^\mu u$. We thus see that if the theory is to be CP invariant $g = g^\star$ in Eq. 2. That is, we need a complex phase in order to violate CP.

The complex phase cannot be trivial.
In other words, it should not be possible to remove the phase with a field redefinition. In the example of Eq. 2, if there are no additional fields, the phase in g is trivial. To see this we write

$$\mathcal{L} = g e^{i\phi} W_\mu^+ \bar{u} \gamma^\mu d + g e^{-i\phi} W_\mu^- \bar{d} \gamma^\mu u \tag{3}$$

and then redefine one of the fields: $d' = e^{i\phi} d$. In terms of the new fields the Lagrangian becomes

$$\mathcal{L} = g W_\mu^+ \bar{u} \gamma^\mu d' + g W_\mu^- \bar{d}' \gamma^\mu u \tag{4}$$

The phase has been rotated away and the theory does not violate CP. Within the standard model, CP violation occurs precisely in interactions of the form Eq. 2. It is now well known that in order to have a non-trivial phase, at least three generations of quarks are needed [5]. The constant g in Eq. 2 is then replaced by a 3×3 unitary matrix with one phase that cannot be rotated away by redefining the quark fields.

Interference is needed to observe CP violation
To observe CP violation one needs a process to which more than one amplitude can contribute, and CP violation will occur in the interference terms. This is

easy to illustrate with an example. Suppose we have a CP violating theory but are looking at the process $i \to f$ to which only one amplitude contributes. The amplitude can be written as $M(i \to f) = e^{i\phi} A_1$ whereas the amplitude for the CP conjugate process would be $M(\tilde{i}^{(CP)} \to \tilde{f}^{(CP)}) = e^{-i\phi} A_1$. The two are different due to the CP violating phase ϕ. However, any observable in the process $i \to f$ will be proportional to $|M(i \to f)|^2$ and similarly for the CP conjugate process. Since the CP violating phase disappears in $|M|^2$, it is unobservable.

If the process receives contributions from at least two amplitudes with different phases (both CP conserving and CP violating phases) then the CP violating phases are observable:

$$M(i \to f) = A_1 e^{i\delta_1} e^{i\phi_1} + A_2 e^{i\delta_2} e^{i\phi_2}$$
$$\bar{M}(\tilde{i}^{(CP)} \to \tilde{f}^{(CP)}) = A_1 e^{i\delta_1} e^{-i\phi_1} + A_2 e^{i\delta_2} e^{-i\phi_2} \qquad (5)$$

A comparison of an observable in the CP conjugate pair of reactions would thus be proportional to

$$|M|^2 - |\bar{M}|^2 \sim \sin(\phi_1 - \phi_2)\sin(\delta_1 - \delta_2) \qquad (6)$$

To obtain this result we have been forced to introduce not only a second amplitude, but a new type (CP conserving) of phase. The notation in Eq. 5 is to use ϕ for a CP violating phase which changes sign in the CP-conjugate process, and to use δ for a CP conserving phase which does not change sign. We have already seen that a CP violating phase changes sign between a pair of CP conjugate processes, these phases appear at the Lagrangian level as in Eq. 2. CP conserving phases, on the other hand, appear as a consequence of unitarity in going beyond the Born approximation. In perturbation theory, they arise when loop diagrams have imaginary parts due to the existence of physical intermediate states. We will discuss this in some detail later on, but for now I can present Figure 2. This figure shows a one-loop contribution to the CP-conjugate pair of processes $K^0 \to \pi^+\pi^-$ and $\bar{K}^0 \to \pi^+\pi^-$. The CP violating phase occurs in the weak vertex, as shown. On the other hand, the CP conserving imaginary part of the amplitude is given by the absorptive part of the two pion intermediate state. Comparing the two diagrams it is intuitively clear that this phase is the same for the two processes.

CP Violation in the Standard Model

In the standard model there is one CP violating phase in the CKM matrix (we ignore the issue of strong CP violation). Recall that quark masses arise via spontaneous symmetry breaking (see for example Ref. [2]). In general, however, this results in non-diagonal mass matrices of the form $M'_{u,d}$

$$\mathcal{L}_M \sim \bar{u}'_L M'_u u'_R + \bar{d}'_L M'_d d'_R \qquad (7)$$

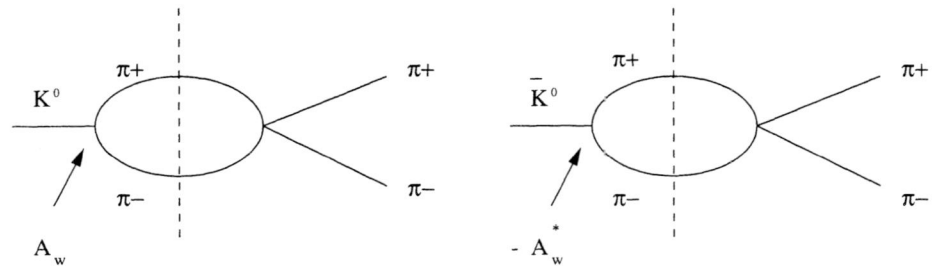

FIGURE 2. One-loop contribution to $K^0 \to \pi^+\pi^-$ and $\bar{K}^0 \to \pi^+\pi^-$. The weak vertex introduces the CP violating phase, whereas the CP conserving phase arises from the cut in the strong re-scattering of $\pi\pi$.

The mass eigenstates are obtained by diagonalizing these mass matrices. In the mass eigenstate basis there are no flavor changing neutral currents [2], and the only remaining effect of the diagonalization appears in the charged current which looks like

$$J^\mu \sim \bar{u}_L \gamma^\mu V d_L \tag{8}$$

with V a unitary 3×3 (CKM) matrix that results from a mismatch between the unitary transformations that diagonalize the left handed up and down type quark mass matrices. A unitary 3×3 matrix has nine arbitrary parameters that can be taken as three angles and six phases. Of these six phases, five are "trivial" in the sense that they can be removed by redefinitions of quark fields, and one is left with a unique CP violating phase [5].

I show here two of the many parameterizations of the CKM matrix that can be found in the literature. The original parameterization of Ref. [5], Eq. 10, and the approximate Wolfenstein [6] parameterization, Eq.11, which is the most commonly used one.

$$V = \begin{pmatrix} V_{ud} & V_{us} & V_{ub} \\ V_{cd} & V_{cs} & V_{cb} \\ V_{td} & V_{ts} & V_{tb} \end{pmatrix} \tag{9}$$

$$V = \begin{pmatrix} c_1 & -s_1 c_3 & -s_1 s_3 \\ s_1 c_2 & c_1 c_2 c_3 - s_2 s_3 e^{i\delta} & c_1 c_2 s_3 + s_2 c_3 e^{i\delta} \\ s_1 s_2 & c_1 s_2 c_3 + c_2 s_3 e^{i\delta} & c_1 s_2 s_3 - c_2 c_3 e^{i\delta} \end{pmatrix} \tag{10}$$

$$V \approx \begin{pmatrix} 1 - \lambda^2/2 & \lambda & A\lambda^3(\rho - i\eta) \\ -\lambda & 1 - \lambda^2/2 & A\lambda^2 \\ A\lambda^3(1 - \rho - i\eta) & -A\lambda^2 & 1 \end{pmatrix} \tag{11}$$

---

[2] This is a property of the standard model, that is imposed in new physics extensions in order to satisfy experimental constraints.

Just from these two forms, you can see that the phase may appear in different entries of the matrix. Since the different forms are related by unobservable field redefinitions, it is convenient to have a parameterization independent measure of CP violation. This was done in Ref. [7] where it was shown that all CP violation in the standard model is proportional to the (Jarlskog) invariant which can be written, for example, as

$$J = \text{Im}(V_{us}V_{cb}V_{ub}^{\star}V_{cs}^{\star}) \qquad (12)$$

In the KM parameterization this is equal to

$$J = c_1 c_2 c_3 s_1^2 s_2 s_3 s_\delta \qquad (13)$$

and from this form it can be deduced immediately that

$$J_{max} = \frac{1}{6\sqrt{3}} \sim 0.1 \qquad (14)$$

In the approximate parameterization of Wolfenstein, on the other hand,

$$J = A^2 \lambda^6 \eta \sim 8 \times 10^{-5} \eta \leq 6 \times 10^{-5} \qquad (15)$$

where the number follows from experimental constraints [9]. A surprising and unexplained result is that the observed CP violation is much smaller than what is allowed by the CKM mechanism. In addition, for the standard model to violate CP, no two quarks of the same charge can be degenerate (because if they are, the phase in the CKM matrix can be removed by a field redefinition). For high energy experiments (say at, or above, the W mass scale), this requirement shows up as an additional suppression in CP odd observables. For kaon and B-meson physics there is no such additional suppression as the requirement is satisfied in the definition of the states.

The unitarity of the CKM matrix has been exploited to find simple descriptions of CP violation in the standard model. For example, $V^\dagger V = 1$ implies that $\sum_i V_{bi}^\dagger V_{id} = 0$ which expanded reads

$$V_{ub}^{\star}V_{ud} + V_{cb}^{\star}V_{cd} + V_{tb}^{\star}V_{td} = 0 \qquad (16)$$

which, in turn, can be interpreted as the equation for a triangle in the complex plane.[3]

The Jarlskog invariant is proportional to the area of the unitarity triangle. The experimental knowledge of the CKM elements at present, can be summarized as follows: $\lambda = 0.2196 \pm -0.0023$ is known from semileptonic kaon decay; $A = 0.819 \pm 0.035$ from $B \to D^\star \ell \nu$. The parameters ρ and η are not precisely known yet. They

---

[3] Detailed discussions on how to measure the different elements of this, and other, unitarity triangles can be found in many review talks such as Ref. [9].

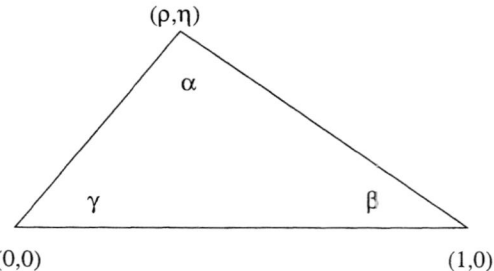

FIGURE 3. bd Unitarity triangle. Recent experimental constraints may be found, for example, in Ref. [9].

are constrained by measurements of CP violation in the kaon system (both ϵ and ϵ'), by oscillations in the B_d and B_s systems and by measurements of V_{ub}. All these results can be found in many review papers such as Ref. [9]. More recently, there has been another constraint from the measurement of the rare decay $K^+ \to \pi^+ \nu \bar{\nu}$. An improved measurement of this mode, and a measurement of its neutral counterpart $K_L \to \pi^0 \nu \bar{\nu}$ [8], will provide two of the cleanest determinations of CKM parameters. For details on the theoretical calculations see for example, Ref. [10] and references therein.

The impact of future measurements of CP violation on the determination of the parameter η can be grouped into two types. Observables such as ϵ'/ϵ and $A(\Lambda_-^0)$ (which we discuss later on) are plagued by hadronic uncertainties and will not provide precise determinations of η in the near future. Observables such as the asymmetry in $B \to \Psi K_s$ (which we also discuss later), and the rate of $K_L \to \pi^0 \nu \bar{\nu}$, on the other hand, are very clean theoretically and should allow us to precisely measure certain combinations of CKM parameters.

NEUTRAL KAON PHENOMENOLOGY

The phenomenology of neutral kaons is a very old subject, and I will not attempt to provide historical details. Discussions similar to the one I give here can be found in many textbooks such as Ref. [1,2]. The neutral kaon flavor eigenstates are K^0 and $\bar{K}^0$. Their valence quark content is $d\bar{s}$ and $\bar{d}s$ respectively. These are the states that are produced in strong interaction reactions, they are eigenstates of strangeness and carry quantum numbers $I = 1/2$ and $J^P = 0^-$. In the phase convention that we use here $CP|K^0> = -|\bar{K}^0>$. The CP eigenstates are the linear combinations

$$K_1^0 = \frac{1}{\sqrt{2}}(K^0 - \bar{K}^0)$$
$$K_2^0 = \frac{1}{\sqrt{2}}(K^0 + \bar{K}^0) \qquad (17)$$

the first one being CP-even and the second one CP-odd. If CP were a good symmetry, these would also be the mass eigenstates observed in nature. Experimentally, the physical states observed are labelled K_L and K_s for "long" and "short" because one lives much longer than the other one. The K_L decays mostly into three pion states and lives 5.2×10^{-8} seconds. The K_s decays mostly into two pion states and lives 0.9×10^{-10} seconds. They both have a mass of approximately 497.7 MeV [11].

Recall that a two pion system with $J = 0$ is a CP even state. Therefore, if $K_{L,s}$ were CP eigenstates, only one of them would decay to two pions. K_s decays predominantly into two pions, but K_L has also been observed to decay into two pions [12]. This is the experimental evidence for CP violation. Noting that $B(K_L \to 3\pi) \sim 20\%$ but $B(K_L \to 2\pi) \sim 2 \times 10^{-3}$, one sees that the observed CP violation in the kaon system is small. Later on we will quantify this statement.

$K^0 - \bar{K}^0$ Mixing

The mass eigenstates $K_{L,s}$ can be written as linear superpositions of the strangeness eigenstates with a two component wave-function as

$$|\Psi(t)>= \begin{pmatrix} a(t) \\ b(t) \end{pmatrix} \equiv a(t)|K^0> + b(t)|\bar{K}^0> \qquad (18)$$

The time evolution of the system is given by a Schrödinger equation with an effective Hamiltonian

$$(M - \frac{i}{2}\Gamma)_{ij} \equiv \frac{1}{2m_K} <K_i^0|H_{eff}|K_j^0> \qquad (19)$$

The factor $1/2m_K$ is needed for the usual normalization of states, and hermiticity of the effective Hamiltonian implies that $M_{12} = M_{21}^*$ and $\Gamma_{12} = \Gamma_{21}^*$.

Assuming that CPT is a good symmetry of the theory one has that

$$<K^0|(CPT)^{-1}(CPT)H_{eff}(CPT)^{-1}(CPT)|K^0> = <\bar{K}^0|H_{eff}|\bar{K}^0> \qquad (20)$$

or, $M_{11} = M_{22}$ and $\Gamma_{11} = \Gamma_{22}$. One can write

$$M - \frac{i}{2}\Gamma = \begin{pmatrix} A & p^2 \\ q^2 & A \end{pmatrix} \qquad (21)$$

with p^2, q^2 complex in general. If CP is a good symmetry of the theory then

$$<K^0|(CP)^{-1}(CP)H_{eff}(CP)^{-1}(CP)|\bar{K}^0> = <\bar{K}^0|H_{eff}|K^0> \qquad (22)$$

so that $p = q$ (and it is real) if CP is conserved. For this reason it is natural to introduce as a measurement of CP violation the quantity

$$\bar{\epsilon} \equiv \frac{p-q}{p+q}$$
$$\frac{q}{p} = \frac{1-\bar{\epsilon}}{1+\bar{\epsilon}} \qquad (23)$$

Diagonalizing the effective Hamiltonian, and identifying the eigenvalues as $m_{L,s} - i/2\Gamma_{L,s}$ leads to the usual expression

$$\frac{q}{p} = 1/2 \frac{\Delta m + \frac{i}{2}\Delta\Gamma}{M_{12} - \frac{i}{2}\Gamma_{12}} \qquad (24)$$

and other equivalent expressions.[4] The notation is $\Delta m \equiv m_L - m_s$, $\Delta\Gamma \equiv \Gamma_s - \Gamma_L$. Similarly, the eigenvectors are

$$|K_L> = \frac{1}{\sqrt{1+|\bar{\epsilon}|^2}} \left(|K_2^0> + \bar{\epsilon}|K_1^0> \right)$$
$$|K_s> = \frac{1}{\sqrt{1+|\bar{\epsilon}|^2}} \left(|K_1^0> + \bar{\epsilon}|K_2^0> \right) \qquad (25)$$

From these expressions one sees, again, that if CP is conserved $\bar{\epsilon} = 0$ and $K_L = K_2$, $K_s = K_1$. With CP violation the long and short states mix with resulting "indirect" ($\Delta S = 2$) CP violation in $K \to \pi\pi$.

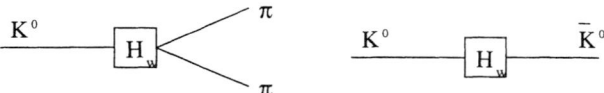

FIGURE 4. Schematic depiction of "direct" and "indirect" CP violation in $K \to \pi\pi$.

Phenomenology of $K \to \pi\pi$

There are six reactions of this type, $K^0 \to \pi^+\pi^-$, $K^0 \to \pi^0\pi^0$ and $K^+ \to \pi^+\pi^0$ and their charge conjugate reactions. Although we can treat the weak $|\Delta S| = 1$ interaction perturbatively, we must deal with the non-perturbative nature of the strong interactions, which among other things, can mix the $\pi^0\pi^0$ and $\pi^+\pi^-$ final states. For this reason it is more convenient to work with final states of definite isospin.

Since the pion has $I = 1$, the two pion state has $I = 1 \oplus 1 = 0, 1, 2$. The two pion states that concern us here originate from the decay of a kaon, and are, therefore, in a $J = 0$ state. This limits the possible isospin states to 0 and 2,

---

[4] I am simply introducing the notation that will be used later, those wanting a detailed derivation of results should consult, for example Refs. [1,2,4].

$$|\pi^+\pi^-> = \sqrt{\frac{2}{3}}|0,0> + \sqrt{\frac{1}{3}}|2,0>$$
$$|\pi^0\pi^0> = \sqrt{\frac{2}{3}}|2,0> - \sqrt{\frac{1}{3}}|0,0> \quad (26)$$

In this isospin basis the final state interaction phases can be extracted from $\pi\pi$ scattering data with the use of Watson's theorem. At the kaon mass energy, pion-pion scattering is elastic (neglecting electroweak interactions) so that $<I|S|I> \equiv e^{i2\delta_I}$ using $|I>$ to denote the two pion state with isospin I. With $S = 1 + iT$, the unitarity of the S matrix can be written as

$$\sum_n <I|T^\dagger|n><n|T|K^0> = i(<I|T^\dagger|K^0> - <I|T|K^0>) \quad (27)$$

where we have introduced a complete set of states $|n>$. For the strong interactions, however, the only state $|n>$ that can connect with $<I|$ is $|I>$ itself. This condition, combined with

$$<I|T^\dagger|K^0> = (<K^0|T|I>)^\star = (<I|T|\bar{K}^0>)^\star, \quad (28)$$

which follows from CPT invariance, then leads to the result

$$e^{-2i\delta_I} <I|T|K^0> = (<I|T|\bar{K}^0>)^\star \quad (29)$$

The reader interested in a more detailed derivation should consult, for example, Refs. [1,4].

The above result, known as Watson's theorem, is what permits us to write the amplitudes in the usual way,

$$<I|T|K^0> = iA_I e^{i\delta_I}$$
$$<I|T|\bar{K}^0> = -iA_I^\star e^{i\delta_I} \quad (30)$$

The strong rescattering phases have been separated from the weak phases, and they are equal to the $\pi\pi$ scattering phases at $\sqrt{s} = m_K$ which can be extracted from experiment (additional details and references can be found in Ref. [2]).

If the weak interaction is either $\Delta I = 1/2, 3/2$ **only**, then

$$A(K^0 \to \pi^+\pi^-) = A_0 e^{i\delta_0} + \frac{A_2}{\sqrt{2}} e^{i\delta_2}$$
$$A(K^0 \to \pi^0\pi^0) = A_0 e^{i\delta_0} - \sqrt{2} A_2 e^{i\delta_2}$$
$$A(K^+ \to \pi^+\pi^0) = \frac{3}{2} A_2 e^{i\delta_2} \quad (31)$$

A fit to the measured rates reveals the "$\Delta I = 1/2$" rule,

$$\omega \equiv |\frac{A_2}{A_0}| \approx 0.045 \approx \frac{1}{22} \quad (32)$$

Measures of CP Violation

If we ignore the small $\Delta I = 3/2$ amplitude in $K \to \pi\pi$ decays, the final state pions are in an $I = 0$ state and we can define one CP violating quantity

$$\epsilon \equiv \frac{A(K_L \to (\pi\pi)_{I=0})}{A(K_s \to (\pi\pi)_{I=0})} \tag{33}$$

with the parameterization of Eq. 31 and the definitions of Eq. 25 one finds after some algebra that

$$\epsilon = \frac{\bar{\epsilon} + i\frac{\mathrm{Im}A_0}{\mathrm{Re}A_0}}{1 + i\bar{\epsilon}\frac{\mathrm{Im}A_0}{\mathrm{Re}A_0}} \tag{34}$$

In the Wu-Yang phase convention, $\mathrm{Im}A_0 = 0$ and $\bar{\epsilon} = \epsilon$. However, this is not the convention usually adopted in the standard model where there are non-zero, small, CP violating phases

$$\xi_I \approx \mathrm{Im}A_I/\mathrm{Re}A_I \tag{35}$$

In the general case, with $A_2 \neq 0$, it is conventional to introduce the quantities

$$\eta_{+-} \equiv \frac{A(K_L \to \pi^+\pi^-)}{A(K_s \to \pi^+\pi^-)} = \epsilon + \epsilon'$$

$$\eta_{00} \equiv \frac{A(K_L \to \pi^0\pi^0)}{A(K_s \to \pi^0\pi^0)} = \epsilon - 2\epsilon' \tag{36}$$

and one finds that

$$\epsilon = \bar{\epsilon} + i\xi_0$$

$$\epsilon' = ie^{i(\delta_2 - \delta_0)} \frac{A_2}{\sqrt{2}A_0}(\xi_2 - \xi_0)$$

$$|\epsilon'| = \frac{\omega}{\sqrt{2}\mathrm{Re}A_0}(\mathrm{Im}A_0 - \frac{1}{\omega}\mathrm{Im}A_2) \tag{37}$$

From these expressions one sees that in order to have a non-zero ϵ' the two weak decay amplitudes A_0 and A_2 must have different phases. One also sees that ϵ'/ϵ is suppressed by the small value of ω and that the importance of $\mathrm{Im}A_2$ is enhanced with respect to that of $\mathrm{Im}A_0$ by the $\Delta I = 1/2$ rule. This last point is the key reason for the importance of electroweak penguins and isospin breaking effects (that we discuss later) for ϵ'.

In practice $\mathrm{Re}\epsilon'/\epsilon$ is extracted experimentally using the relation $\mathrm{Re}\epsilon'/\epsilon = 1/6(|\eta_{+-}/\eta_{00}|^2 - 1)$. The theoretical calculation of ϵ'/ϵ is quite involved and has been described in detail in reviews such as Ref. [10,13]. In the next section I will present a sketch of the ingredients that go into the calculation emphasizing only the basics and leaving out all technical details.

$K \to \pi\pi$ IN THE STANDARD MODEL- A SKETCH

For simplicity lets look at the mode $\bar{K}^0 \to \pi^+\pi^-$. In terms of quark model quantum numbers the transition is $(\bar{d}s) \to (\bar{d}u)(\bar{u}d)$ for a net flavor change $s \to d$. At tree-level, the Feynman diagram responsible for this transition is shown in Figure 5a which can be easily evaluated using the Feynman rules for the Standard Model. Noting that a typical momentum transfer through the W boson propagator is much smaller than the W mass, one can write the amplitude as

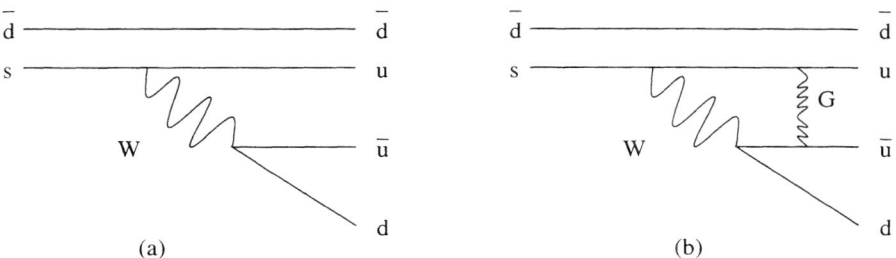

FIGURE 5. a) Tree-level contribution to the process $s \to u\bar{u}d$. b) Example of a QCD correction.

$$A_{tree} = -i\frac{G_F}{\sqrt{2}}V_{us}V_{ud}^\star \bar{u}\gamma_\mu(1-\gamma_5)s\bar{d}\gamma^\mu(1-\gamma_5)u \tag{38}$$

The usual analysis is to note that this amplitude can also be derived from an effective low energy Hamiltonian of the form $H_{eff} = G_F/\sqrt{2}V_{us}V_{ud}^\star C_A O_A$, where C_A is a coefficient (that is equal to one at this level) and O_A is the local four-quark operator $O_A = \bar{u}\gamma_\mu(1-\gamma_5)s\bar{d}\gamma^\mu(1-\gamma_5)u$.

This expression is purely CP conserving since $V_{us}V_{ud}^\star$ is real. This is in agreement with the statement that all three generations must be involved in order to generate CP violation within the standard model.

An obvious problem in the expression for H_{eff} is that it contains quark fields and yet we need to find the matrix element of the operator between meson states. This is the main source of theoretical uncertainty in the calculation. The form of O_A (looks like the factorized product of two quark bilinears) suggests factorization as a first attempt to compute the matrix element $< \pi^+\pi^- |O_A| K^0 >$. Anticipating the fact that naive factorization will not give the correct result, we, nevertheless, describe it next.

Introduction to Chiral Perturbation Theory

Once again, I will only have time to sketch the basics. The interested reader should consult, for example, Ref. [14] or text books such as Ref. [2]. The QCD Lagrangian can be written in the form

$$\mathcal{L} = -1/4 G^{a\mu\nu} G^a_{\mu\nu} + \bar{\Psi}_L i\gamma_\mu D^\mu \Psi_L + \bar{\Psi}_R i\gamma_\mu D^\mu \Psi_R + m\bar{\Psi}_L \Psi_R + m\bar{\Psi}_R \Psi_L \quad (39)$$

For the light quarks u, d, s, one can think of the mass terms as perturbations to a Lagrangian that is invariant under separate $SU(3)$ rotations of the left and right-handed fermion fields. This is the chiral $SU(3) \times SU(3)$ symmetry. At low energy, the spectrum consists of light mesons, the pions and kaons, in an $SU(3)$ octet. The chiral symmetry is broken spontaneously to $SU(3)_V$ and there appear eight (pseudo)-Goldstone bosons, the light mesons. That they are not massless is a consequence of the explicit breaking of chiral symmetry by the quark masses, and this is dealt with in perturbation theory.

The basic degrees of freedom for the low energy theory are thus written as a matrix,

$$\phi = \frac{1}{\sqrt{2}} \begin{pmatrix} \pi^0/\sqrt{2} + \eta/\sqrt{6} & \pi^+ & K^+ \\ \pi^- & -\pi^0/\sqrt{2} + \eta/\sqrt{6} & K^0 \\ K^- & \overline{K^0} & -2\eta/\sqrt{6} \end{pmatrix} \quad (40)$$

and a non-linear Lagrangian is written in terms of the matrix $U = \exp(2i\phi/f_\pi)$ which transforms under chiral symmetry as $U \to RUL^\dagger$. L, R are elements of $SU(3)_{L,R}$ respectively. The strong interaction Lagrangian is constructed from chiral invariants and organized as a series of operators with increasing number of derivatives. Since $U^\dagger U = 1$, the leading order Lagrangian contains only one term

$$\mathcal{L}^{(2)} = \frac{f_\pi^2}{4} \text{Tr}(\partial_\mu U \partial^\mu U^\dagger) \quad (41)$$

The normalization constant $f_\pi \sim 93$ MeV is measured in $\pi^+ \to e^+ \nu$, for example. The effective low energy Lagrangian contains an infinite number of terms. For example, with four derivatives one can write

$$\mathcal{L} = \frac{f_\pi^2}{4} \frac{1}{\Lambda^2} \left(\text{Tr}(\partial_\mu U \partial^\mu U^\dagger) \right)^2 \quad (42)$$

The theory is, nevertheless, useful and manageable if the higher dimension operators are suppressed. This is the case, for example, if one studies processes at low energies compared to the scale Λ. Amplitudes induced by Eq. 42 are then suppressed with respect to those induced by Eq. 41 by E^2/Λ^2. Experimentally, the scale Λ is about 1 GeV.

To move on to weak processes we need to construct currents, and this is most easily done by promoting the global chiral symmetry to a gauge symmetry,

$$\mathcal{L}_{QCD} \to \mathcal{L}_{QCD} + \bar{\Psi}\gamma^\mu(v_\mu + \gamma_5 a_\mu)\Psi - \bar{\Psi}(s - ip\gamma_5)\Psi \quad (43)$$

The v_μ, a_μ, s, p are external fields, hermitian 3×3 matrices (in flavour space), and color singlets. Under a gauge $SU(3)_L \times SU(3)_R$ symmetry they transform as $(\ell_\mu, r_\mu = v_\mu \mp a_\mu)$

$$\ell_\mu \to L\ell_\mu L^\dagger + iL\partial_\mu L^\dagger$$
$$r_\mu \to Rr_\mu R^\dagger + iR\partial_\mu R^\dagger$$
$$s + ip \to R(s + ip)L^\dagger \tag{44}$$

When one includes these sources, there is one additional term that can be written for the leading order strong chiral Lagrangian

$$\mathcal{L}^{(2)} = \frac{f_\pi^2}{4}\Big(\mathrm{Tr}(D_\mu U D^\mu U^\dagger) + \mathrm{Tr}(\xi U^\dagger + U\xi^\dagger)\Big) \tag{45}$$

where now

$$D_\mu U = \partial_\mu - ir_\mu U + iU\ell_\mu$$
$$\xi = 2B(s + ip) \tag{46}$$

A new constant, B has been introduced and it relates the meson masses to the quark masses because explicit chiral symmetry breaking can now be introduced by setting $s = diag(m_u, m_d, m_s)$, $p = 0$. This leads to relations such as $m_\pi^2 = (m_u + m_d)B$ and to the Gell-Mann-Okubo mass relation. The electroweak currents can also be introduced, by identifying

$$r_\mu = eQ(A_\mu - \tan\theta_W Z_\mu)$$
$$\ell_\mu = r_\mu + \frac{eQ_L}{\sin\theta_W}Z_\mu + \frac{e}{\sqrt{2}\sin\theta_W}(VW_\mu^+ + V^\dagger W_\mu^-) \tag{47}$$

with the matrices $Q = diag(2/3, -1/3, -1/3)$, $Q_L = diag(1, -1, 1)$ and $V_{12} = V_{ud}$, $V_{13} = V_{us}$ (all other elements of V being zero). These are, of course, the usual electroweak charges of the three light quarks.

In general, currents can be obtained by usual methods. For example,

$$J_L^\mu \equiv \bar\Psi_L \gamma^\mu \Psi_L = \frac{\delta\mathcal{L}_{QCD}}{\delta\ell_\mu} \tag{48}$$

can be "bosonized" by applying the same operation to the effective chiral Lagrangian. At lowest order,

$$J_L^\mu = \frac{\delta\mathcal{L}^{(2)}}{\delta\ell_\mu} = \frac{i}{2}f_\pi^2(D^\mu U^\dagger)U$$
$$= f_\pi D^\mu \phi - \frac{i}{\sqrt{2}}(\phi D^\mu \phi - D^\mu \phi \phi) + \cdots \tag{49}$$

The first term describes, for example, $\pi \to \ell\nu$ decays from which f_π can be extracted. The second term describes semileptonic $K_{\ell 3}$ decays, and so on.

If one requires better precision than that provided by the leading order calculation, one constructs higher order chiral Lagrangians and repeats the process.

For example, the next to leading order chiral Lagrangian has been constructed by Gasser and Leutwyler [14]. It contains ten new constants, all of which have been determined. Higher order calculations are a bit more involved because a consistent power counting requires that the leading order Lagrangian be treated at the one-loop order (for next-to-leading order). We will not have time to discuss this in further detail.

Unfortunately our job does not end here for we need to calculate non-leptonic weak decays. At tree-level we saw that they were induced by an effective Hamiltonian that can be written as

$$\mathcal{H}_{eff} = 4\frac{G_F}{\sqrt{2}} V_{ud}^\star V_{us} J_{L13}^\mu J_{\mu L21} \qquad (50)$$

The naive factorization approximation consists of replacing the currents in Eq. 50 with their bosonized form of Eq. 49. This leads to the predictions

$$A_0 = -\frac{G_F}{\sqrt{2}} V_{ud}^\star V_{us} \frac{2\sqrt{2}}{3} f_\pi (m_K^2 - m_\pi^2)$$
$$A_2 = -\frac{G_F}{\sqrt{2}} V_{ud}^\star V_{us} \frac{2}{3} f_\pi (m_K^2 - m_\pi^2) \qquad (51)$$

so that $A_0/A_2 = \sqrt{2}$ which is far from the measured $A_0/A_2 \sim 22$. In more detail, the prediction for A_0 is too small by about a factor of 8 whereas the prediction for A_2 is large by a factor of 2.

A discussion in terms of symmetry arguments alone leads to a chiral Lagrangian for the weak interactions directly [15], and this approach has been detailed, for example, in Ref. [16]. It is not useful for the purposes of computing quantities such as ϵ' because the couplings are not known. Let us instead go back to the discussion in terms of quark fields to describe some of the additional effects that have not been incorporated so far.

The first thing one thinks of doing is to incorporate perturbative QCD effects, such as the one depicted in Figure 5b. This results in several effects:

- The coefficient C_A in the effective Hamiltonian is changed.

- New operators appear. In this case $O_B = \bar{u}_L \gamma_\mu u_L \bar{d}_L \gamma^\mu s_L$.

- The figure shows how QCD "connects" the two currents so that one does not expect factorization to hold.

- At this level the amplitude is still CP conserving as there is no phase.

- The relative size of this loop diagram relative to the tree diagram is about $g_s^2/(16\pi^2) \log(M_W^2/\mu^2)$. For a renormalization scale $\mu \sim 1$ GeV where $\alpha_s \sim 0.4$ this is about 30%. The QCD corrections are much larger than what one might have anticipated, they are enhanced by large logarithms, and this must be handled correctly with renormalization group methods. Again, we will not have time to discuss this in further detail, but it can be found, for example, in Ref. [10]. Some of the early papers on this subject are Refs. [17,18].

To get an imaginary part that violates CP we need to involve all three generations. This happens at the one-loop level through the so-called "penguin" diagrams like the one in Figure 6. The main features of this diagram are

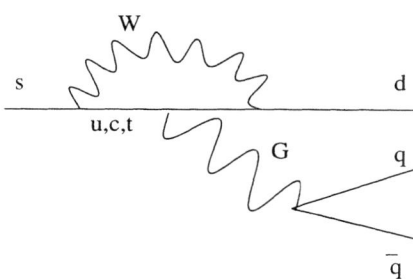

FIGURE 6. So-called "penguin" diagram contribution to the quark-level process $s \to dq\bar{q}$.

- It generates a new operator, $\bar{s}_L q_R \bar{q}_R d_L$, which looks like the product of (pseudo)scalar densities.

- All three generations are involved as intermediate states in the loop and therefore a CP violating phase is possible. In fact, a complete numerical analysis [10] results in a dominant penguin contribution to an operator O_6 with coefficient $C_6 = -0.011 + 0.08 A^2 \lambda^4 (1 - \rho + i\eta)$, where we can see the Jarlskog invariant appearing in the imaginary part.

- This operator is completely (mostly) $\Delta I = 1/2$ and its effect goes in the right direction to explain the $\Delta I = 1/2$ rule [17]. However, its magnitude is too small for an explanation of this effect [19].

- In factorization, it is possible to obtain a bosonized version of this operator in a similar manner as was described for O_A. It can be found in detail, for example, in Ref. [4].

Finally there is another distinct class of diagram. These are the so-called "electroweak penguins" where the gluon line in Figure 6 is replaced by a photon or a Z boson. The main features of these contributions are

- All three generations are involved generating additional CP violation.

- Unlike the gluonic penguin, electroweak penguins contribute to both A_0 and A_2.

- Although the coefficient of this operator is suppressed with respect to that of the gluon penguin operator by a ratio of α/α_s, its contribution to A_2 is quite important in the calculation of ϵ'/ϵ where it is enhanced by a factor of 22 relative to the phase of A_0 as in Eq. 37.

Once again, a complete analysis can be found in Ref. [10], where the effective Hamiltonian is written as a sum of at least seven operators and the coefficients are calculated in detail.

It is instructive to present an approximate result from Ref. [20]

$$\frac{\epsilon'}{\epsilon} \approx 13 A^2 \lambda^5 \eta [B_6^{1/2}(1 - \Omega_{\eta,\eta'}) - 0.4 B_8^{3/2}] \qquad (52)$$

In this equation $B_6^{1/2}$ is the matrix element of the penguin operator between a K^0 and the $I = 0$ two pion state normalized to the factorization result. Similarly, $B_8^{3/2}$ is the matrix element of the electroweak penguin operator between a K^0 and an $I = 2$ two pion state normalized to its value in factorization. The one ingredient in the result that we have not discussed is the quantity $\Omega_{\eta,\eta'}$. This quantity arises from the increased importance of $\mathrm{Im} A_2$ (enhanced by 22 with respect to $\mathrm{Im} A_0$) that makes isospin violating effects important. In short, $K^0 \to \pi^0 \pi^0$ receives additional contributions from η, η' intermediate states that mix with a π^0 when there is isospin violation. We show a sketch of this type of contribution in Figure 7. A detailed discussion can be found in Ref. [21].

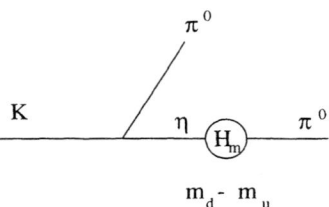

FIGURE 7. Isospin breaking contribution to $K^0 \to \pi^0 \pi^0$ induced by $\pi^0 - \eta$ mixing proportional to $(m_d - m_u)$.

Finally, how do theory and experiment compare? Much has been made of the fact that this year's results from KTeV, $\epsilon'/\epsilon = (28.0 \pm 4.1) \times 10^{-4}$ [22] and NA48, $\epsilon'/\epsilon = (18.5 \pm 7.3) \times 10^{-4}$ [23] are not in agreement with the "central" value of the theoretical range, around 7×10^{-4} [20]. In my opinion, it is premature to reach any conclusion in this regard as the theoretical result contains at least three ($B_6^{1/2}$, $B_8^{3/2}$, $\Omega_{\eta,\eta'}$) non-perturbative parameters that cannot really be calculated reliably at present. I say at least three, because others enter indirectly when one inputs values for CKM parameters extracted from experiment.

CP VIOLATION IN HYPERON DECAY

In Figure 8 we show a sketch of the reaction $\Lambda^0 \to p\pi^-$ in which the Λ initial state is assumed to be polarized with polarization $\vec{\omega}$ and the final pion has momentum q. There are several reactions of this type, but experimentally, this is the only one being actively investigated by Fermilab experiment E871 [24]. The isospin of the final state is $I = 1/2$ or $3/2$, corresponding to a $\Delta I = 1/2$ or $3/2$ weak transition respectively. There are also two possibilities for the parity of the final state. They are the s-wave, $l = 0$, parity odd state (thus reached via a parity violating amplitude); and the p-wave, $l = 1$, parity even state reached via a parity conserving amplitude.

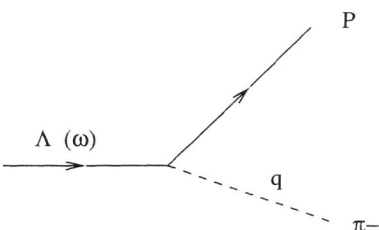

FIGURE 8. Kinematics for $\Lambda \to p\pi^-$. The Λ^0 has a polarization vector $\vec{\omega}$ and the pion has momentum $\vec{q}$.

The amplitude can be written as

$$A(B_i \to B_f \pi) = s + p\vec{\omega} \cdot \vec{q} \tag{53}$$

In terms of these quantities one can compute the decay distribution, and the total decay rate. One finds that the decay is characterized by three independent observables: the total decay rate and two parameters that determine the angular distribution. The total decay rate is given by

$$\Gamma = \frac{|\vec{q}|(E_P + M_P)}{4\pi M_\Lambda} G_F^2 m_\pi^4 \left(|s|^2 + |p|^2\right). \tag{54}$$

whereas the angular distribution is proportional to

$$\frac{d\Gamma}{d\Omega} \sim 1 + \alpha \hat{q} \cdot \vec{\omega} \tag{55}$$

where

$$\alpha \equiv \frac{2\mathrm{Re}\, s^* p}{|s|^2 + |p|^2}, \tag{56}$$

A third observable is possible if the polarization of the final baryon is observed. For simplicity we will ignore this here.

To construct CP-odd observables we compare the reactions $\Lambda^0 \to p\pi^-$ and $\overline{\Lambda}^0 \to \overline{p}\pi^+$ in terms of the two independent observables. CP symmetry predicts that

$$\overline{\Gamma} = \Gamma$$
$$\overline{\alpha} = -\alpha \qquad (57)$$

and, therefore, one can construct the following CP-odd observables: [25]

$$\Delta \equiv \frac{\Gamma - \overline{\Gamma}}{\Gamma + \overline{\Gamma}}$$
$$A \equiv \frac{\alpha\Gamma + \overline{\alpha}\overline{\Gamma}}{\alpha\Gamma - \overline{\alpha}\overline{\Gamma}} \approx \frac{\alpha + \overline{\alpha}}{\alpha - \overline{\alpha}} + \Delta \qquad (58)$$

To discuss the final state interaction phases it is convenient to decompose the final pion-nucleon system in terms of isospin and parity eigenstates. In that way we can make use of Watson's theorem again to handle the strong phases. The pion-nucleon system with fixed isospin and parity is an eigenstate of the strong interaction. Furthermore, at an energy equal to the Λ mass, there are no other states with the same quantum numbers. The pion-nucleon system will then re-scatter due to the strong interactions into itself, and in the process pick up a phase δ_ℓ^I [1].

Writing the amplitudes as

$$S(\Lambda_-^0) = S_1 e^{i(\delta_1 + \phi_1^s)} + S_3 e^{i(\delta_3 + \phi_3^s)}$$
$$P(\Lambda_-^0) = P_1 e^{i(\delta_{11} + \phi_1^p)} + P_3 e^{i(\delta_{33} + \phi_3^p)} \qquad (59)$$

it is possible to construct approximate expressions based on the fact that there are three small parameters in the problem:

- The strong rescattering phases which are measured to be [26] $\delta_1^s \approx 6.0°$, $\delta_1^p \approx -1.1°$ and the $I = 3/2$ phases even smaller.

- The $\Delta I = 3/2$ amplitudes are much smaller than the $\Delta I = 1/2$ amplitudes. This $\Delta I = 1/2$ rule is stronger here than in kaon decay and it is also not understood.

- The CP violating phases are presumed to be small.

One finds [25],

$$\Delta(\Lambda_-^0) = \sqrt{2}\frac{S_3}{S_1} \sin(\delta_3^s - \delta_1^p) \sin(\phi_3^s - \phi_1^s)$$
$$A(\Lambda_-^0) = -\tan(\delta_1^p - \delta_1^s) \sin(\phi_1^p - \phi_1^s) \qquad (60)$$

These expressions illustrate that Δ arises dominantly from an interference of $\Delta I = 1/2$ and $\Delta I = 3/2$ s-wave amplitudes. It is therefore suppressed by all three small factors. On the other hand A arises from the interference of s and p-wave amplitudes with $\Delta I = 1/2$ and is, therefore, not suppressed by the $\Delta I = 1/2$ rule.

To calculate the weak phases in the standard model we start from the same effective weak Hamiltonian discussed for $K \to \pi\pi$

$$H_W^{eff} = \frac{G_F}{\sqrt{2}} V_{ud}^* V_{us} \sum_i c_i(\mu) Q_i(\mu) \tag{61}$$

Recall that the coefficients have been calculated [10] so that the problem is, once again, the calculation of the matrix elements of the four-quark operators. In this case we write

$$\langle p\pi| H_w^{eff} |\Lambda^0 \rangle |_\ell^I = \mathrm{Re} M_\ell^I + i \mathrm{Im} M_\ell^I, \tag{62}$$

and adopt the strategy of extracting the real part from experiment (assuming no CP violation) and of estimating the imaginary part in a model like factorization. This is in fact the same strategy adopted in the calculation of ϵ'.

There is a main difference between this calculation and the one for ϵ'. Because this calculation is dominated by $\Delta I = 1/2$ amplitudes, only the dominant penguin operator Q_6 is important in the estimate of the imaginary parts. There is no subtle cancellation against the electroweak penguins as there could be in ϵ'. On the other hand a detailed knowledge of the dynamical difference between the s and p-waves is required and it is unlikely that factorization will be accurate. I will not describe the factorization calculation, which can be found in the literature, but simply quote the result [25]

$$A(\Lambda_-^0) = (-3 \pm 2.6) \times 10^{-5} \tag{63}$$

which is just shy of the reach of E871.

CP VIOLATION IN B-DECAY

CP violation in B-decay is extremely interesting, and will be the subject of intense experimental scrutiny within the next few years. For this reason it is imperative that I say something about it. Given the time constraints, however, it will be impossible to do any justice to this field. I will therefore have to refer you to the literature and concentrate here on one example that illustrates some of the basic features of CP violation that we have been discussing in these lectures. CP violation in hyperon decays, discussed in the previous section, is an example of a CP asymmetry arising from the interference of two final state amplitudes, in contrast, I will use the B system to discuss a CP asymmetry that arises through mixing. This corresponds to two different paths to reach the same final state. An early paper in this field is Ref. [27], and a good recent review is Ref. [28].

The formalism to describe $B - \bar{B}$ mixing is identical to the one introduced for $K - \bar{K}$ mixing, but the phenomenology of the two cases is completely different. Just as in the kaon case, we have here,

$$(M - \frac{i}{2}\Gamma)_{ij} = \frac{<B_i^0|H_{eff}|B_j^0>}{2m_B} \qquad (64)$$

From this expression we see that the origin of Γ_{12} is in intermediate states that are common decay products of B and $\bar{B}$. For the kaon case this is depicted in Figure 9. In that case the two-pion state is not only common to K^0 and $\bar{K}^0$, but it is also the

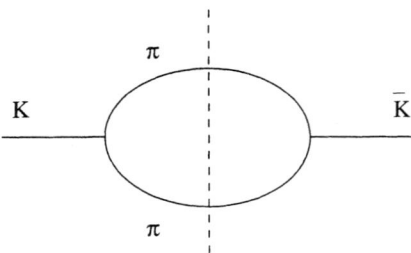

FIGURE 9. Diagram giving rise to Γ_{12} in the neutral kaon system.

dominant decay mode. This implies that $\Delta\Gamma \sim \Gamma$, as verified experimentally. For the B system there are many modes that can appear as intermediate states in a diagram such as Figure 9 (not only two particle states but more as well), and they all have small rates. This leads one to expect that $\Delta\Gamma << \Gamma$ or Γ_{12} is small. A theoretical argument based on inspection of the box diagrams responsible for the mixing in the standard model [2] tells us that $\Gamma_{12} << M_{12}$ as well.

As a consequence of Γ_{12} being negligible, p/q is, to a very good approximation, a pure phase $p/q \equiv exp(2i\phi_M)$ (see Eq. 24).

The time evolution of the two mass eigenstates $B_{H,L}$ (for heavy and light respectively) is given by the usual expression

$$A(B_{H,L}(t)) = A(B_{H,L}(0))e^{-(\Gamma/2 + iM_{H,L})t} \qquad (65)$$

One can then write the flavor eigenstates B^0 and $\bar{B}^0$ in terms of the mass eigenstates to determine their time evolution. For example, a state that was purely B_0 at $t=0$ will evolve as

$$|B(t)> = g_+(t)|B^0> + g_-(t)|\bar{B}^0> \qquad (66)$$
$$g_+(t) = e^{-\Gamma t/2}e^{-iMt}\cos(\Delta Mt/2)$$
$$g_-(t) = e^{-\Gamma t/2}e^{-iMt}i\sin(\Delta Mt/2) \qquad (67)$$

Now let us consider the example of decays into CP eigenstates. The usual framework is to define the amplitudes

$$A \equiv <f_{cp}|H_w|B^0>$$
$$\bar{A} \equiv <f_{cp}|H_w|\bar{B}^0> \qquad (68)$$

If CP is conserved, then $\bar{A} = \pm A$. Defining $\lambda \equiv q/p\bar{A}/A$ one can easily show that a measurement of the time dependent asymmetry

$$a_{f_{cp}}(t) \equiv \frac{\Gamma(B^0 \to f) - \Gamma(\bar{B}^0 \to f)}{\Gamma(B^0 \to f) + \Gamma(\bar{B}^0 \to f)} \tag{69}$$

permits an extraction of the CP violating quantity $\mathrm{Im}\lambda$. The most interesting example is the special case in which both $|q/p| = 1$ and $|A/\bar{A}| = 1$, which leads to an asymmetry $a_{fcp}(t) = -\mathrm{Im}\lambda \sin(\Delta M t)$, with no hadronic uncertainty. We have already seen that in the B system, $|q/p| = 1$ to a very good approximation. The question is then to find the conditions under which $|A/\bar{A}| = 1$ as well. In general, if several amplitudes contribute to a process we can write (in a manner similar to the decomposition used for hyperon decay)

$$A = \sum_i A_i e^{i\delta_i} e^{i\phi_i}$$
$$\bar{A} = \sum_i A_i e^{i\delta_i} e^{-i\phi_i} \tag{70}$$

From these expressions it is easy to see that if all the weak phases ϕ_i are the same, then $\bar{A}/A = exp(-2i\phi_i)$ and the condition for no hadronic uncertainty is satisfied. In practice, there is always more than one amplitude, and they have different weak phases. The practical question is then whether there are some processes in which one amplitude is so dominant that the condition is satisfied to a very good approximation.

In order to quantify this discussion some, let us assume that there are two amplitudes that contribute to a given process with different weak phases but that one of them is larger, say $A_1 > A_2$. We can then write

$$A = A_1 e^{i\delta_1} e^{i\phi_1} + A_2 e^{i\delta_2} e^{i\phi_2}$$
$$\bar{A} = A_1 e^{i\delta_1} e^{-i\phi_1} + A_2 e^{i\delta_2} e^{-i\phi_2} \tag{71}$$

From this we can write an expansion for $|\bar{A}/A|$,

$$|\frac{\bar{A}}{A}| = 1 + 2\frac{A_2}{A_1}\sin(\delta_2 - \delta_1)\sin(\phi_2 - \phi_1) + \cdots \tag{72}$$

notice how this expression is reminiscent of that which occurred in hyperon decay. It is the second term and beyond that contain the hadronic uncertainty, as one is forced to calculate the A_i for them.

Let us now briefly discuss two modes that are usually identified as measuring the angles α and β of the unitarity triangle, Figure 3.

$$B_d \to \Psi K_s$$

This is the "golden mode" that will measure the angle β of Figure 3 without hadronic uncertainties. At the quark level one can think of two main types of diagrams contributing to the process. A tree-level diagram as in Figure 10a and a penguin diagram as in Figure 10b.

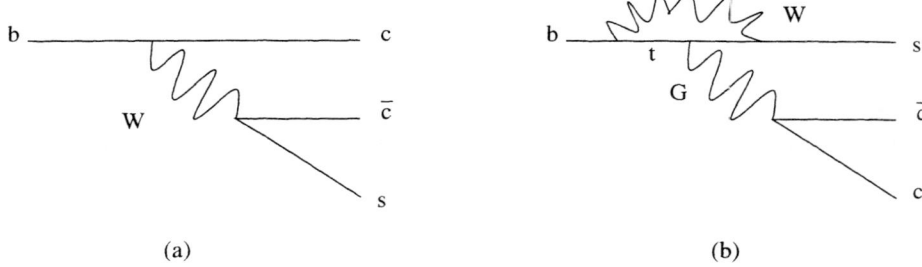

FIGURE 10. Tree-level (a), and penguin (b) diagrams contributing to the quark-level transition $b \to c\bar{c}s$.

The weak phase of the tree-level diagram is the phase of $V_{cb}V_{cs}^*$, approximately equal to $A^2\lambda^4\eta$ which is very small. The weak phase of the penguin diagram with the top-quark intermediate state is instead that of $V_{tb}V_{ts}^*$ or approximately zero. Notice that the penguin diagram is dominated by the top quark intermediate state, and that the charm-quark intermediate state has the same weak phase as the tree amplitude. From this we can estimate how much $\bar{A}/A$ will differ from a pure phase using Eq. 72. The result is,

$$\frac{A_P}{A_T} \sim \frac{\alpha_s(m_b)}{4\pi} A^2 \lambda^4 \eta \qquad (73)$$

which is completely negligible. Note that I have ignored the strong phases since they can only make the ratio smaller. We conclude that this mode is, to a very good approximation, free of hadronic uncertainty.

By examining the box diagrams responsible for B_d mixing and for K mixing (which enters this mode in the final state) one finds that [28] $\text{Im}\lambda = -\sin 2\beta$ for this mode.

$$B_d \to \pi\pi$$

This mode is often mentioned as one to measure the angle α of Figure 3. Again, at the quark level we can think of at least two contributions from the tree-level and penguin diagrams of Figure 11.

In this case the weak phase of the tree diagram is that of $V_{ub}V_{ud}^*$ which is approximately $-\eta/\rho$. The weak phase of the penguin diagram with a top-quark intermediate state, on the other hand, is that of $V_{tb}V_{td}^*$ or approximately $\eta/(\rho-1)$. The same type of naive dimensional counting that we used above can be used to find that

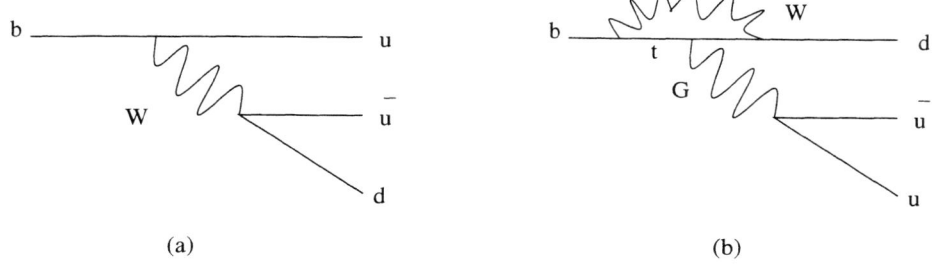

FIGURE 11. Tree-level (a), and penguin (b) diagrams contributing to the quark-level transition $b \to u\bar{u}d$.

$$\frac{A_P}{A_T} \sim \frac{\alpha_s(m_b)}{4\pi} \qquad (74)$$

In this case the weak phase difference can be of order one, the strong phase difference is not known and the suppression is just due to one of the amplitudes being of higher order in the QCD coupling constant. This, as you might expect, is a problem because QCD effects can not really be computed reliably at the moment. This has spun an enormous number of papers dealing with the question of whether it is really possible to extract α from this mode and what would it take. I refer you to the literature for this discussion [29]. [5]

ACKNOWLEDGEMENTS

This work was supported in part by DOE under contract number DE-FG02-92ER40730. I thank the theory group at Brookhaven National Lab for their hospitality while these lectures were prepared. I thank the organizing committee and in particular Miguel Angel Perez for their kind invitation to present these lectures and for their hospitality during the workshop.

REFERENCES

1. T. D. Lee. PARTICLE PHYSICS AND INTRODUCTION TO FIELD THEORY. Harwood Academic, 1981. 865p. (Contemporary Concepts in Physics, v. 1)
2. John F. Donoghue, Eugene Golowich, Barry R. Holstein. DYNAMICS OF THE STANDARD MODEL. Cambridge Univ. Press, 1992. 540p.
3. G. Valencia, hep-ph/9411441. Lectures given at Theoretical Advanced Study Institute in Elementary Particle Physics (TASI 94). Boulder, CO, 29 May - 24 Jun

---

[5] This is sometimes referred to as "penguin pollution", a term to which our friend Carlos Garcia Canal objects, correctly pointing out that, penguins are the victims rather than the cause of pollution in the southern ocean!

1994; L. Littenberg and G. Valencia, *Ann. Rev. Nucl. Part. Sci.* **43**, 729 (1993) [hep-ph/9303225].

4. E. de Rafael, hep-ph/9502254. Lectures given at Theoretical Advanced Study Institute in Elementary Particle Physics (TASI 94). Boulder, CO, 29 May - 24 Jun 1994.
5. M. Kobayashi and T. Maskawa, *Prog. Theor. Phys.* **49**, 652 (1973).
6. L. Wolfenstein, *Phys. Rev. Lett.* **51**, 1945 (1983).
7. C. Jarlskog, *Phys. Rev.* **D35**, 1685 (1987).
8. L. S. Littenberg, *Phys. Rev.* **D39**, 3322 (1989).
9. S. Mele, hep-ph/9808411.To be published in the proceedings of Workshop on CP Violation, Adelaide, Australia, 3-8 Jul 1998.
10. G. Buchalla, A. J. Buras and M. E. Lautenbacher, *Rev. Mod. Phys.* **68**, 1125 (1996) [hep-ph/9512380].
11. C. Caso et al., *Eur. Phys. J.* **C3** (1998) 1.
12. J. H. Christenson, J. W. Cronin, V. L. Fitch and R. Turlay, *Phys. Rev. Lett.* **13** (1964) 138.
13. S. Bertolini, M. Fabbrichesi and J. O. Eeg, hep-ph/9802405.
14. J. Gasser and H. Leutwyler, *Nucl. Phys.* **B250**, 465 (1985).
15. J. A. Cronin, *Phys. Rev.* **161**, 1483 (1967).
16. J. Kambor, J. Missimer and D. Wyler, *Nucl. Phys.* **B346**, 17 (1990).
17. A. I. Vainshtein, V. I. Zakharov and M. A. Shifman, *JETP Lett.* **22**, 55 (1975).
18. F. J. Gilman and M. B. Wise, *Phys. Rev.* **D20**, 2392 (1979).
19. R. S. Chivukula, J. M. Flynn and H. Georgi, *Phys. Lett.* **B171**, 453 (1986).
20. S. Bosch, A. J. Buras, M. Gorbahn, S. Jager, M. Jamin, M. E. Lautenbacher and L. Silvestrini, hep-ph/9904408.
21. J. F. Donoghue, E. Golowich, B. R. Holstein and J. Trampetic, *Phys. Lett.* **B179**, 361 (1986); A. J. Buras and J. M. Gerard, *Phys. Lett.* **B192**, 156 (1987); S. Gardner and G. Valencia, *Phys. Lett.* **B466**, 355 (1999). hep-ph/9909202.
22. A. Alavi-Harati et al. [KTeV Collaboration], *Phys. Rev. Lett.* **83**, 22 (1999) [hep-ex/9905060]
23. V. Fanti et al. [NA48 Collaboration], *Phys. Lett.* **B465** (1999) 335 [hep-ex/9909022].
24. C. G. White et al., *Nucl. Phys. Proc. Suppl.* **71**, 451 (1999).
25. J. F. Donoghue and S. Pakvasa, *Phys. Rev. Lett.* **55**, 162 (1985); J. F. Donoghue, X. He and S. Pakvasa, *Phys. Rev.* **D34**, 833 (1986); X. He and G. Valencia, *Phys. Rev.* **D52**, 5257 (1995) [hep-ph/9508411].
26. L. D. Roper, R. M. Wright and B.Feld, *Phys. Rev.* **138**, 190 (1965).
27. A. B. Carter and A. I. Sanda, *Phys. Rev.* **D23**, 1567 (1981).
28. Y. Nir and H. R. Quinn, *Ann. Rev. Nucl. Part. Sci.* **42**, 211 (1992).
29. M. Gronau and D. London, *Phys. Rev. Lett.* **65**, 3381 (1990); N. G. Deshpande and X. He, *Phys. Rev. Lett.* **74**, 26 (1995) [hep-ph/9408404].

TOP AND HIGGS AT THE TEVATRON:
measurements, searches, prospects

J. KONIGSBERG

University of Florida, Department of Physics, Gainesville, FL 32611, USA
E-mail: konigsberg@phys.ufl.edu

Abstract. In this paper we summarize the status of Top Quark Physics and of searches for the Standard Model Higgs at the Tevatron. Results from both the CDF and D0 experiments are discussed and the prospects for the upcoming Run 2, in the year 2001, are outlined. Much work has been performed on these topics and due to the nature of these proceedings only a brief explanation can be offered here. For more details the reader should turn to the excellent sources listed in the reference section.

I TOP QUARK PHYSICS

A Introduction

The announcement of the discovery of the top quark was made in March 1995 [1,2]. Since then the two Tevatron experiments, CDF and D0, have analyzed all their data taken during the 1992-1996 "Run 1" period and have been upgrading their detectors to match the high-luminosity running conditions that the Tevatron collider will provide during "Run 2". The results presented here come from datasets corresponding to a total integrated luminosity of about 110 pb^{-1} for CDF and about 125 pb^{-1} for D0.

At the Tevatron top quarks are produced mainly in pairs, via $q\bar{q}$ anihilation [3], with a cross section of about 5 pb. Therefore only about 600 $t\bar{t}$ pairs, per experiment, were created during the entire Run 1. When geometrical acceptances and detection efficiencies are folded in, the datasets available for studies are rather small. Nonetheless, as it will be shown here, significant information has been extracted about the identity and behaviour of this newest quark.

During Run 2 the luminosity is expected to increase by about a factor of twenty, to 2 fb^{-1}, and the Tevatron center-of-mass energy is expected to reach 2 TeV. We expect to collect datasets that are about thirty times larger than those in Run 1 which will help improve significantly our knowledge about the top quark [4].

In the Standard Model (SM) the top quark decays effectively 100% of the time to a W boson and a b quark. The top decay width is approx. 1.5 Gev (at $M_{top} \sim$ 175 GeV) and the corresponding lifetime is 4×10^{-25} sec. In this short time there is no hadronization and the top quark decays as a free quark. No hadronic states with top can be formed and there is no toponium spectroscopy to be studied. The signatures for $t\bar{t}$ pair production depend exclusively on the decays of the two W bosons from each of the $t \to W, b$ decays. The channels are classified according to the number of leptons appearing in the final state. Excluding final states with tau leptons, which are more difficult to detect, the "dilepton" channel ($t\bar{t} \to \ell\bar{\nu}b\ell\nu\bar{b}$) has a branching fraction of 5%, the "lepton+jets" channel ($t\bar{t} \to q\bar{q}b\ell\nu\bar{b}$) of 30% and the "all-jets" channel ($t\bar{t} \to q\bar{q}bq'\bar{q}'\bar{b}$) of 40%. Significant excess of events has been observed in all these channels and the results reported below come from the analyses of these channels.

B Measurements

1 Event selection

The large mass of the top quark allows for an event selection biased towards high transverse energy, or momentum, objects. This helps reduce backgrounds significantly [5–8]. For leptons, jets, and missing transverse energy ($\not{E}_t$) the requirement is $E_T(P_T) > 20$ GeV approx. These objects are also required to be isolated from other objects in the event and in the central region of the detector with pseudorapidity $|\eta| < 2$ approx. Also due to the large M_{top}, global event kinematical variables are useful in separating signal from backgrounds. The aplanarity and sphericity of the event and the sum of transverse energies of all objects have been used for this purpose. Neural nets are also useful in separating further the top signal from backgrounds.

A key feature of $t\bar{t}$ events is the production of two b quarks and the ability to tag these objects has proven essential in top quark physics. The tagging of long-lived heavy flavor quarks can be performed by means of secondary vertex detectors which can be $\sim 50\%$ efficient in finding at least one tag in a $t\bar{t}$ event (CDF) and by finding the softer leptons in semi-leptonic b decays with $\sim 20\%$ efficiency (D0 and CDF). The secondary vertex method works best because of the relatively low fake rate.

In the dilepton channel (including taus), after the selection cuts CDF (D0) finds 13 (9) events, with a signal purity of $\sim 66\%$ ($\sim 90\%$). The dominant backgrounds come from Drell-Yan, $Z \to \tau\tau$, WW and fake lepton processes [9,10]. In the lepton+jets channels CDF (D0) finds 74 (30) events with a purity of $\sim 57\%$ ($\sim 63\%$). The dominant backgrounds are $W + jets$ production and QCD multi-jet processes [11,12]. The all-jets channel is dominated by QCD multi-jet productions and b-tagging is essential, CDF (D0) finds 344 (41) events with a purity of about $\sim 24\%$ ($\sim 39\%$) [13]. It is with these relatively small samples of events that the measurements described below are performed. In the lepton+jets channel the experiments have been able to reconstruct the invariant mass of the hadronic W decay and the Jacobian peak for the leptonic W decay. Additionally the kinematical distributions for all examined variables, including the invariant mass of the $t\bar{t}$ system (which could reveal a resonant production) match very well those expected from SM $t\bar{t}$ production.

2 Production cross section

The $t\bar{t}$ production cross section has been measured separately in the channels described above and all channels yield consistent results. A final result is obtained by combining the individual channel cross sections. CDF measures $6.5^{+1.7}_{-1.4}$ pb^{-1} and D0 5.9 ± 1.7 pb^{-1} [14,7]. The uncertainty is of the order of 25%. The theoretical predictions range between 4.7 and 5.5 pb^{-1} depending on assumptions about gluon resumation. In Run 2 the experimental uncertainty is expected to go down to less than 10% and will be dominated by systematics associated with $t\bar{t}$ Monte Carlo event simulations and Luminosity measurements.

From the ratios of observed number of events in the different channels CDF has extracted a measurement of the branching fractions for top to decay leptonically: $BR(t \to eX, or\, t \to \mu X) = 0.094 \pm 0.024$

3 Top Quark mass

In the lepton+jets channel, events with four or more jets are selected and all objects are measured precisely [15,5]. The jet energies are corrected to what, on average, parton energies would have been. A 2C kinematic fit is performed for each combinatorial assignment of objects to the $t\bar{t}$ system. The b-tagging helps reduced the combinatorics. For each event, the solution with the best χ^2 is chosen and the reconstructed mass distribution is compared to a Monte-Carlo mix of $t\bar{t}$ events at a given top mass and background. The top mass that results in the best fit is chosen as the central value and the statistical uncertainty is taken as the range in which the χ^2 changes by ± 0.5 (see Fig. I B 3). The systematics are dominated by the understanding of the jet energy scale in the detectors and of initial and final state radiation. The CDF and D0 results are: $175.9 \pm 4.8(stat.) \pm 5.3(syst.)$ and $173.3 \pm 5.6(stat.) \pm 5.5(syst.)$, respectively.

In the all-jets channel the kinematics are also over-constrained and a 3C-fit is performed using the same technique as in the lepton+jets channel [13]. CDF measures $186.0 \pm 10.0(stat.) \pm 5.7(syst)$. For the dilepton events, the kinematical fit is under-constrained and the event reconstruction is performed hypothesizing a given M_{top} and performing a scan over M_{top}. For each value of M_{top} the likelihood that the reconstructed kinematics compare well with those expected from $t\bar{t}$ production is measured and the reconstructed M_{top}^{best} for the maximum likelihood is chosen. The distribution of M_{top}^{best} is then compared with Monte-Carlo to find

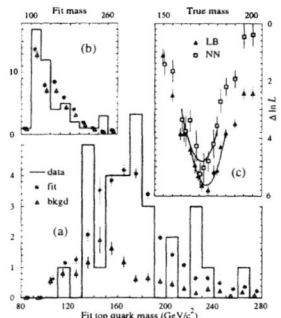

FIGURE 1. D0's top mass in the lepton+jets channel.

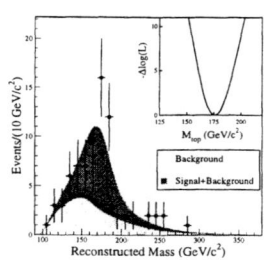

FIGURE 2. CDF's top mass in the lepton+jets channel.

the best value of M_{top} [15,10]. D0 gets $168.4 \pm 12.3(stat.) \pm 3.6(syst.)$ and CDF $167.4 \pm 10.3(stat.) \pm 4.8(syst.)$. For the all-jets and the dilepton channels the systematics are also dominated by the jet energy scale and knowledge of initial and final state radiation.

The lepton+jets channel yields the measurement with the smallest statistical uncertainty while the systematic uncertainties are comparable in all channels. The CDF and D0 combined result is: $M_{top} = 174.3 \pm 3.2(stat.) \pm 4.0(syst.)$ GeV/c^2 or 174.3 ± 5.1, combining in quadrature the systematic and statistical uncertainties. In Run 2 the systematics, dominated by jet selection and jet energy measurements, will be controlled better through higher statistics control samples and better understood Monte Carlo simulations. It is estimated that in Run 2 M_{top} can be measured with a total uncertainty of about 2 GeV [4].

4   Measurement of $|V_{tb}|$

CDF uses the taggable fraction of top decays, $R_b \equiv \frac{BR(t \to Wb)}{BR(t \to Wq)}$ as a way to measure $|V_{tb}|$. R_b is obtained from measuring the ratio of events with 0,1, or 2 b-taggs. In a 3-generation Standard Model $R_b = \frac{|V_{tb}|^2}{|V_{tb}|^2+|V_{ts}|^2+|V_{td}|^2}$. The result: $R_b = 0.99 \pm 0.29 (> 0.58 \ at \ 95\% \ c.l.)$ and $|V_{tb}| = 0.99 \pm 0.15 (> 0.76 \ at \ 95\% \ c.l.)$ [16].

5   Measurement of the W helicity in top decays

The helicity of the W boson in top decays carries information on the $W - t - b$ coupling. In the Standard Model a large fraction of top decays are expected to produce a W with helicity=0. CDF uses the lepton P_T distributions in the lab frame to measure the W-helicity distribution in the top rest frame. The P_T of left-handed W bosons is expected to be softer than that of zero-helicity $W's$. The fit of these two distributions to the data results in $F_o = 0.91 \pm 0.37(stat.) \pm 0.13(syst.)$, consistent with the theoretical expectation of $F_o = (70.6 \pm 1.6)\%$. In Run 2, with more statistics a $\sim 6\%$ or better measurement will be possible.

C Searches

1 Single top production

The production of single top at the Tevatron is expected to occur via Wg-fusion and via W^* with cross sections of about 1.7 and 0.7 pb, respectively. these processes suffer from copious backgrounds from $W+2-jet$ QCD production. CDF has used the secondary vertex tagger to reduce the backgrounds and has performed fits to kinematical distributions that reconstruct the single top mass and include the rapidity distribution of jets in order to set limits to the single top production cross section. The obtained limits are: $\sigma(Wg) < 15.4$ pb at 95% c.l. and $\sigma(W^*) < 15.8$ pb at 95% c.l.. These limits are still a factor of 10-20 above the theoretical expectations and will improve considerably in Run 2, perhaps to the point of actually observing this process.

2 Rare top decays

CDF has looked for the rare FCNC decays $t \to \gamma q$ and $t \to Zq$, with $q = u, c$. Only one event is observed in each channel, consistent with the expected backgrounds. The corresponding 95% c.l. limits on the branching fractions are [17]: $BR(t \to \gamma q) < 3.2\%$ and $BR(t \to Zq) < 33\%$. These sensitivities should improve about one order of magnitude in Run 2.

II HIGGS SEARCHES

A Introduction

At the Tevatron energies the processes with the highest cross section for Standard Model Higgs production are [18]: $gg \to H$ and $q\bar{q} \to WH, ZH$. The first process, which has the largest cross section, is very difficult to detect because of copious backgrounds, so searches have focused on the later processes (associated production") with high P_T leptons to help reduce the background. For $M_H \sim 120$ GeV the cross sections are about 0.1 and 0.2 pb for WH and ZH, respectively. These are about 25 smaller than the $t\bar{t}$ cross section!

For $M_H \sim 100$ GeV the decay $H \to b\bar{b}$ dominates ($\sim 90\%$); for $M_H \geq \sim 130$ GeV $H \to WW$ starts to contribute significantly and dominates after $M_H \geq \sim 160$ GeV. Excellent b-tagging is therefore crucial for Higgs searches.

B Results

Both CDF and D0 have performed searches in the $WH \to \ell\nu b\bar{b}$, $ZH \to \ell^+\ell^- b\bar{b}$ and $ZH \to \nu\bar{\nu} b\bar{b}$ channels [18]. CDF uses b-tagging with secondary vertices and soft leptons and D0 uses the soft lepton taggs with a more relaxed kinematical selection. Both experiments find the data consistent with well known backgrounds and set limits for the production cross section times the branching fraction ($\sigma \cdot BR$) in the range $\sim 90 < M_H < \sim 140$ GeV. The enhanced b-tagging capability of CDF allows for better limits and also enables the search in the $VH \to q\bar{q}b\bar{b}$ channels [18]. The limits on $\sigma \cdot BR$ are still more than one order of magnitude higher than the Standard Model prediction but show that these searches are possible and limited only by luminosity, provided the experiments have excellent b-tagging and b-jet energy resolution. Both CDF and D0 have improved their detectors with precisely these capabilities in mind. Figure II B show the CDF limits and the expected sensitivity for Higgs searches as a function of integrated luminosity in Run 2 and beyond when CDF and D0 are combined' For 10 fb^{-1} the Higgs can be observed at the 3σ level up to $M_H \sim 130$ GeV and excluded up to 190 GeV at the 95% c.l. This is definitely a goal the Tevatron should aim for.

The lightest Higgs In supersymmetric theories, where five Higgs bosons are expected ($h^o, A^o, H^0, H^{\pm}$), the mass of the lightest one, h^o, is expected to be $< \sim 130$ GeV, and therefore accessible in Run 2. In some regions of the SUSY parameter space (large $\tan\beta$) the production cross-sections can be very large. CDF has already set preliminary mass limits, of about 120 GeV, for h^o and A^o, for $\tan\beta > \sim 50$ in the $p\bar{p} \to b\bar{b}\phi^o \to b\bar{b}, b\bar{b}$ channel ($\phi^c = h^o, A^o$). Both CDF and D0 will likely be able to discover the light neutral SUSY Higgs in Run 2 if indeed its mass is where predicted. The CDF and D0 charged Higgs searches in Run 1 have also yielded mass limits (as a function of $\tan\beta$); it can be excluded below about 130 GeV for $\tan\beta > \sim 100$ and $\tan\beta < \sim 0.2$. A much larger fraction of the parameter space is expected to be covered in Run 2.

III CONCLUSIONS

The Tevatron at Fermilab, for the next several years, remains the frontier of high energy physics. In Run 2, and beyond, with the upgraded CDF and D0 detectors and the improved accelerator complex, it has the capability of producing very important and exciting physics. The top quark will be studied with much more precision and a window of opportunity exists, before the LHC takes over the energy frontier, where the the Higgs boson(s) could be found if enough luminosity is gathered.

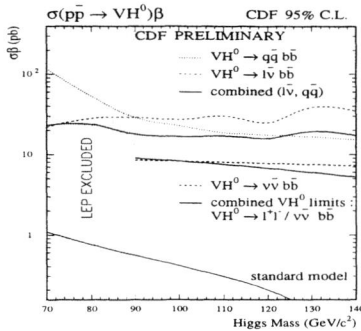

FIGURE 3. CDF SM Higgs limits.

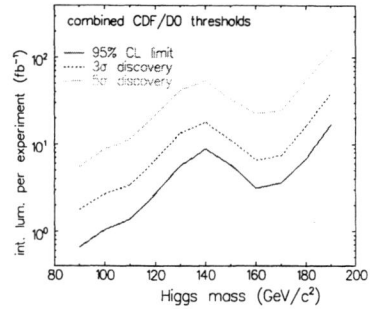

FIGURE 4. Tevatron sensitivity to SM Higgs.

ACKNOWLEDGMENTS

The author thanks the workshop organizers for the invitation to participate in such well organized and successfull workshop. The author also thanks his colleagues in the CDF and DO collaborations.

REFERENCES

1. F. Abe et al., Phys. Rev. Lett. **74**, 2626 (1995)
2. A. Abachi et al., Phys. Rev. Lett. **74**, 2632 (1995)
3. Laenen, E.J. et al., 1994, Phys. Lett. **B321**, 254; Berger, E.L. et al., 1996, Phys. Rev. D **54**, 3085; Catani, S. et al., 1996, Phys. Lett. **B378**, 329.
4. "Thinkshop" at FNAL, Top Quark Physics for Run 2: http://lutece.fnal.gov/thinkshop/.
5. F. Abe et al., Phys. Rev. Lett. **80**, 1197 (1998)
6. F. Abe et al., Phys. Rev. Lett. **79**, 3585 (1997)
7. S. Abachi et al., Phys. Rev. Lett. **79**, 1203 (1997)
8. B. Abbott et al., submitted to Phys. Rev. D, Fermilab-Pub-98/130-E (hep-ex/9808034).
9. F. Abe et al., Phys. Rev. Lett. **80**, 2779 (1998); Phys. Rev. Lett. 79, 3585 (1997)
10. B. Abbott et al., Phys. Rev. Lett. **80**, 2063 (1998)
11. F. Abe et al., Phys. Rev. Lett. **80**, 2767 (1998);
12. S. Abachi et al., Phys. Rev. Lett. **79**, 1197 (1997); B. Abbott et al., Phys. Rev. D **58**, 52001 (1998)
13. F. Abe et al., Phys. Rev. Lett. **79**, 1992 (1997);
14. F. Abe et al. Phys. Rev. Lett. 80, 2773 (1998).
15. F. Abe et al., Phys. Rev. Lett. 82, 271 (1999)
16. F. Tartarelli, Proceedings of "International Europhysics Conference on High Energy Physics", Jerusalem, Israel, August 19-26, 1997.
17. F. Abe et al., Phys. Rev. Lett. 80, 2525 (1998).
18. Higgs and SUSY workshop: http://fnth37.fnal.gov/susy.html

Trilinear Gauge Boson Couplings and Vector Boson Pair Production

A. Sánchez-Hernández

(for the DØ Collaboration)
Depto. de Física, CINVESTAV
Apdo. postal 14-740, 07000 México, D.F.

Abstract. The trilinear couplings appear as the three gauge boson vertices and can be measured by studying the gauge boson pair production processes. The measurement of the coupling parameters is one of the few remaining crucial tests of the Standard Model. DØ has studied $W\gamma$, $Z\gamma$, WW, and WZ production and found no evidence of anomalous production. In this paper we review all the current results from DØ data.

INTRODUCTION

Over the past two decades, experiments have beautifully confirmed the predictions of the Standard Model (SM). However two crucial sectors remain poorly tested: The symmetry breaking sector, and the self-interactions of gauge bosons. The self-couplings of the gauge bosons are completely fixed by the $SU(2) \times U(1)$ symmetry of the SM. The trilinear couplings appear as the three gauge boson vertices and can be measured by studying the gauge boson pair production processes. Deviations of the couplings parameters values from the SM ones signal new physics.

The WWV ($V = \gamma$ or Z) vertices are described by a generalized effective Lagrangian [1] with two overall couplings parameters ($g_{WW\gamma} = -e$ and $g_{WWZ} = -e \cdot \cot\theta_W$) and six dimensionless coupling parameters g_1^V, κ_V and λ_V, where $V = \gamma$ or Z, after imposing C, P and CP invariance. Furthermore g_1^γ is restricted to unity by electromagnetic gauge invariance. The general Lagrangian is reduced to the SM Lagrangian by setting $g_1^\gamma = g_1^Z = \kappa_V = 1$ ($\Delta\kappa_V \equiv \kappa_V - 1 = 0$) and $\lambda_V = 0$. The amplitudes for gauge boson pair production with the non-SM coupling parameters grows with energy ($\hat{s}$). In order to avoid unitarity violation, the coupling parameters are modified by form factors with a cutoff scale Λ; $\lambda_V(\hat{s}) = \frac{\lambda_V}{(1+\hat{s}/\Lambda^2)^2}$ and $\Delta\kappa_V(\hat{s}) = \frac{\Delta\kappa_V}{(1+\hat{s}/\Lambda^2)^2}$. Λ is physically interpreted as the mass scale where the new phenomenon which is responsible for the anomalous couplings would be directly observed.

In an analogous manner, the $Z\gamma V$ ($V = \gamma$ or Z) vertices are described by a general vertex function [2] with eight dimensionless coupling parameters h_i^V ($i =$

$1, 4; V = \gamma$ or Z). In the SM, all h_i^V's are zero. The form factors for these vertices, which are required to constrain the cross sections amplitudes within the unitarity limit, are $h_i^V(\hat{s}) = \frac{h_{i0}^V}{(1+\hat{s}/\Lambda^2)^n}$, where $n = 3$ for $i = 1, 3$ and $n = 4$ for $i = 2, 4$.

The DØ collaboration has performed several searches for anomalous trilinear gauge boson couplings. In this paper we review all measurements of trilinear gauge boson couplings based on the direct observation of diboson final states produced in $p\bar{p}$ collisions at $\sqrt{s} = 1.8$ TeV during the 1992-1996 data taking period using the DØ detector at Fermilab. Limits on the anomalous coupling parameters were obtained at a 95% CL from the following processes: $p\bar{p} \to Z\gamma + X \to l\bar{l}\gamma + X$ ($l = e, \mu, \nu$), $p\bar{p} \to W\gamma + X \to l\nu\gamma + X$ ($l = e, \mu$), $p\bar{p} \to WW/WZ + X \to l\nu jj + X$ ($l = e, \mu$), $p\bar{p} \to WW + X \to l\nu l\nu + X$ ($l = e, \mu$), and $p\bar{p} \to WZ + X \to l\nu ll + X$ ($l = e, \mu$). Combined limits with LEP experiments have also been obtained.

$W\gamma$ ANALYSIS

The DØ collaboration has studied $W\gamma$ production from two decay modes of the W boson: $W \to e\nu$ and $W \to \mu\nu$, reported in Ref. [3]. In each case the photon was required to have a minimum transverse momentum of 10 GeV/c and to be spatially separated from the charged lepton by at least 0.7 units of $\mathcal{R}_{l\gamma}$, $\mathcal{R} \equiv \sqrt{(\Delta\eta)^2 + (\Delta\phi)^2}$ ($\eta = -\log\tan(\theta/2)$). We have observed $84.4^{+12.3}_{-11.3} \pm 8.7$ signal events from $\sim 89\text{pb}^{-1}$ of data taken during 1992-1993 and 1993-1995 Tevatron collider runs. The asymmetrical error is the 1σ uncertainty due to Poisson statistics, and the second error is due to the uncertainties in the background estimates.

From this observation we calculated the $W\gamma$ cross section times branching ratio of W bosons to leptons, for our photon requirements, to be: $\sigma(p\bar{p} \to W\gamma + X) \times \text{BR}(W \to l\nu) = 11.3^{+1.7}_{-1.5} \pm 1.6(syst)$ pb. This is in agreement with the SM prediction of 12.5 ± 1.0 pb. A combined likelihood analysis of the p_T^γ spectra from the individual $W(e\nu)$ and $W(\mu\nu)$ analyses allowed us to set 95% CL limits on the anomalous $WW\gamma$ coupling parameters of $-0.98 < \Delta\kappa < 0.94$ and $-0.31 < \lambda < 0.29$. These are the 95% CL limits when only one of the couplings is allowed to vary at a time.

$Z\gamma$ ANALYSIS

Measurements of $Z\gamma$ production through the $ee\gamma$, $\mu\mu\gamma$, and $\nu\nu\gamma$ decay channels with the DØ detector were previously reported in Ref. [4]. Here we briefly describe those analyses. The measurements of $ee\gamma$, and $\mu\mu\gamma$ channels are based on ~ 100 pb^{-1} of data collected in 1993-1995 Tevatron collider run, while the measurement of the $\nu\nu\gamma$ production is based on 13.5 pb^{-1} of data collected in the 1992-1993 run. Event selection for the $ee\gamma(\mu\mu\gamma)$ analysis required two electrons (muons) with high E_T and a photon with $E_T > 10$ GeV. We additionally required that the photon was separated from either electron (muon) by at least 0.7 units in $\eta - \phi$ space.

These channels are dominated by $Z + j$ and multijet production with jets faking the photon or electrons (muons). This background was derived from data. The observed yield events agree well with the SM predictions and background estimates.

For the $\nu\nu\gamma$ analysis, we required a much tighter cut on the photon energy: $E_T > 40$ GeV which was forced by a dominant background from $W \to e\nu$ decays with the electron being misidentified for a photon due to inefficiency of the central tracker. Additional cuts were applied to the shape of the photon EM shower in transverse and longitudinal directions to ensure that it was consistent with a photon originating from a real vertex. The residual background, which had roughly equal contributions from $W \to e\nu$ decays and *bremsstrahlung* photons from cosmic and beam halo muons, was derived from data. The observed yield is consistent with the SM prediction and brackground estimates. Combined limits on anomalous couplings were set at 95% CL by the E_T^γ fit: $|h_{10,30}^Z| < 0.36$, $|h_{10,30}^\gamma| < 0.37$, and $|h_{20,40}^V| < 0.05$ using a cutoff scale of $\Lambda = 750$ GeV. This represents the most stringent limits available today.

WW ANALYSIS

DØ has searched for W pair production in the dilepton decay modes: $e\nu e\nu$, $e\nu\mu\nu$, and $\mu\nu\mu\nu$ [5]. The analyses require two isolated leptons plus missing transverse energy. In order to remove the background coming from top quark pair production, DØ require the vector sum of the E_T from hadrons to be less than 40 GeV. This cut reduces the this background by a factor of more than four, while is 95% efficient for SM W^+W^- events. A cut in the transverse missing energy in introduced to avoid backgrounds from $Z \to \tau^+\tau^-$ and Drell–Yan processes $\gamma/Z \to e^+e^-, \mu^+\mu^-$. Events are also rejected if the transverse missing energy vector points along or opposite the direction of a lepton. Also, events with a dilepton mass greater than 75 GeV/c² or less than 110 GeV/c² are rejected. 5 events pass the above selection criteria while the estimated background is 3.1 ± 0.4 events. This leds to an upper limit on the cross section for $p\bar{p} \to W^+W^-$ of 37.1 pb at the 95% CL. Using a binned likelihood to the measured p_T spectra of the two leptons the limits at 95% CL and a cutoff scale of 1.5 TeV are: $-0.62 < \Delta\kappa < 0.77$ and $-0.52 < \lambda < 0.56$ varing only one coupling at a time.

WZ ANALYSIS

WZ production have also been studied using the $e\nu ee$ and $\mu\nu ee$ decay modes at DØ [6]. In that analysis we searched for unusual signature of three charged high–E_T leptons and the missing transverse energy due to the high–E_T neutrino. We use about 92 pb⁻¹ of data. One event was found which passed the selection criteria (an $e\nu ee$ candidate). The SM prediction for both channels combined was found to be 0.25 ± 0.02 events, with a estimated background of 0.50 ± 0.17 events. Based on these the 95% CL upper limit on the cross section was 47 pb, consistent with the SM.

Since no excess of events, which would be an indication of non-SM WWZ couplings, was seen, DØ set limits on anomalous couplings. The analysis is most sensitive to the λ_Z and g_1^Z parameters because the helicity amplitudes have larger factors multiplying λ_Z and g_1^Z compared with $\Delta\kappa_Z$. Due to WZ production is sensitive only to the WWZ couplings, the results are independent on any assumption on $WW\gamma$ couplings. Using a cutoff scale of $\Lambda = 1.0$ TeV, the one-dimensional 95% CL limits were: $|\lambda_Z| < 1.42$ and $|\Delta g_1^Z| < 1.63$.

WW/WZ ANALYSIS

The WW/WZ candidates were selected by searching for events containing an isolated electron (muon) with high E_T, large missing transverse energy $\not{E}_T$ and two high E_T jets. The transverse mass of the electron and neutrino system was required to be consistent with a W boson decay ($M_T > 40$ GeV/c^2). The invariant mass of the two jet system was required to be $50 < m_{jj} < 110$ GeV/c^2, as expected for a W or Z decay. Additionally for the electron channel, it was also required that the p_T of the two gauge bosons was balanced ($|p_T(jj) - p_T(e\nu)| < 40$ GeV/c) as expected for WW/WZ production. The number of events that satisfied all of the requirements were 483 for the electron channel and 224 for the muon one. There were two major sources of background for these processes, QCD multijet events with a jet misidentified as an electron or muon and W boson production with two associated jets. Total number of background events was estimated to be 463 ± 40 and 224 ± 55 respectively. The SM predicts 21 ± 3 events and 5 ± 1 respectively for the above requirements and thus no significant deviation from the SM prediction was seen.

A maximum likelihood fit to the p_T^W spectrum, calculated from the E_T of electron (muon) and missing E_T, was performed to set limits on the anomalous couplings. Using a cutoff scale of $\Lambda = 2.0$ TeV, the 95% confidence level (CL) limits were: $-0.43 < \Delta\kappa < 0.59$ (with $\lambda = 0$) and $-0.33 < \lambda < 0.36$ (with $\Delta\kappa = 0$) for the electron channel, and $-0.60 < \Delta\kappa < 0.74$ (with $\lambda = 0$) and $-0.43 < \lambda < 0.44$ (with $\Delta\kappa = 0$) for the muon channel. These results were reported in Ref. [6,7].

COMBINED ANALYSIS

The DØ experiment has performed combined limits on the parameters of the $WW\gamma$ and WWZ couplings using a simultaneous fit to the p_T distribution in the $W\gamma$ data, the lepton $p_T^{(l\nu)}$ distribution in the $WW \to l\nu l'\nu'$, and $WW/WZ \to l\nu jj$ data, and the number of observed event in $WZ \to l\nu ll$. This exercise is reported in Ref. [6]. There, the limits on the $WW\gamma$ and WWZ coupling parameters are extracted from that fit. Correlations between the uncertainties due the integrated luminosity, the selection efficiencies and the background estimates are properly taken into account. In these exercise $\Delta\kappa$, λ, and g_1^Z parameters are used, as well as the LEP ones: $\alpha_{B\phi}$, $\alpha_{W\phi}$, and α_W.

The results obtained on $WW\gamma$ and WWZ coupling parameters have comparable sensitivity to those from the LEP experiments. The limits on the $\alpha_{B\phi}$, α_W parameters obtained by DØ are also the most stringent constraints. However, the LEP measurements are most sensitive to $\alpha_{W\phi}$. The LEP limits are complementary to the Tevatron ones because they are obtained from a different process ($e^+e^- \rightarrow W^+W^-$) exploiting the behavior of the angular distributions of the decay products. Table 1 shows limits on λ, $\Delta\kappa$, and where applicable on Δg_1^Z, $\alpha_{B\phi}$, $\alpha_{W\phi}$, and α_W, for $\Lambda = 1.5$ and 2.0 TeV.

Couplings	$\Lambda = 1.5$ TeV	$\Lambda = 2.0$ TeV
$\lambda_\gamma = \lambda_Z (\Delta\kappa_\gamma = \Delta\kappa_Z = 0)$	-0.20, 0.20	-0.18, 0.19
$\Delta\kappa_\gamma = \Delta\kappa_Z = 0 (\lambda_\gamma = \lambda_Z = 0)$	-0.27, 0.42	-0.25, 0.39
λ_γ (HISZ [8])($\Delta\kappa_\gamma = 0$)	-0.20, 0.20	-0.18, 0.19
$\Delta\kappa_\gamma$ (HISZ) ($\lambda_\gamma = 0$)	-0.31, 0.56	-0.29, 0.53
λ_Z (SM $WW\gamma$)($\Delta\kappa_Z = \Delta g_1^Z = 0$)	-0.26, 0.29	-0.24, 0.27
$\Delta\kappa_Z$ (SM $WW\gamma$)($\lambda_Z = \Delta g_1^Z = 0$)	-0.37, 0.55	-0.34, 0.51
Δg_1^Z (SM $WW\gamma$)($\lambda_Z = \Delta\kappa_Z = 0$)	-0.39, 0.62	-0.37, 0.57
λ_γ (SM WWZ) ($\Delta\kappa_\gamma = 0$)	-0.27, 0.25	-0.25, 0.24
$\Delta\kappa_\gamma$ (SM WWZ) ($\lambda_\gamma = 0$)	-0.57, 0.74	-0.54, 0.69
$\alpha_{B\phi}$ ($\alpha_{W\phi} = \alpha_W = 0$)	-0.73, 0.59	-0.67, 0.56
$\alpha_{W\phi}$ ($\alpha_{B\phi} = \alpha_W = 0$)	-0.19, 0.38	-0.18, 0.36
α_W ($\alpha_{B\phi} = \alpha_{W\phi}$)	-0.20, 0.20	-0.18, 0.19
Δg_1^Z ($\alpha_{B\phi} = \alpha_W = 0$)	-0.25, 0.49	-0.23, 0.47

TABLE 1. One-dimensional limits at 95% CL. from a simultaneos fit to the DØ $WW\gamma$, $WW \rightarrow$ dilepton, $WW/WZ \rightarrow l\nu jj$, and $WZ \rightarrow$ trilepton data samples.

REFERENCES

1. K. Hagiwara, R.D. Peccei, D. Zeppenfeld and K. Hikasa, *Nucl. Phys.* **B 282**, 253 (1987).
2. U. Baur and E. L. Berger, *Phys. Rev.* **D 41**, 1476 (1990).
3. S. Abachi, *et.al.*, (DØ Collaboration), *Phys. Rev. Lett.* **75**, 1034 (1995); *Phys. Rev. Lett.* **78**, 3634 (1997).
4. S. Abachi, *et.al.*, (DØ Collaboration), *Phys. Rev. Lett.* **75**, 1028 (1995); *Phys. Rev. Lett.* **78**, 3640 (1997); B. Abbott, *et.al.*, (DØ Collaboration), *Phys. Rev.* **D 57**, 3817 (1998).
5. S. Abachi, *et.al.*, (DØ Collaboration), *Phys. Rev. Lett.* **75**, 1023 (1995); B. Abbott, *et.al.*, (DØ Collaboration), *Phys. Rev.* **D 58**, 051101 (1998).
6. B. Abbott, *et.al.*, (DØ Collaboration), *Phys. Rev.* **D 60**, 072002 (1999).
7. S. Abachi, *et.al.*, (DØ Collaboration), *Phys. Rev. Lett.* **77**, 3303 (1996); B. Abbott, *et.al.*, (DØ Collaboration), *Phys. Rev. Lett.* **79**, 1441 (1997).
8. K. Hagiwara, S. Ishihara, R. Szalapski, and D. Zeppenfeld, Phys. Rev. **D 48**, 2182, (1993). They parametrize WWZ couplings in terms of the $WW\gamma$ couplings: $\Delta\kappa_Z = \Delta\kappa_\gamma(1 - tan^2\theta_W)/2$, $\Delta g_1^Z = \Delta\kappa_\gamma/(2cos^2\theta_W)$ and $\lambda_Z = \lambda_\gamma$.

A short course in effective Lagrangians

José Wudka

Physics Department, UC Riverside
Riverside CA 92521-0413, USA

Abstract. These lectures provide an introduction to effective theories concentrating on the basic ideas and providing some simple applications.

I INTRODUCTION

When studying a physical system it is often the case that there is not enough information to provide a fundamental description of some of its properties. In such cases one must parameterize the corresponding effects by introducing new interactions with coefficients to be determined phenomenologically. Experimental limits or measurement of these parameters then (hopefully) provides the information needed to provide a more satisfactory description.

A standard procedure for doing this is to first determine the dynamical degrees of freedom involved and the symmetries obeyed, and then construct the most general Lagrangian, the *effective Lagrangian* for these degrees of freedom which respects the required symmetries. The method is straightforward, quite general and, most importantly, *it works!*

In following this approach one must be wary of several facts. First it is clear that the relevant degrees of freedom can change with scale (e.g. mesons are a good description of low-energy QCD, but at higher energies one should use quarks and gluons); in addition, physics at different scales may respect different symmetries (e.g. mass conservation is violated at sufficiently high energies). It follows that the effective Lagrangian formalism is in general applicable only for a limited range of scales. It is often the case (but no always!) that there is a scale Λ so that the results obtained using an effective Lagrangian are invalid for energies above Λ.

The formalism has two potentially serious drawbacks. First, effective Lagrangian has an infinite number of terms suggesting a lack of predictability. Second, even though the model has an UV cutoff Λ and will not suffer from actual divergences, simple calculations show that is a possible for this type of theories to generate radiative corrections that grow with Λ, becoming increasingly important for higher and higher order graphs. Either of these problems can render this approach useless. It is also necessary to verify that the model is unitary.

I will discuss below how these problems are solved, an provide several applications of the formalism. The aim is to give a flair of the versatility of the approach, not to provide an exhaustive review of all known applications.

II FAMILIAR EXAMPLES

A Euler-Heisenberg effective Lagrangian

This Lagrangian summarizes QED at low energies (below the electron mass) [1]. At these energies only photons appear in real processes and the effective Lagrangian will be then constructed using the photon field A_μ, and will satisfy a $U(1)$ gauge and Lorenz invariances. Thus it can be constructed in terms of the field strength $F_{\mu\nu}$ or the loop variables $\mathcal{A}(\Gamma) = \oint_\Gamma A \cdot dx$. The latter are non-local, so that a local description would involve only F, namely [1]

$$\mathcal{L}_{\text{eff}} = \mathcal{L}_{\text{eff}}(F)$$
$$= aF^2 + bF^4 + c(F\tilde{F})^2 + dF^2(F\tilde{F}) \cdots \qquad (1)$$

One can arbitrarily normalize the fields and so choose $a = -1/4$. The constants b, c and d have units of mass^{-2}.

Note that the term $\propto d$ violates CP. Though we know QED respects C and P, it is possible for other interactions to violate these symmetries, there is nothing in the discussion above that disallows such terms and, in fact, weak effects will generate them. For this system we are in a privileged position to know the underlying physics, and so we can calculate b, c, d, …. The leading effects come form QED which yields $b, c \sim 1/(4\pi m_e)^2$ at 1 loop [1]. The parameters b and c summarize all the leading *virtual* electron effects (see Fig. 1). Forgetting about this underlying structure we could have simply *defined* a scale M and taken $b, c \sim 1/M^2$ (so that $M = 4\pi m_e$), and while this is perfectly viable, M is not relevant phenomenologically speaking as it does not corresponds to a physical scale. In order to extract information about the physics underlying the effective Lagrangian from a measurement of b and c we must be able to at least estimate the relation between these constants and the underlying scales.

In addition we also know that $d \sim \xi/(4\pi v)$ with $v \sim 246$GeV and ξ is a very small constant proportional to the Jarlskog determinant [2]. The effective Lagrangian can hold terms with radically different scales and limits on some constants cannot, in general, translate to others. In this case the terms are characterized by different CP transformation properties, and it is often the case that such global symmetries are useful in differentiating terms in the effective Lagrangian. The point being that a term violating a given global symmetry at scale Λ will generate all terms in the effective Lagrangian with the same symmetry properties through radiative corrections. The caveat in the argument being that the underlying theory might

---

[1]) There are no $F\tilde{F}$ terms since it is a total derivative.

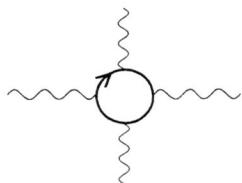

FIGURE 1. Graph generating the leading terms in the Euler-Heisenberg effective Lagrangian

have some additional symmetries not apparent at low energies which might further segregate interactions and so provide different scales for operators with the same properties under all low energy symmetries.

When calculating with the effective Lagrangian the effects produced by the new terms proportional to b, c are suppressed by a factor $\sim (E/4\pi m_e)^2$, where E is the typical energy on the process and $E \ll m_e$. Thus the effects of these terms are tiny, yet they are noticeable because they generate a *new* effect: $\gamma - \gamma$ scattering.

B (Standard) Superconductivity

This is a brief summary of the very nice treatment provided by Polchinski [3]. The system under consideration has the electron field ψ as its only dynamical variable (the phonons are assumed to have been integrated out, generating a series of electron self-interactions), it respects $U(1)$ electromagnetic gauge invariance, as well as Galilean invariance and Fermion number conservation.

Assuming a local description, the first few terms in the effective Lagrangian expansion are (neglecting those containing photons for simplicity)

$$\mathcal{L}_{\text{eff}} = \int_k \psi_{\mathbf{k}}^* [i\partial_t - e_{\mathbf{k}} + \mu] \psi_{\mathbf{k}} + \int \psi_{\mathbf{k}}^* \psi_{\mathbf{l}} \psi_{\mathbf{q}} \psi_{\mathbf{p}}^* \delta(\mathbf{k} - \mathbf{l} - \mathbf{q} + \mathbf{p}) V_{\mathbf{klq}} + \cdots \quad (2)$$

In this equation the relation $e_{\mathbf{k}} = \mu$ determines the Fermi surface, while $V \sim \frac{\text{(electron-photon coupling)}^2}{\text{(phonon mass)}^2}$ summarizes the virtual phonon effects. In order to determine the importance of the various terms we need the dimensions of the field ψ. A vector $\mathbf{k}$ lies on the Fermi Surface (FS) if $e_{\mathbf{k}} = \mu$, if $\mathbf{p}$ is near the FS one can write $\mathbf{p} = \mathbf{k} + \ell \hat{\mathbf{n}}$ (with $e_{\mathbf{k}} = \mu$). Scaling towards the FS implies $\ell \to s\ell$ with $s \to 0$. Then assuming $\psi \to s^d \psi$ the quadratic terms in the action will be scale invariant provided $d = -1/2$. The quartic terms in the action then scales as s and becomes negligible near the FS *except* when the pairing condition $\mathbf{q} + \mathbf{l} = 0$ is obeyed. In this case the quartic term scales as s^0 and cannot be ignored. In fact this term determines the most interesting behavior of the system at low temperatures (see [3] for full details).

C Electroweak interactions

Again I will follow the general recipe. I will concentrate only on the (low energy) interactions involving lepton fields, which are then the degrees of freedom. Since I assume the energy to be well below the Fermi scale, the only relevant symmetries are $U(1)$ gauge and Lorenz invariances. In addition there is the question whether the heavy physics will respect the discrete symmetries C, P or CP; using perfect hindsight I will retain terms that violate these symmetries.

Assuming a local description I have [1]

$$\mathcal{L}_{\text{eff}} = \sum \bar{\psi}_i (i\slashed{D} - m_i)\psi_i + \sum f_{ijkl} \left(\bar{\psi}_i \Gamma^a \psi_j\right)\left(\bar{\psi}_k \Gamma_a \psi_l\right) + \cdots \qquad (3)$$

where the ellipsis indicate terms containing operators of higher dimension, or those involving the electromagnetic field. The matrices Γ are to be chosen among the 16 independent basis $\Gamma^a = \{1, \gamma_\mu, \sigma_{\mu\nu}, \gamma_\mu \gamma_5, \gamma_5\}$

The coefficients for the first two terms are be fixed by normalization requirements. While a SM calculation gives $f \sim g^2/m_W^2 = 1/v^2$ ($v \simeq 246\text{GeV}$) and is generated by tree-level graphs (see Fig. 2) because of this the scale $1/\sqrt{f}$ is, in fact, the scale of the heavy physics and so the model is applicable at energies well below v. The four fermion interactions summarize the leading virtual gauge boson effects. The contributions of the four-fermion operators to processes with typical energy E are suppressed by a factor E^2/v^2. These can be observed (or bounded) despite the $E \ll v$ condition because they generate *new* effects: C and P (and some of them chirality) violation.

FIGURE 2. Standard model processes generating four fermion interactions at low energies (e.g.. Bhaba scattering)

D Strong interactions at low energies

In this case we are interested in the description of the interactions among the lightest hadrons, the meson multiplet. The most convenient parameterization of these degrees of freedom is in terms of a unitary field [4] U such that $U = \exp(\lambda_a \pi^a / F)$ where π^a denote the eight meson fields, λ^a the Gell-Mann matrices and F is a constant (related to the pion decay constant). The symmetries obeyed by the system are chiral $SU(3)_L \times SU(3)_R$, Lorenz invariance, C and P.

With these constraints the effective Lagrangian takes the form

$$\mathcal{L}_{\text{eff}} = a\,\text{tr}\partial U^\dagger \cdot \partial U + \left[b\,\text{tr}\partial_\mu U^\dagger \partial_\nu U \partial^\mu U^\dagger \partial^\nu U + \cdots\right] + \cdots \tag{4}$$

I can set $a \sim F^2$ by properly normalizing the fields. In this case the leading term in the effective Lagrangian will determine all (leading) low-energy pion interactions in terms of the single constant F. The effects form the higher-order terms that have been measured and the data requires $b \sim 1/(4\pi)^2$. This result is also predicted by the consistency of this approach which requires that radiative corrections to a, b, etc. should be at most of the same size as their tree-level values.

III BASIC IDEAS ON THE APPLICABILITY OF THE FORMALISM

Being a model with an intrinsic cutoff there are no actual ultraviolet divergences in most effective Lagrangian computations. Still there are interesting renormalizability issues that arise when doing effective Lagrangian loop computations.

Imagine doing a loop calculation including some vertices terms of (mass) dimension higher than the dimension of space-time. These must have coefficients with dimensions of mass to some negative power. The loop integrations will produce in general terms growing with Λ, the UV cutoff, which are polynomials in the external momenta [2] and will preserve the symmetries of the model [5]. Hence these terms which may *grow* with Λ correspond to vertices appearing in the most general effective Lagrangian and can be absorbed in a renormalization of the corresponding coefficients. They have no observable effects (though they can be used in naturality arguments [6]).

Effective theories will also be unitary *provided* one stays within the limits of their applicability. Should one exceed them new channels will open (corresponding to the production of the heavy excitations) and unitarity violating effects will occur. This is *not* produced by real unitarity violating interactions, but due to our use of the model beyond its range of applicability (e.g. the typical energy of the process under consideration reaches or exceeds Λ). One can, of course, *extend* the model, but this necessarily introduces *ad-hoc* elements and will dilute the generality gained using effective theories.

For example consider WWZ interactions with an effective Lagrangian of the form

$$\mathcal{L}_{\text{eff}} = \lambda(p,k) W_{\mu\nu}(k) W^{\nu\rho}(p) Z_\rho{}^\mu(-p-k) + \cdots ; \tag{5}$$

(where $V_{\alpha\beta} = \partial_\alpha V_\beta - \partial_\beta V_\alpha$). One can then choose λ to insure unitarity is preserved (at least in some processes), for example [7]

$$\lambda(p,k) = \frac{\lambda_0}{(p \cdot k + \Lambda)^n} \tag{6}$$

---

[2] Since a graph can be rendered convergent by taking sufficient number of derivatives with respect to the external momenta.

which, for n sufficiently large insures that the cross section for the reaction $e^+e^- \to Z \to WW$ is unitary, since it behaves as s^{2-2n} for a CM energy= $s \gg \Lambda^2$. But the very same effective vertex also modifies other reactions such as, for example $u\bar{d} \to W \to ZW$ where the cross section now has a factor $(s - \Lambda^2)^{-2n}$ and will exhibit resonant behavior if $s \sim \Lambda^2$. If one requires $s \ll \Lambda^2$ (as required by the consistency of the formalism) there are neither unitarity violations nor resonance effects. If, however, one uses the above Ansatz to extend the range of applicability to $s \sim \Lambda^2$ and beyond then very clear resonances should be observed in hadron colliders. Given that these have not been observed one *must* use for Λ a value significantly larger than the average CM energy for the hard W pair production cross section.

IV USING EFFECTIVE LAGRANGIANS

Effective Lagrangians provide an efficient way of summarizing some (perhaps very complex) interactions. The idea is simply to include all the effective vertices produces by those excitations which are not directly observed.

For example given a real scalar field ϕ and assume that all Fourier components above a scale Λ are not directly observable (i.e. the available energies lie all below Λ), then the effective Lagrangian is obtained by integrating over the variables observable at energies $\geq \Lambda$; writing $\phi = \phi_0 + \phi_1$, with

$$\phi_0(\mathbf{k}): \ |\mathbf{k}| < \Lambda \qquad \phi_1(\mathbf{k}): \ \Lambda \leq |\mathbf{k}| < \Lambda_1 \tag{7}$$

then by definition

$$e^{iS_{\text{eff}}} = \int [d\phi_1] e^{iS(\phi_0, \phi_1)}, \qquad S_{\text{eff}} = \int d^n x \mathcal{L}_{\text{eff}} \tag{8}$$

where $\mathcal{L}_{\text{eff}}$ is obtained by expanding S_{eff} in powers of Λ which gives an infinite tower of local operators.

Another common situation where effective Lagrangians appear occurs when some heavy excitations are integrated out. This can be illustrated by the following toy model [3]

$$S = \int d^n x \left[\bar{\psi}(i\slashed{\partial} - m)\psi + \frac{1}{2}(\partial\phi)^2 - \frac{1}{2}\Lambda^2 \phi_1^2 + f\phi\bar{\psi}\psi \right] \tag{9}$$

where ϕ is heavy. A simple calculation gives

$$S_{\text{eff}} = \int d^n x \left[\bar{\psi}(i\slashed{\partial} - m)\psi + \frac{1}{2} f^2 \bar{\psi}\psi \frac{1}{\Box + \Lambda^2} \bar{\psi}\psi \right] \tag{10}$$

and

---

[3] I'm cheating in order to get a closed form for the effective action, a more realistic model should include a term $\propto \phi_1^4$.

$$\mathcal{L}_{\text{eff}} = \bar{\psi}(i\not{\partial} - m)\psi + \frac{f^2}{2\Lambda^2} \sum_{l=1}^{\infty} \bar{\psi}\psi \left(\frac{\Box}{\Lambda^2}\right)^n \bar{\psi}\psi \qquad (11)$$

Note that terms with large number of derivatives will be suppressed by a large power of the small factor (E/Λ), if we are interested in energies $E \sim \Lambda$ the *whole* infinite set of vertices must be included in order to reproduce the ϕ pole.

A How to parameterize ignorance

If one knows the theory, we can, in principle, calculate $\mathcal{L}_{\text{eff}}$ (or do a full calculation). Yet there are many cases where the underlying theory is not known. In these cases an effective theory is obtained by writing *all* possible interactions among the light excitations. The model then has an infinite number of terms each with an unknown parameter, and these constants then parameterize *all* possible underlying theories. The terms which dominate are those usually called renormalizable (or, equivalently, marginal or relevant). The other terms are called non-renormalizable, or irrelevant, since their effects become smaller as the energy decreases

This recipe for writing effective theories must be supplemented with some symmetry restrictions. The most important being that the all the terms in the effective Lagrangian must respect the local gauge invariance of the low-energy physics (more technically, the one respected by the renormalizable terms in the effective action) [8]. The reason is that the presence of a gauge variant term will generate *all* gauge variant interactions thorough renormalization group evolution.

Using a simple argument it is possible to turn any theory into a gauge theory [9] and so it appears that the requirement of gauge invariance is empty. That this is not the case is explained here. I first describe the trick which grafts gauge invariance onto a theory and then discuss the implications.

Consider an arbitrary theory with matter fields (spin 0 and 1/2) and vector fields V_μ^n, $n = 1, \ldots N$. Then

- Choose a (gauge) group G with N generators $\{T^n\}$. Define a covariant derivative $D_\mu = \partial_\mu + V_\mu^n T^n$ and *assume* that the V_μ^n are gauge fields.

- Invent a unitary field U transforming according to the fundamental representation of G and construct the gauge invariant composite fields

$$\mathcal{V}_\mu^n = -\text{tr} T^n U^\dagger D_\mu U \qquad (12)$$

Taking $\text{tr} T^n T^m = -\delta_{nm}$, it is easy to see that in the unitary gauge $U = 1$, $\mathcal{V}_\mu^n = V_\mu^n$.

Thus, if we simply replace $V \to \mathcal{V}$ in the original theory we get a gauge theory. Does this mean that gauge invariance is irrelevant since it can be added at will? In my opinion this is not the case.

In the above process *all matter fields are assumed gauge singlets* (none are minimally coupled to the gauge fields).In the case of the standard model , for example, the universal coupling of fermions to the gauge bosons would be accidental in this approach. In order to recover the full predictive power commonly associated with gauge theories, the matter fields must transform non-trivially under G which can be done only if there are strong correlations among some of the couplings. It is *not* trivial to say that the standard model group is $SU(3) \times SU(2) \times U(1)$ with left-handed quarks transforming as $(3, 2, 1/6)$, left-handed leptons as $(1, 2, -1/2)$, etc., as opposed to a $U(1)^{12}$ with all fermions transforming as singlets [10].

B How to estimate ignorance

A problem which I have not addressed so far is the fact that effective theories have an infinite number of coefficients, with the (possible) problem or requiring an infinite number of data points in order to make any predictions. On the other hand, for example, if this is the case why is it that the Fermi theory of the weak interactions is so successful?

The answer to this question lies in the fact that not all coefficients are created equal, there is a *hierarchy* [4,10]. As a result, given any desired level of accuracy, only a finite number of terms need to be included. Moreover, even though the effective Lagrangian coefficients cannot be calculated without knowing the underlying theory, they can still be *bounded* using but a minimal set of assumptions about the heavy interactions. It is then also possible to estimate the errors in neglecting all but the finite number of terms used.

As an example consider the standard model at low energies and calculate two processes: Bhaba cross section and the anomalous magnetic moment of the electron. For Bhaba scattering there is a contribution due the Z-boson exchange (see Fig. 2)

$$e^+e^- \to Z \to e^+e^- \quad \text{generates} \quad \mathcal{O} = \frac{1}{2m_Z^2} \left(\bar{e}\Gamma\gamma^\mu e\right)\left(\bar{e}\Gamma\gamma_\mu e\right) \quad (13)$$

where $\Gamma = g_V + g_A\gamma_5$. The coefficient of the effective operator $\mathcal{O}$ is then $\sim$ (coupling/physical mass)$^2 \sim 1/v^2$

The electron anomalous magnetic moment receives contributions from virtual W, Z and H exchanges (see Fig. 3). The corresponding low-energy operator is

$$\mathcal{O} = \bar{e}\sigma_{\mu\nu}eF^{\mu\nu} \quad (14)$$

In this case the coefficient $\sim \{\text{coupling}/[4\pi(\text{physical mass})]\}^2 \sim 1/(4\pi v)^2$ [4].

The point of this exercise is to illustrate the fact that, for weakly coupled theories, loop-generated operators have smaller coefficients than operators generated at tree level. Leading effects are produced by operators which are generated at tree level.

---

[4]) In addition the coefficient is suppressed by a factor of m_e since it violates chirality.

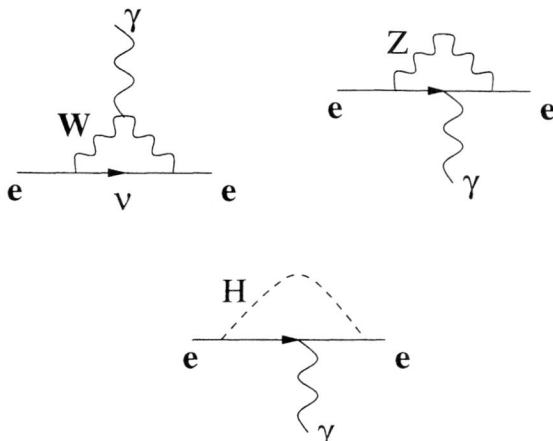

FIGURE 3. Weak contributions to the electron anomalous magnetic moment

C Coefficient estimates

In this section I will provide arguments which can be used to estimate (or, at least bound) the coefficients in the effective Lagrangian. These are order of magnitude calculations and might be off by a factor of a few; it is worth noting that no single calculation has provided a significant deviation from these results.

The estimate calculations should be done separately for weakly and strongly interacting theories. I will characterize the first as those where radiative corrections are smaller than the tree-level contributions. Strongly interacting theories will have radiative corrections of the same size at any order [5].

1 Weakly interacting theories

In this case leading terms in the effective Lagrangian are those which can be generated at tree level by the heavy physics. Thus, the dominating effects are produced by operators which have the lowest dimension (leading to the smallest suppression from inverse powers of Λ) and which are tree-level generated (TLG) [11].

When the heavy physics is described by a gauge theory it is possible to obtain all TLG operators [11]. The corresponding vertices fall into 3 categories, symbolically

- vertices with 4 fermions
- vertices with 2 fermions and k bosons; $k = 2, 3$

---

[5] Should the radiative corrections increase with the order of the calculation, it is likely that the dynamic variables being used are not appropriate for the regime where the calculation is being done.

- vertices with n bosons; $n = 4, 6$.

A particular theory may not generate one or more of these vertices, the only claim is that there is *a* gauge theory which does.

In the case of the standard model with lepton number conservation the leading operators have dimension 6 [12,11]. Subleading operators are either dimension 8 and their contributions are suppressed by an additional factor $(E/\Lambda)^2$ in processes with typical energy E. Other subleading contributions are suppressed by a loop factor $\sim 1/(4\pi)^2$. Note that it is possible to have situations where the only two effects are produced by either dimension 8 TLG operators or loop generated dimension 6 operators. In this case the former dominates only when $\Lambda > 4\pi E$.

The terms in the electroweak effective Lagrangian which describe the interaction of the W and Z bosons generated by some heavy physics underlying the standard model has received considerable attention recently [13]. In terms of the $SU(2)$ and $U(1)$ gauge fields W and B and the scalar doublet ϕ these interactions are

$$\mathcal{L}_{\text{eff}} = \frac{1}{\Lambda^2} (\alpha_W \mathcal{O}_W + \alpha_{BW} \mathcal{O}_{BW})$$
$$\mathcal{O}_W = \epsilon_{IJK} W^I_{\mu\nu} W^{J\nu}{}_\lambda W^{K\lambda\mu}$$
$$\mathcal{O}_{WB} = \phi^\dagger \tau^I \phi W^I_{\mu\nu} B^{\mu\nu} \tag{15}$$

The above arguments inly that there is no TLG operator containing three gauge bosons. This means that all effective contributions to the WWZ and $WW\gamma$ interactions are loop generated, so their coefficients *necessarily* take the form $\prod(\text{coupling constants})/(16\pi^2)$. Thus the parameters κ and λ commonly used to parameterize these interactions are of order 5×10^{-3}. *Experiments providing limits significantly above this value provide <u>no</u> information about the heavy physics.*

2 Strongly interacting theories

I will imagine a theory containing scalars and fermions which interact strongly. Gauge couplings are assumed to be small and will be ignored. This calculation is useful for low energy chiral theories but not for low energy QCD [14,15,4].

A generic effective operator in this type of theories takes the form

$$\mathcal{O}_{abc} \sim \lambda \Lambda^4 \left(\frac{\phi}{\Lambda_\phi}\right)^a \left(\frac{\psi^{3/2}}{\Lambda_\psi}\right)^b \left(\frac{\partial}{\Lambda}\right)^c \tag{16}$$

Then the condition that these dynamic variables appropriately describe the physics below Λ implies that radiative corrections to the couplings are at most as large as the tree-level values, namely $\delta_{\text{rad}} \lambda \leq \lambda$. A straightforward estimate (including a factor of $1/(16\pi^2)$ for each loop) shows that this condition is satisfied only if

$$\Lambda_\psi = \frac{1}{(4\pi)^{2/3}} \Lambda, \quad \Lambda_\phi = \frac{1}{4\pi} \Lambda, \quad \lambda = \frac{1}{16\pi^2} \tag{17}$$

In terms of $U \sim \exp(\phi/\Lambda_\phi)$, the operators take the form

$$\mathcal{O}_{abc} = \frac{1}{(4\pi)^{2-b}} \Lambda^{4-c-3b/2} \partial^c U^{a'} \psi^b \tag{18}$$

In particular the coefficient of the two derivative operators $\text{tr}\partial U^\dagger \partial U$ is $\propto \Lambda_\phi^2$.

For the case where ϕ represents the interpolating field for the lightest mesons PCAC implies $\Lambda_\phi = f_\pi$ [14,4]. Then

$$\psi^4 \propto \frac{1}{f_\pi^2} \qquad \partial^4 U^4 \propto \frac{1}{16\pi^2} \qquad \psi^2 \partial^2 U^2 \propto \frac{1}{4\pi f_\pi} \tag{19}$$

(note that these are upper bounds). The extensive data on low energy meson reactions can be used to gauge the validity of these predictions, they are indeed satisfied. In particular the $(\partial U)^4$ terms have coefficients $\sim 1/(16\pi^2)$.

For the case of the standard model the field U can be used to provide masses for the W and Z bosons without a physical Higgs being present (the price is that the model breaks down at energies $\sim 4\pi v = 3\text{TeV}$). In this case the gauge fields are introduced minimally and it is the term $(DU)^2$ that gives a mass to the W and Z, which fixes $\Lambda_\phi = v = 246\text{GeV}$, $\Lambda = 3\text{TeV}$; as before, the model makes no sense beyond this scale [6]. In addition, when the gauge fields are reintroduced, the terms with 4 derivatives will generate triple-vector boson couplings, again leading to the estimates $\lambda, \kappa \sim 5 \times 10^{-3}$ [10].

D Radiative corrections

Despite the presence of higher-dimensional operators radiative corrections can be calculated in the usual way. As an example imagine calculating the corrections to the cross section for the reaction $e^+ e^- \to e^+ e^-$ using the standard model with the addition of a 4-fermion interaction

$$\mathcal{L}_{\text{eff}} = \mathcal{L}_{\text{eff}}^{\text{SM}} + \frac{f}{\Lambda^2} \left(\bar\psi \gamma^\mu \psi\right) \left(\bar\psi \gamma_\mu \psi\right) + \cdots \tag{20}$$

where ψ denotes the electron field.

The calculation is illustrated in Fig. 4 where the loops involving the 4-fermion operator are cut-off at a scale Λ. The SM and new physics (NP) contributions are, symbolically,

$$\begin{aligned}\text{SM:} & \frac{1}{v^2}\left[1 + \frac{g^2}{16\pi^2} + \cdots\right] \\ \text{NP:} & \frac{f}{\Lambda^2}\left[1 + \frac{f}{16\pi^2} + \cdots\right]\end{aligned} \tag{21}$$

---

[6] Tough it is conceivable that a full non-perturbative calculation would show that the theory cures itself and can be extended beyond this scale, there is no indication that this miracle occurs.

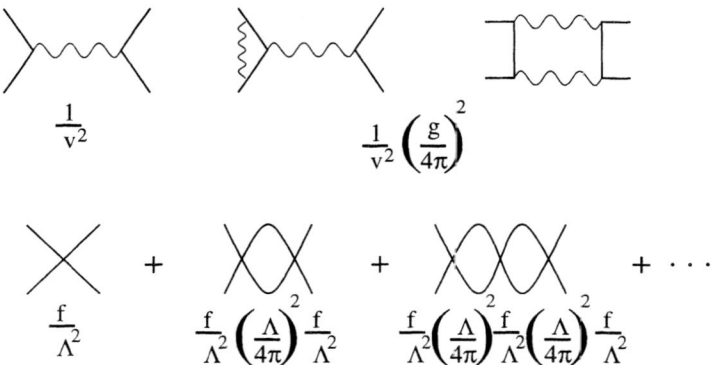

FIGURE 4. Radiative corrections to Bhaba scattering in the presence of a 4-fermion interaction.

Note that this consistent behavior (that the new physics effects disappear as $\Lambda \to \infty$) results form having the physical scale of new physics Λ in the coefficient of the operator. Had we used f'/v^2 instead of f/Λ^2, the new physics effects would appear to be enormous, and growing with each new loop. It is not that the use of f'/v^2 is wrong, it is only that it is misleading to believe f' can be of order one; it *must* be suppressed by the small factor $(v/\Lambda)^2$.

Using these results we see that this reaction is sensitive to Λ provided $f(v/\Lambda)^2 >$sensitivity. If the sensitivity is, say 1% this corresponds to $\Lambda/\sqrt{f} > 2.5\text{TeV}$ [16].

This perturbative calculation is manageable provided $f < 16\pi^2$, otherwise the underlying physics is strongly coupled. It is still possible in that case to provide estimates of the new physics contributions, though these are less reliable; these estimates imply that $1 + f/(4\pi)^2 + \cdots \sim 1$ when $f \sim 16\pi^2$.

V APPLICATIONS TO ELECTROWEAK PHYSICS

With the above results one can determine, for any given process, the leading contributions (as parameterized by the various effective operator coefficients). Using then the coefficient estimates one can provide the expected magnitude of the new physics effects with only Λ as an unknown parameter, and so estimate the sensitivity to the scale of new physics.

It is important to note that this is sometimes a rather involved calculation as all contributing operators must be included. For example, in order to determine the heavy physics effects on the oblique parameters one must calculate not only those affecting the vector boson polarization tensors, but also those which modify the

Fermi constant, the fine structure constant, etc. as these quantities are used when extracting S, T and U from the data [18].

A Effective lagrangian

In the following I will assume that the underlying physics is weakly coupled and derive the leading operators that can be expected form the existence of heavy excitations at scale Λ.

The complete list of dimension 6 operators was cataloged a long time ago for the case where the low energy spectrum includes a single scalar doublet [12] [7]. It is then straightforward to determine the subset of operators which can be TLG, they are [11]

- Fermions: $\left(\bar{\psi}_i \Gamma^a \psi_j\right)\left(\bar{\psi}_k \Gamma^a \psi_l\right)$
- Scalars: $|\phi|^6$, $(\partial|\phi|^2)^2$
- Scalars and fermions: $|\phi|^2 \times$ Yukawa term
- Scalars and vectors: $|\phi|^2|D\phi|^2$, $|\phi^\dagger D\phi|^2$
- Fermions, scalars and vectors: $\left(\phi^\dagger T^n D^\mu \phi\right)\left(\bar{\psi}_i T^n \gamma_\mu \psi_j\right)$

where T denotes a group generator and Γ a product of a group generator and a gamma matrix.

Observables affected by the operators in this list provide the highest sensitivity to new physics effects provided that the standard model effects are themselves small (or that the experimental sensitivity is large enough to observe small deviations). I will illustrate this with two (incomplete) examples.

B b-parity

This is a proposed method for probing new flavor physics [19]. Its virtue lies in the fact that it is very simple and sensitive (though it does not provide the highest sensitivity for all observables). The basic idea is based on the observation that the standard model acquires an additional global $U(1)_b$ symmetry in the limit $V_{ub} = V_{cb} = V_{td} = V_{ts} = 0$ (given the experimental values $0.002 < |V_{ub}| < 0.005$, $0.036 < |V_{cb}| < 0.046$, $0.004 < |V_{td}| < 0.014$, $0.034 < |V_{ts}| < 0.046$, deviations from exact $U(1)_b$ invariance will be small). Then for any standard model interaction a reaction of the type

$$n_i \, b-\text{jet} + X \to n_f \, b-\text{jet} + Y \qquad (22)$$

will obey

---

[7] More complicated scalar sectors have also been studied [17], though not exhaustively.

$$(-1)^{n_i} = (-1)^{n_f} \qquad (23)$$

to very high accuracy. The number $(-1)^{\#\text{ of }b\text{ jets}}$ defines the **b-parity** of a state (it being understood that the top quarks have decayed).

The standard model is then b-parity even, and the idea is to consider a lepton collider [8] and simply count the number of b jets in the final state; new physics effects will show up as events with odd number of b jets.

The standard model produces no measurable irreducible background, yet there are significant *reducible* backgrounds which reduced the sensitivity to Λ. To estimate these effects I define

- $\epsilon_b = b -$ jet tagging efficiency
- $t_c = c -$ jet **mis**tagging efficiency (probability of mistaking a $c -$ jet jet for a $b -$ jet)
- $t_j =$ light-jet **mis**tagging efficiency (probability of mistaking a light-jet for a $b -$ jet)

so that the *measured* cross section with k-b-jets is

$$\bar{\sigma}_k = \sum_{u+v+w=k} \left[\binom{n}{u}\epsilon_b^u(1-\epsilon_b)^{n-u}\right]\left[\binom{m}{v}t_c^v(1-t_c)^{m-v}\right]\left[\binom{\ell}{w}t_j^w(1-tj)^{\ell-w}\right]\sigma_{nm\ell} \qquad (24)$$

where $\sigma_{nm\ell}$ denotes the cross section for the final state with n b-jets, m c-jets, and ℓ light jets. Note that $\left[\binom{n}{u}\epsilon_b^u(1-\epsilon_b)^{n-u}\right]$ is the probability of tagging u and missing $n-u$ b-jets out of the n available.

As an example consider

$$\mathcal{L}_{\text{eff}} = \mathcal{L}_{\text{sm}} + \frac{f_{ij}}{\Lambda^2}\left(\bar{\ell}\gamma^\mu\ell\right)\left(\bar{q}_i\gamma_\mu q_j\right) \qquad (25)$$

where $i \neq j$ denote family indices. Taking $m_H = 100\text{GeV}$ $|f| = 1$ $t_c = t_j = 0$ the sensitivity to Λ is summarized by the following table

Limits from $e^+e^- \to t\bar{c} + \bar{t}c + b\bar{s} + bs \to 1b-\text{jet} + X$				
$\sqrt{s}$	L	$\epsilon_b = 50\%$	$\epsilon_b = 60\%$	$\epsilon_b = 70\%$
200 GeV	2.5 fb^{-1}	1.4 TeV	1.5 TeV	1.6 TeV
500 GeV	75 fb^{-1}	5.0 TeV	5.2 TeV	5.5 TeV
1000 GeV	200 fb^{-1}	9.5 TeV	10.0 TeV	10.7 TeV

These results are promising yet they will be degraded in a realistic calculation. First one must include the effects of having $t_{c,j} \neq 0$. In addition there are complications in using inclusive reactions such as $e^+e^- \to b + X$ since the contributions from events with large number of jets can be very hard to evaluate (aside from the calculational difficulties there are additional complications when *defining* what a jet is). A more realistic approach is to restrict the calculation to a sample with a fixed number of jets (2 and 4 are the simplest) and determine the sensitivity to Λ for various choices of ϵ_b and t_j using this population only.

---

8) In hadron colliders there are sea-b quarks which foul-up the argument.

C CP violation

Just as for b-parity the CP violating effects are small within the standard model and so precise measurements of CP violating observable might be very sensitive to new physics effects.

In order to study CP violations it is useful to first define what the CP transformation *is*. In order to do this in general denote the Cartan group generators by H_i and the root generators by E_α, then it is possible to find a basis where *all* the group generators are real and, in addition, the H_i are diagonal [20]. Define then CP transformation by

$$\psi \to C\psi^* \text{ (fermions)}$$
$$\phi \to \phi^* \text{ (scalars)}$$
$$A_\mu^{(i)} \to -A_\mu^{(i)}, \; (i: \text{ Cartan generator})$$
$$A_\mu^{(\alpha)} \to -A_\mu^{(-\alpha)}, \; (\alpha: \text{ root})$$

it is easy to see that the field strengths and currents transform as A_μ, while $D\phi \to (D\phi)^*$. It then follows that in this basis the whole gauge sector of *any* gauge theory is CP conserving; CP violation can arise *only* in the scalar potential and fermion-scalar interactions using this basis.

In order to apply this to electroweak physics I will need the list of TLG operators of dimension 6 which violate CP, they are given by [9]

$$\left(\bar{\ell}e\right)\left(\bar{d}q\right) - \text{h.c.} \quad (\bar{q}u)\,\varepsilon\,(\bar{q}d) - \text{h.c.} \quad \left(\bar{q}\lambda^A u\right)\varepsilon\left(\bar{q}\lambda^A d\right) - \text{h.c.}$$
$$\left(\bar{\ell}e\right)\varepsilon\,(\bar{q}u) - \text{h.c.} \quad \left(\bar{\ell}u\right)\varepsilon\,(\bar{q}e) - \text{h.c.} \quad |\phi|^2\left(\bar{\ell}e\phi - \text{h.c.}\right)$$
$$|\phi|^2\left(\bar{q}u\tilde{\phi} - \text{h.c.}\right) \quad |\phi|^2\left(\bar{q}d\phi - \text{h.c.}\right) \quad |\phi|^2\partial_\mu\left(\bar{\ell}\gamma^\mu\ell\right)$$
$$|\phi|^2\partial_\mu\left(\bar{e}\gamma^\mu e\right) \quad |\phi|^2\partial_\mu\left(\bar{q}\gamma^\mu q\right) \quad |\phi|^2\partial_\mu\left(\bar{u}\gamma^\mu u\right)$$
$$|\phi|^2\partial_\mu\left(\bar{d}\gamma^\mu d\right)$$
$$\mathcal{O}_1 = \left(\phi^\dagger\tau^I\phi\right)D_\mu^{IJ}\left(\bar{\ell}\gamma^\mu\tau^J\ell\right)$$
$$\mathcal{O}_2 = \left(\phi^\dagger\tau^I\phi\right)D_\mu^{IJ}\left(\bar{q}\gamma^\mu\tau^J q\right)$$
$$\mathcal{O}_3 = \left(\phi^\dagger\varepsilon D_\mu\phi\right)(\bar{u}\gamma^\mu d) - \text{h.c}$$

All operators except $\mathcal{O}_{1,2,3}$ violate chirality and their coefficients are strongly bounded by their contributions to the strong CP parameter θ; in addition some chirality violating operators contribute to meson decays (which again provide strong bounds for fermions in the first generation) and, finally, in natural theories some contribute radiatively to fermion masses and will be then suppressed by the smaller of the corresponding Yukawa couplings. For these reasons I will not consider them

---

[9] The notation is the following: ℓ and q denote the left-handed lepton and quark doublets; u, d and e denote the right handed quark and charged lepton fields. λ^A denote the Gell Mann matrices, τ the Pauli matrices, and $\epsilon = i\tau^2$. D_μ represents the covariant derivatives and ϕ the scalar doublet.

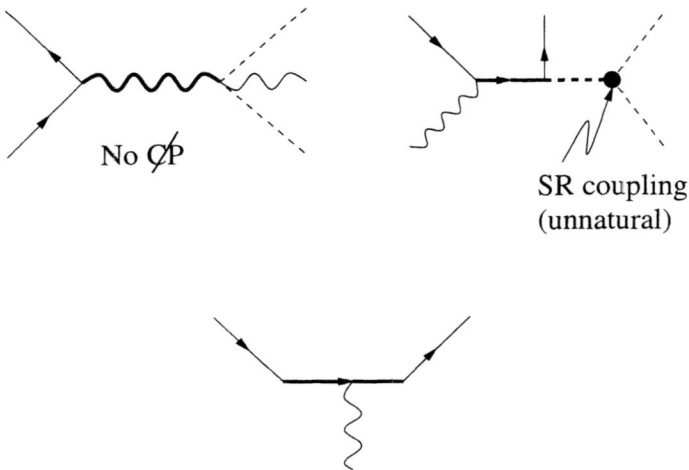

FIGURE 5. Heavy physics contributing to CP violating operators. Wavy lines denote vectors, solid lines fermions, and dashed ones scalars. Heavy lines denote heavy excitations.

further. Moreover, since I will be interested in limits that can be obtained using current data, I will ignore operators whose only observable effects involve Higgs particles.

With these restrictions only $\mathcal{O}_{1,2,3}$ remain; their terms not involving scalars are

$$\mathcal{O}_1 \to -\frac{igv^2}{\sqrt{2}} \left(\bar{\nu}_L \slashed{W}^+ e_L - \text{h.c.} \right)$$

$$\mathcal{O}_2 \to -\frac{igv^2}{\sqrt{2}} \left(\bar{u}_L \slashed{W}^+ d_L - \text{h.c.} \right)$$

$$\mathcal{O}_3 \to -\frac{igv^2}{\sqrt{8}} \left(\bar{u}_R \slashed{W}^+ d_R - \text{h.c.} \right)$$

The contributions from $\mathcal{O}_{1,2}$ can be absorbed in a renormalization of standard model coefficients whence only $\mathcal{O}_3$ produces observable effects, corresponding to a right-handed quark current. Existing data (from τ decays and m_W measurements) implies $\Lambda \gtrsim 500 \text{GeV}$.

One can also determine the type of new interactions which might be probed using these operators [11]. The heavy physics which can generate $\mathcal{O}_3$ at tree level is described in Fig. 5. If the underlying theory is natural we conclude that there will be no super-renormalizable couplings; in this case $\mathcal{O}_3$ will be generated by heavy fermion exchanges only [10]

---

[10] It is true that vertices involving light fermions, light scalars and heavy fermions produce mixings between the light and heavy scales, but this occurs at the one loop level. In contrast cubic terms of order Λ in the scalar potential would shift v at tree level.

Note finally that these arguments are only valid for weakly coupled heavy physics. For strongly coupled theories other CP violating operators can be important, e.g.

$$\frac{f}{\Lambda^2} B^{\mu\nu} (\bar{e}\gamma_\mu D_\nu e - \text{h.c}) \tag{26}$$

since $|f| \sim 1$.

VI OTHER APPLICATIONS

The effective Lagrangian approach can be applied in many other situations such as gravity and high temperature field theory. I will briefly consider the latter.

A Large temperatures

It is a well-known fact that the thermodynamics of a system with Hamiltonian H can be derived form the partition function $\text{tr} e^{-\beta H}$. This resembles closely the (trace of the) quantum evolution operator e^{-iHt} hence we can obtain the thermodynamics of a system by the replacement $-it \to \beta$: non-zero temperature field theory corresponds to Euclidean field theory on a cylinder of perimeter$= \beta$, I will denote the corresponding Euclidean time by τ [21]

Since the time direction is finite the fields are expanded in a Fourier series. For bosons one obtains

$$\phi = \sum_{n=-\infty}^{\infty} \int \frac{d^3 k}{(2\pi)^3} \phi_n(\mathbf{k}) e^{i(2n\pi T \tau + \mathbf{k} \cdot \mathbf{r})} \tag{27}$$

and the corresponding free propagator is given by

$$\frac{1}{(2n\pi T)^2 + \mathbf{p}^2 + m^2} \tag{28}$$

The field is periodic in τ due to the commutativity of the variables in the functional integral (there is a much more physical reason, called the Kubo-Martin-Schwinger condition) [21].

Note that the $n \neq 0$ modes become heavy as $T \to \infty$ so that in this limit only the $n = 0$ modes remain and the theory reduces to a 3-D Euclidean field theory (there might be some subtleties involved, see below).

For fermions the expansion is in odd Fourier modes since the corresponding integration variables anticommute. Explicitly

$$\psi = \sum_{n=-\infty}^{\infty} \int \frac{d^3 k}{(2\pi)^3} \psi_n(\mathbf{k}) e^{i(2n+1)\pi T \tau + \mathbf{k} \cdot \mathbf{r}} \tag{29}$$

with free propagator

$$\frac{1}{[i(2n+1)\pi T + \mu]\gamma^0 - \mathbf{k}\cdot\gamma - m} \tag{30}$$

which shows that all modes become heavy as $T \to \infty$. There will be then no fermions in the spectrum at very large temperatures. Note that this occurs independently of the fermion mass [21].

Despite the absence of heavy fermions and scalars (effective mass $\sim T$) at large temperatures, we can still ask what is their effect on the scalar modes that survive in this regime. To this end we can construct the corresponding effective theory. I will illustrate the procedure using a simple example.

Consider the following scalar theory

$$\mathcal{L}^{(4)} = \frac{1}{2}\left(\partial\phi^2\right) - \frac{1}{2}m^2\phi^2 - \frac{\lambda}{4!}\phi^4 \tag{31}$$

Then the excitations which survive at large T are

$$\varphi(\mathbf{x}) = \sqrt{T} \int_0^\beta d\tau\, \phi(\mathbf{x}, \tau) \tag{32}$$

where φ is the dynamical variable of a 3 dimensional Euclidean field theory (in 3 dimensions the scalar fields have units of $\sqrt{\text{mass}}$ which explains the $\sqrt{T}$ factor). The only symmetry (aside form Euclidean invariance) is the reflection symmetry $\varphi \to -\varphi$. The scale of the new theory is set by $\Lambda = T$, but in this case the model is supposed to describe physics *above* Λ

With these considerations we can write the effective theory for φ,

$$\mathcal{L}_{\text{eff}} = \frac{1}{2}(\nabla\varphi)^2 + \frac{1}{2}a\varphi^2 + \frac{1}{4!}b\varphi^4 + \frac{c}{6!}\varphi^6 + O(1/T) \tag{33}$$

note that b is a super-renormalizable coupling and may lead to infrared problems.

The coefficients a, b, c, etc. can be calculated from the original theory. At one loop one obtains

$$a = \frac{\lambda T^2}{24}, \qquad b = -\frac{m}{2\pi}\left(\frac{\lambda T}{4m}\right)^2, \qquad c = \frac{1}{4\pi}\left(\frac{\lambda T}{4m}\right)^3.$$

But this calculation has some potential problems. Consider the $2k$ point function at zero external momentum; the corresponding graphs are given in Fig. 6 A simple estimate (verified by explicit calculation) shows that

$$\text{Graph} \propto \underbrace{\frac{\lambda^k}{m^{2k-4}}}_{\text{prefactors+dim. analysis}} \times \underbrace{\left(\frac{T}{m}\right)^{k+1}}_{\text{integral+sums}} \tag{34}$$

which corresponds to the operator

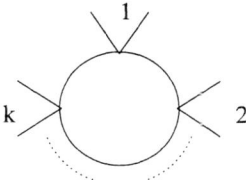

FIGURE 6. Graphs exhibiting interesting infrared behavior at high temperatures

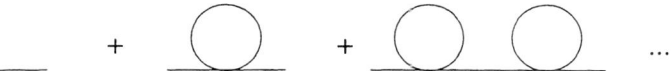

FIGURE 7. Radiative corrections to the $n = 0$ propagator which cure the infrared divergences in the effective coefficients.

$$\mathcal{O}^{(k)} \sim \frac{\lambda^k}{m^{2k-4}} \left(\frac{T}{m}\right)^{k+1} \left(\sqrt{T}\,\varphi\right)^{2k} \frac{1}{T} = m^3 \left(\frac{\sqrt{\lambda}\,T\,\varphi}{m^{3/2}}\right)^{2k} \qquad (35)$$

whose coefficient has *positive* powers of Λ ($= T$) and are not suppressed at large temperatures. In fact, should this be correct the, effective theory expansion would be useless.

The solution to this infrared problem (diverging effective coefficients as $m \to 0$ is well known for this type of theories [21]: the propagator for the $n = 0$ mode gets dressed and in so doing the m^2 gets shifted by an amount $\propto T^2$. Explicitly, the graphs in Fig. 6 shift

$$m^2 \to m^2 + \frac{\lambda}{24} T^2 \qquad (36)$$

so that the previous expression for the effective operator coefficient becomes

$$\mathcal{O}^{(k)} \sim \frac{\lambda^{(3-k)/2}}{T^k} \varphi^{2k} \qquad (37)$$

which vanishes as $T \to \infty$. Note that there is still a remnant of the infrared properties of the theory in that the coefficients still diverge as $\lambda \to 0$.

The previous arguments can be applied to the case of gauge theories. Just as for the scalar field, the gauge field is periodic in β and can be expanded in Fourier modes. At high temperatures, all but the $n = 0$ modes are heavy with masses $\sim T$. The remaining light modes are

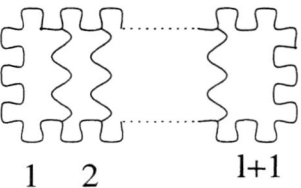

FIGURE 8. Some radiative corrections to the QCD free energy

$$\mathbf{A}^A_{n=0} \equiv \mathbf{a}^A \qquad A^{0A}_{n=0} \equiv \varphi^A \qquad (38)$$

leaving a 3-D Euclidean $SU(3)$ model with gauge fields $\mathbf{a}^A$ and with a scalar octet (the φ^A). The 3-D gauge coupling constant is $g\sqrt{T}$ (where g denotes the QCD gauge coupling)

The simplest infrared divergences are cured by the dressing the gluon propagator at one loop [21]; the ϕ^A propagator at large T then becomes

$$\frac{1}{p^2} \to \frac{1}{p^2 + cg^2T^2} \qquad (39)$$

for some numerical constant c. But this effect is not extended to the $\mathbf{a}^A$ for the corresponding vacuum polarization obeys $\Pi_{ii}(p \to 0) \to 0$ [21].

The fact that the $\mathbf{a}$ remain massless leads to various interesting problems. For example the higher order corrections to the free energy, provided by graphs in Fig. 8. Suppose that the gauge bosons have a (dynamically generated) mass m. In this case a graph with ℓ loops behaves as [21]

$$g^6 T^4 (g^2 T/m)^{\ell-3} \qquad (\ell > 3) \qquad (40)$$

For the case where internal lines correspond to A^0 (or φ^A) $m \sim gT$ and the graph is well behaved, $\sim g^{\ell+3}T^4$. On the other hand when the internal lines represent A^i (or, equivalently, $\mathbf{a}^A$) propagator a problem will arise unless $m \sim g^2 T$ is generated (we already know there is no $O(g)$ correction to m). This so-called magnetic mass has not been obtained perturbatively though it is widely believed to be generated.

Additional problems arise since the gauge coupling constant in the 3-d theory has dimensions of $\sqrt{\text{mass}}$ leading to super-renormalizable interactions with the related infrared divergences [22].

VII CONCLUSIONS

In these lectures I have provided a review of some of the very many aspects and properties of effective theories, as well as some of their application. Despite this drawback I hope it does give a flair for the strength of the approach.

Effective theories will be used in deriving the implications of new data on the properties of the physics which underlies the standard model, but in addition it can be applied to a wide variety of phenomena ranging form QCD to superconductivity. It is this flexibility which makes the formalism so attractive.

REFERENCES

1. C. Itzykson and J.-B. Zuber, *Quantum field theory* (McGraw-Hill, N.Y., 1980).
2. C. Jarlskog, Phys. Rev. Lett. **55**, 1039 (1985).
3. J. Polchinski, hep-th/9210046.
4. J.C. Collins, *Renormalization* (Cambridge U. Press, N.Y., 1984).
5. C. Arzt, M. B. Einhorn and J. Wudka, Phys. Rev. **D49**, 1370 (1994) [hep-ph/9304206].
6. U. Baur and D. Zeppenfeld, Phys. Lett. **B201**, 383 (1988).
7. M. Veltman, Acta Phys. Polon. **B12**, 437 (1981).
8. C. P. Burgess and D. London, hep-ph/9203215.
9. H. Georgi, *Weak interactions and modern particle theory* (Benjamin/Cummings Pub. Co., Menlo Park, Calif., 1984); Nucl. Phys. **B363**, 301 (1991); Nucl. Phys. **B361**, 339 (1991).
10. J. Wudka, Int. J. Mod. Phys. **A9**, 2301 (1994) [hep-ph/9406205].
11. C. Arzt, M. B. Einhorn and J. Wudka, Nucl. Phys. **B433**, 41 (1995) [hep-ph/9405214].
12. C. J. Burges and H. J. Schnitzer, Nucl. Phys. **B228**, 464 (1983). C. N. Leung, S. T. Love and S. Rao, Z. Phys. **C31**, 433 (1986). W. Buchmuller and D. Wyler, Nucl. Phys. **B268**, 621 (1986).
13. J. Ellison and J. Wudka, hep-ph/9804322.
14. J. Gasser and H. Leutwyler, Nucl. Phys. **B250**, 465 (1985); Annals Phys. **158**, 142 (1984).
15. A. Manohar and H. Georgi, Nucl. Phys. **B234**, 189 (1984). H. Georgi Phys. Lett. **B298**, 187 (1993) [hep-ph/9207278].
16. B. Grzadkowski and J. Wudka, Phys. Lett. **B364**, 49 (1995) [hep-ph/9502415].
17. M. A. Perez, J. J. Toscano and J. Wudka, Phys. Rev. **D52**, 494 (1995) [hep-ph/9506457].
18. G. Sanchez-Colon and J. Wudka, Phys. Lett. **B432**, 383 (1998) [hep-ph/9805366].
19. S. Bar-Shalom and J. Wudka, hep-ph/9904365.
20. M. B. Einhorn and J. Wudka, in preparation.
21. J. I. Kapusta, *Finite-temperature field theory* (Cambridge U. Press, N.Y., 1989). M. Le Bellac, *Thermal field theory*,(Cambridge U. Press, Cambridge, N.Y., 1996). A. L. Fetter and J. D. Walecka, *Quantum theory of many-particle systems* (McGraw-Hill, San Francisco, 1971).
22. E. Braaten and A. Nieto, Phys. Rev. **D51**, 6990 (1995) [hep-ph/9501375]; Phys. Rev. Lett. **76**, 1417 (1996) [hep-ph/9508406].

Experimental Techniques

Jürgen Engelfried

*Institito de Física, Universidad Autónoma de San Luis Potosí,
Álvaro Obregón 64, Zona Centro, 78000 San Luis Potosí, México
jurgen@ifisica.uaslp.mx, http://www.ifisica.uaslp.mx/~jurgen*

Abstract. In this course we will give examples for experimental techniques used in particle physics experiments. After a short introduction, we will discuss applications in silicon microstrip detectors, wire chambers, and single photon detection in Ring Imaging Cherenkov (RICH) counters. A short discussion of the relevant physics processes, mainly different forms of energy loss in matter, is enclosed.

INTRODUCTION

In this course we will not try to reproduce standard text books about detectors (see for example [1,2]) and descriptions of interaction with matter (a good summary can be found in [3]), covering in great details all aspects of experimental high energy physics. Due to the time restrictions (3×50 min were assigned by the organizers for this course) we will rather discuss examples on the use of some detector families. The selection is highly biased, since the author decided to use examples he knows best, e.g. he either worked on some of the detectors directly, or they were part of an experiment he participated in.

In a particle physics experiments detectors of various kinds are placed downstream (some even upstream) of the fixed target or surrounding a collision point of colliding beams. In general different detectors and electronic apparatus are required to perform the different tasks most experiment require: tracking, momentum analysis, particle identification, neutral particle detection, triggering, and data acquisition. Clearly a very important part is also the analysis of data, which comes in two parts: the event reconstruction, using all the detector information available (which requires for example a good alignment and calibration of all parts), and the final physics analysis. All these pieces cannot live by them self, to perform a successful experiment requires that all components, be it hardware or software, work all together to reach the final goal: A good and significant physics result.

Usually resources (both money and person-power) are limited when designing, building and operating an experiment. Careful consideration is necessary to decide, if some fancy or expensive detector is really necessary to obtain the physics goals or if a simpler, less expensive version would be sufficient, and the free resources can be applied to more essential parts of the experiments. Sometimes there are also political constraints that make a clear technical decision more complicated. All these arguments hold for the hardware as well as for the software part of the experiment.

In the following we will discuss generically MWPCs, the properties of a silicon microstrip array used in the SELEX experiment, and a longer part about RICH detectors used in WA89, SELEX, and CKM.

MULTIWIRE PROPORTIONAL AND DRIFT CHAMBERS

This section should be seen as an introduction to energy loss in matter and its use for particle detection, with these devices as examples.

Long time ago Geiger invented his counter: A charged particle will ionize gas, and the liberated electrons drift under the influence of an electric field to a thin counting wire. Close to the wire the field is strong enough to accelerate the electrons to ionize again further gas atoms or molecules (gas multiplication). In typical applications multiplication gains of 10^5 or more can be obtained. The electrons drift to the wire, producing there a fast signal which is usually not resolved by the preamplifier, but the backwards drifting ions induce an additional, slower, signal, which can be easily detected.

Without any further consideration this will actually lead to a spark, usually something not welcome (but in the 60's "spark chambers" where used for particle detection), since in the multiplication avalanche not only ions, but also excited atoms or molecules are produced, which will de-excite with the emission of a (UV)-photon, which can again ionize. Two tricks already used by Geiger help to avoid this: 1) A sufficiently large resistor is included in the HV line, so that the current drawn will lead to a voltage drop. 2) The addition of so-called quencher gases, usually some alcohol, to the detector gas, who will absorb the UV-photons without being ionized.

In the 60's several groups, most famous the group around Charpak at CERN, started to develop counters later known as MWPC: Instead of one wire, a lot of wires are stretch parallel, with the electrons drifting to the *nearest* wire. This allows construction of larger area detectors, something necessary in the time when people tried to develop electronic detectors to replace the bubble chambers. The resolution of the detector is given by the wire distance, and space points can be obtained be putting MWPCs under different angles. In practice the wire distance is limited to about 1 mm in small size chambers, and even more in larger areas for two reasons: The wires have to be stretched, supported by a strong frame, and, even stretched, electrostatic deflection has to be taken into account.

In the mid-60's, a new idea came up, first realized in Heidelberg by Heintze and Walenta [4]: If the time a particle passes the counting gas is known, e.g. by using for example an scintillator somewhere in the experiment, the drift time of the electrons from the point of ionization to the wire contains also space information. Putting the wires further apart, and forming with cathodes a homogeneous electric field (with exception close to the wire), a constant drift velocity in the order of several cm/μsec is observed. The resolution, limited by the diffusion of the electrons, obtained with drift chambers can be well below 100 μm, even with drifts of 10's of cm. The advantage is clearly the reduced number of wires and readout channels, with the additional cost of the need of measuring the drift time with some sort of TDC.

The energy loss of moderately charged particles (other than electrons) in matter is primarily ionization. The mean rate of energy loss is given by the Bethe-Bloch equation:

$$-\frac{dE}{dx} = Kz^2\frac{Z}{A}\frac{1}{\beta^2}\left[\frac{1}{2}\ln\frac{2m_ec^2\beta^2\gamma^2T_{\max}}{I^2} - \beta^2 - \frac{\delta}{2}\right] \quad (1)$$

Here K is some constant, ze the charge of the particle, Z and A the atomic number and charge of the medium, $T_{\max}$ the maximum energy of a free electron after one collision, I the mean excitation energy of the medium, and δ is a correction factor; the other symbols have there usual obvious meaning. dx is measured here in units of g/cm^2. The Bethe-Bloch formula only describes the mean energy loss; for finite path lengths, there are significant fluctuations in the actual energy loss. The distribution is skewed towards high values, described by the Landau distribution. Only for a thick layer the distribution is nearly Gaussian.

As seen from eq. 1, for $\beta\gamma \gtrsim 3$ the energy loss has a so-called "relativistic raise". If the momentum of the particle is known, this can be used to identify the particle. The problem is that one has to sample the energy loss several times to be able to extract the average loss (Landau!). This is explicitly done in so-called "jet chambers". An most up-to-date example is the OPAL central jet chamber [5-7], where a normal track gets measured at 159 points, using 4 m long wires spaced by 1 cm. The 3-dim space information (r: wire number, ϕ: drift time, z: charge division) gets used to measure the momentum (a magnetic field is present) and the total charge information helps to identify the particle. Another application is a TPC (Time Projection Chamber); the whole drift volume contains only gas with parallel electrical and magnetic field (to reduce the diffusion), and the electrons drift to the end plates, where wire or pads are used to obtain space information.

SILICON MICROSTRIP DETECTORS

In the 1980's, silicon microstrip detectors became used heavily in HEP. They are absolutely necessary to measure properties of particles containing charm and beauty quarks. Examples for very successful experiments using this kind of detectors include E691 at Fermilab, WA82 at CERN, and, in colliders, CDF, the 4 LEP experiments (Aleph, DELPHI, L3, OPAL), and the HERA experiments. Today there are a lot of experiments using silicon microstrips, with channel counts up to 1 million or more.

The detector allows to measure with a precision of down to a few μm the one-dimensional position of a passing charged track. Newer devices, the so-called pixel detectors, measure a two-dimensional position. The detector uses as basic detection device a pn-junction, shown in fig. 1 left, a diode which is operated in blocking direc-

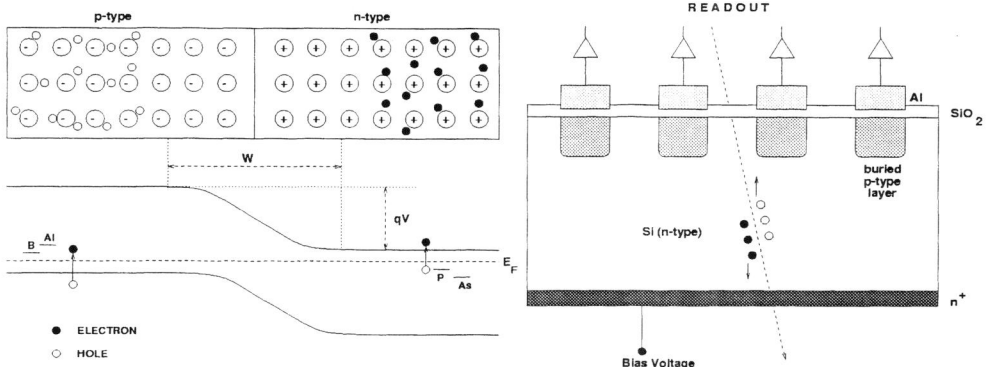

FIGURE 1. Left: Model of a pn-junction in a semiconductor. Right: Schematic drawing of silicon strip detector. Both figures taken from [8].

tion with a sufficiently high voltage so that the entire device is depleted, e.g. there are no free electron-hole pairs (also called charge carriers). In one device several (up to several thousand) of these pn-junctions are operated, arranged in parallel strips. Should a charged particle pass through the detector (see fig. 1 right), new electron-hole pair are created and one of the carrier types will drift towards the nearest strip. In Silicon, the energy loss $dE/dx \approx 3.8$ MeV/cm, and the energy needed to create one electron-hole pair is 3.6 eV[1], so in a typically 300 μm thick detector about $3 \cdot 10^4$ pairs will be created.

The construction of the detector itself seems to be under control today. There are several companies available which will produce the silicon detector with a well understood process. The smallest strip distance used today is 10 μm, so that the structure is actually much simpler than the achieved sub-micron structures in todays semiconductor chips. The real challenge in these detectors is the readout: Imagine a 5 cm × 5 cm detector with 10 μm strip distance: 5000 strips with there small signals have to be readout. Every single strip needs a preamplifier, and some kind of signal detection like a discriminator, otherwise noise will overwhelm the data acquisition. To reduce the number of cables (anyway, how to have a cable every 10 μm?) it would be nice to chain several channels together, at best even all 5000.

---

[1] The band gap in Silicon is only 1.1 eV, but Silicon is an indirect semiconductor.

The chips should then be clever enough only to send a strip number to the data acquisition, e.g. the signal gets digitized and zero suppressed already at the detector.

A system like this, called SVX [9], was developed about 10 years ago by LBL for collider experiments (CDF), and also used in WA89 [10] and SELEX [8,11]. The basic layout of the SVX system is shown in fig. 2. The current is integrated onto a ca-

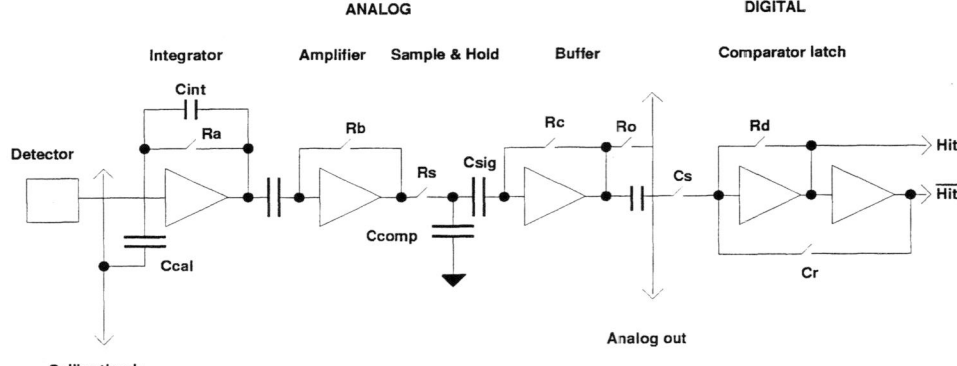

FIGURE 2. Schematic of a single SVX channel. One SVX1 chips contains 128 channels. Figure taken from [8].

pacitor as long as R_a is closed. The charge is then transfered via a sequence of switch operations and compared to a pre-stored charge. If the signal charge is bigger, the channel offers its number to the readout. Up to 64 chips with 128 channels each can be chained for readout. Since this chip was developed for collider, the fact that the chip is integrating is not very important, since a clear cycle can be performed before every collision. In fixed target operation it is not known when an interaction will happen, so in general the chip will integrate several interactions until a clear cycle is performed, which has to be closely coupled with the trigger of the experiment since during a clear the detector is not sensitive. Depending on the beam rate, the number of interactions and the sensitivity of the experiment to out-of-time tracks, the ratio of integration and clear time has to be optimized for the experiment.

The layout of a typical fixed target vertex detector is shown in fig. 3. Tracks originating from the targets are transversing the silicon planes oriented in 4 different orientations (rotated by 45°) to allow the reconstruction of tracks in space. They eventually get fit to form a vertex, and the obtained resolution is shown in fig. 4. At high momentum the resolution is limited by the strip distance, but at lower momentum multiple scattering becomes more and more important. Nevertheless, the fit takes all error contributions correctly into account, as seen from a constant $\chi^2 = 1$ for all momenta. This is another lesson to learn: more detectors is not always good.

Another nice example is the silicon drift detector. The device developed for ALICE was presented in this workshop by E. Crescio [12].

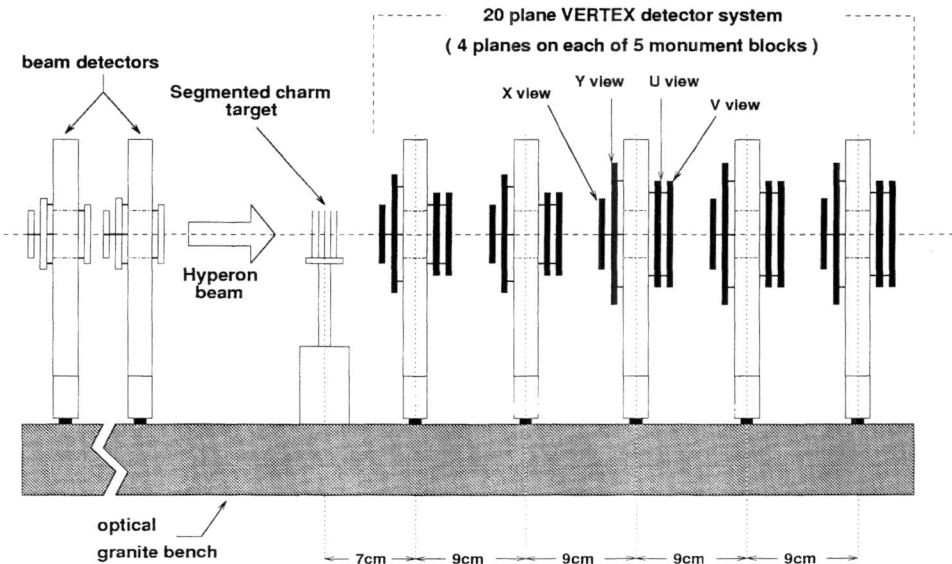

FIGURE 3. Layout of the SELEX vertex detector. After the target is a total of 20 planes with 20 μm and 25 μm strip distance in 4 orientations. Figure taken from [8].

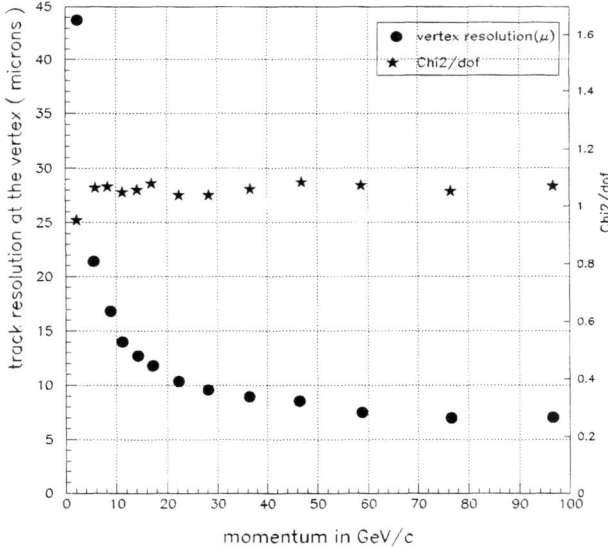

FIGURE 4. Mean χ^2/dof and vertex track resolution as a function of momenta. Figure taken from [8].

RING IMAGING CHERENKOV COUNTERS

Introduction

Even though the basic idea of determining the velocity of charged particles via measuring the Cherenkov angle was proposed in 1960 [13], and in 1977 a first prototype was successfully operated [14], it was only during the last decade that Ring Imaging Cherenkov (RICH) Detectors were successfully used in experiments. A very useful collection of review articles and detailed descriptions can be found in the proceedings of three international workshops on this type of detectors, which were held in 1993 (Bari, Italy) [15], 1995 (Uppsala, Sweden) [16], and 1998 (Ein Gedi, Israel) [17], respectively.

Charged particles with a velocity v larger than the speed of light in a medium with refractive index n will emit Cherenkov radiation under an angle θ, given by [18]

$$\cos\theta = \frac{1}{\beta n} \qquad (2)$$

with $\beta = v/c$, c being the speed of light in vacuum. The number of photons N emitted per energy interval dE and path length dl is given by [19]

$$\frac{d^2N}{dEdl} = \frac{\alpha}{\hbar c}\left(1 - \frac{1}{(\beta n)^2}\right) = \frac{\alpha}{\hbar c}\sin^2\theta \qquad (3)$$

or, expressed for a wavelength interval $d\lambda$,

$$\frac{d^2N}{d\lambda dl} = \frac{2\pi\alpha}{\lambda^2}\sin^2\theta \qquad (4)$$

By measuring the Cherenkov angle θ one can in principle determine the velocity of the particle, which will, together with the momentum p obtained via a magnetic spectrometer, lead to the determination of the mass and therefor to the identification of the particle[2].

Neglecting multiple scattering and energy loss in the medium, all the Cherenkov light (in one plane) is parallel, and can therefor be focused (for small θ) with a spherical mirror (radius R) onto a point, as shown in fig. 5. Since the emission is symmetrical in the azimuthal angle around the particle trajectory, this leads to a ring of radius r in the focus, which is itself a sphere with radius $R/2$. The radius r is given by

$$r = \frac{R}{2}\tan\theta \qquad (5)$$

The dependence of the ring radius on the momentum for different particles is shown in fig. 6

---

[2] For this reason Cherenkov detectors are usually described under the chapter "particle identification" in particle detectors books.

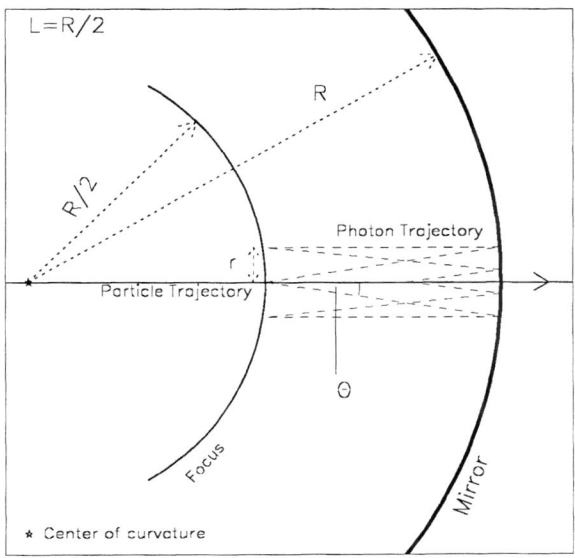

FIGURE 5. Schematic of a RICH detector

Since the number of photons is $\propto \lambda^{-2}$, most of the light is emitted in the VUV range. To fulfill equation 2, the refractive index has to be $n > 1$, so there will be no Cherenkov radiation in the x-ray region. Also it is very important to remember that n is a function of the wavelength ($n = n(\lambda)$, chromatic dispersion) and most materials have a absorption line in the VUV region, where $n \to \infty$, as shown in fig. 7, using Neon as example. Since usually the wavelength of the emitted photon is not measured, this leads to a smearing of the measured ring radius, and one has to match carefully the wavelength ranges which one wishes to use: Lower wavelengths gives more photons, but larger chromatic dispersion.

A very useful formula is obtained by integrating eq. 4 over λ (or E), taking into account all efficiencies etc., obtaining a formula for the number of detected photons N_{ph} [14]:

$$N_{\text{ph}} = N_0 L \sin^2 \theta \tag{6}$$

where N_0 is an overall performance measure (quality factor) of the detector, containing all the details (sensitive wavelength range, efficiencies), and L is the path length of the particle within the radiator. A " very good" RICH detector has $N_0 = 100 \, \text{cm}^{-1}$, which gives typically around 10 to 15 detected photons (N_{ph}) per $\beta = 1$ ring.

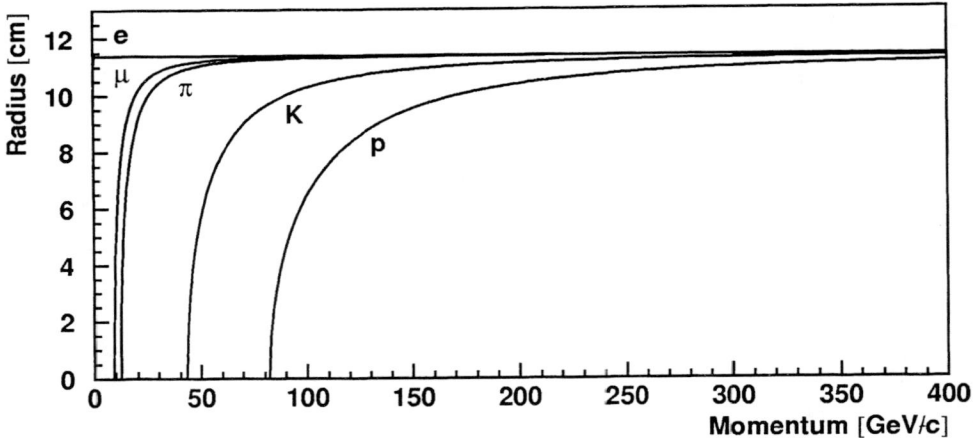

FIGURE 6. Ring radii for different particles as a function of momentum, in the case of the SELEX RICH detector ($R = 1982$ cm, radiator Neon).

The usual construction of a RICH detector is to use a radiator length of $L = R/2$, e.g. equal to the focal length; but any other configuration, like folding the light path with additional (flat) mirrors is possible.

All the presented arguments and the drawing in fig. 5 only work for small θ, which is always fulfilled in gases, since n only differs little from 1. Also important is the fact, that, should the particles not pass through the common center of curvature of mirror(s) and focal spheres, the ring gets deformed to an ellipse or, in more extreme cases, to a hyperbola. If the photon detector is able to resolve this, and the resolution is needed for the measurement, these deviations from a perfect circle have to be taken into account in determining the velocity β. In general this effect can be neglected, and all parallel particles (with the same β) will give the same ring in the focal surface, due to the fact that all emitted Cherenkov light is parallel. The position of the ring center is determined by the angle of the tracks, not by their position.

In the following, we will describe two RICH detectors used in experiments, and a new application for RICH detectors for a new, proposed experiment. The author works or worked on all of them, so the selection is clearly biased. Even so, we feel that they are good examples for the use of this kind of detectors.

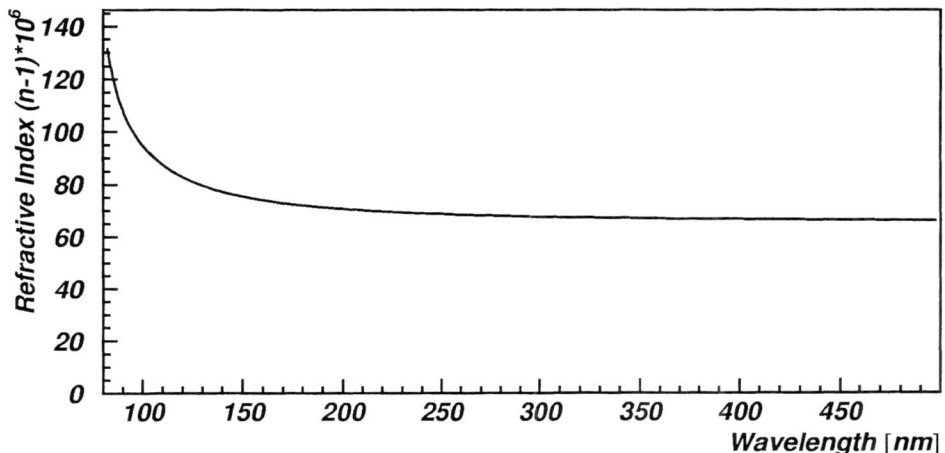

FIGURE 7. Refractive index $n-1$ for Neon as a function of wavelength at STP [20].

The CERN Omega-RICH

In the middle of the 1980's, first attempts were made to apply the prototype results obtained by Séguinot and Ypsilantis [14] to experiments in a larger scale. One of these attempts was performed at the CERN Omega facility in the West Hall. Experiments WA69 and WA82 tried to use this detector for there analysis, but only succeeded partly. An overview about this history can be found in [21]. When in 1987 a new experiment, later named WA89, was proposed [22,23], an important part was a necessary upgrade of this detector for the use by this new experiment. Two main parts where changed: New photon detectors using TMAE as photo sensitive component, and new mirrors to perform the focusing. Details about the detector can be found in [21,24–27].

As seen in the overall layout of the detector (fig. 8), a RICH detector is basically a simple device: a big box, some mirrors at the end, and photon-sensitive detectors at the entrance. The real challenge is to combine all the parameters together to obtain a perfect match for the overall system.

The size of the radiator box and the photon detector is given by the angular distribution of tracks which have to be identified at the location of the detector. Since usually this detector is placed behind a magnetic spectrometer, and the momentum spectrum of the interesting tracks depends on the physics goals of the experiment, the surfaces to cover have to be determined for every setup and experiment, usually with Monte Carlo simulations during the design phase of the experiment [24]. In the case of WA89, the mirror surface needed was about 1 m × 1.5 m, much smaller than the 4 m × 6 m covered by the original Omega-RICH. It was therefor decided to replace only the central mirrors with

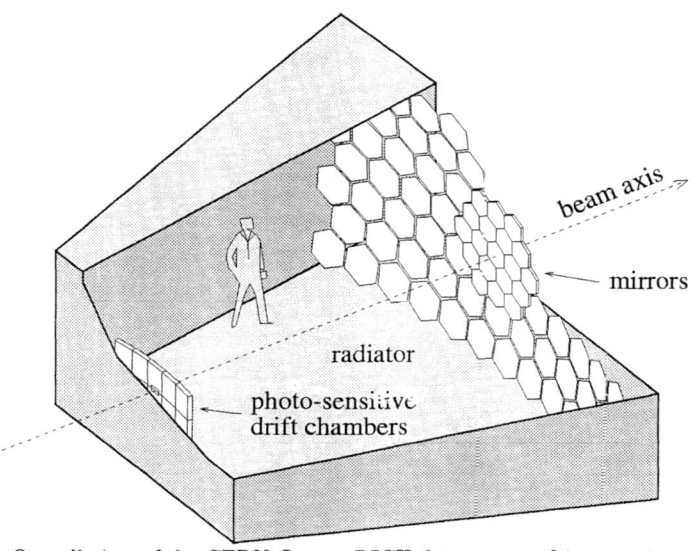

FIGURE 8. Overall view of the CERN Omega RICH detector, used in experiment WA89. The radiator gas is pure Nitrogen, so usually there is no person within the radiator box.

smaller (as seen in fig. 8), higher surface quality mirrors to obtain better resolution.

The detector surface was calculated to be $1.6\,\text{m} \times 0.8\,\text{m}$, with a spatial resolution of a few millimeters for every detected photon. The pixel size could therefor not be much bigger than also a few millimeters, leading to about 100000 pixels in the detector plane. The solution was to build drift chamber (TPC) modules, shown in fig. 9, covering an area of $35\,\text{cm} \times 80\,\text{cm}$, and approximating the focal sphere with a polygon of 5 modules. After passing a 3 mm thick quartz window (to reduce absorption), the photons hit TMAE (see fig. 10) molecules, converting via photo-effect into an single electron. TMAE is present with a concentration of about 0.1 % with the driftgas, which is otherwise pure ethan. The use of a quartz window together with TMAE as photo-sensitive gas leads that the detector is only sensitive in a small wavelength range between $165\,\text{nm} < \lambda < 230\,\text{nm}$, as demonstrated in fig. 11. TMAE has a very low vapor pressure, so that at ambient temperatures the molecule is saturated within a gas. To obtain a short enough conversion length of around 1 cm (otherwise the conversion would occur to far away from the focal plane and lead to an addition contribution to the resolution), the drift gas (ethane) is led through a bubbler, containing TMAE liquid at $30°\,\text{C}$. This means that everything after the bubbler, e.g. the whole detector including radiator box, had to be heated to $40°\,\text{C}$ to avoid condensation. Other unpleasant properties of TMAE include a high reactivity with Oxygen, producing highly electro-negative oxides, which will

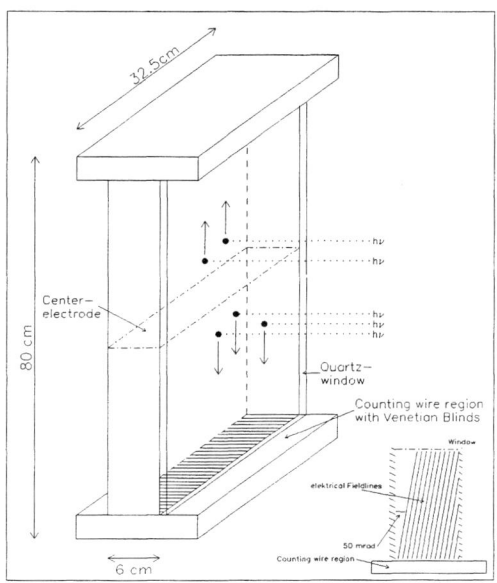

FIGURE 9. Schematic of the Omega-RICH photon detector module. $h\nu$ denotes Cherenkov photons. At the central electrode a voltage of $-40\,\text{kV}$ was applied. The inset shows a side view, demonstrating the electrical field lines (figure taken from [24]).

attach an electron easily, change the drift velocity by a factor of several thousand, leading to a loss of electrons. Since the signal is a single electron (photo-effect!) this is catastrophic. The counting gas had an Oxygen contents of < 1 ppm.

Mostly due to the presence of TMAE, the operation of this detector was not trivial. All parameters were monitored electronically, and hardware limits on some critical parameters (like temperature, Oxygen content of Ethane) lead to a automatic shutdown of the detector, waiting for an expert to arrive in the experimental hall.

Once the electron was released, it was drifting under the influence of an electric field of $1\,\text{kV/cm}$ (drift velocity $5.4\,\text{cm}/\mu\text{m}$) upwards or downwards over maximal $40\,\text{cm}$ towards $6\,\text{cm}$ long counting wires (gold-coated tungsten, $15\,\mu\text{m}$ diameter), spaced by $2.54\,\text{mm}$. The two-dimensional spatial information about the conversion point of the photon is obtained with the position of the wire and the drift time of the electron. In total, 1280 wires were used in the detector. An additional complication was that the charged particles itself where passing through the chambers, leaving a dE/dx signal of several hundred electrons, which is to be compared to the single electron which is our signal. This leads to increased requirements for

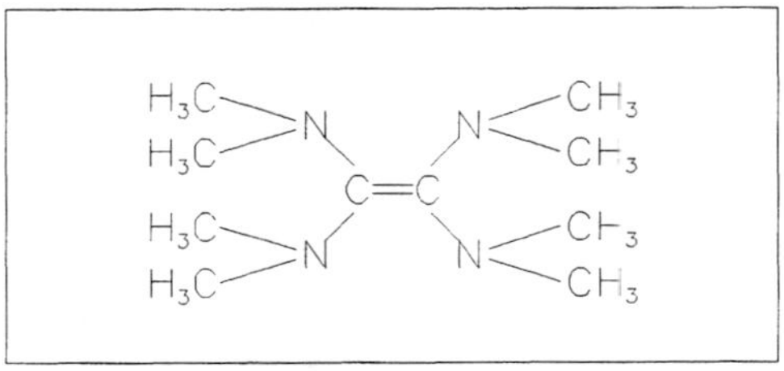

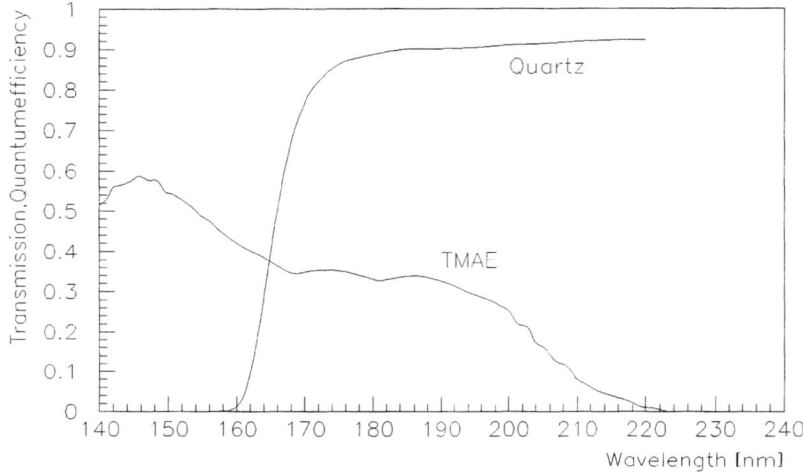

FIGURE 10. Structural formula for Tetrakis(dimethylamino)-ethylen (TMAE). This derivate of Ammonium is the molecule with the lowest ionization energy (5.3 eV).

FIGURE 11. Transmission of a quartz window actually used in the Omega-RICH, together with the quantum efficiency (probability of releasing an electron via photo-effect) of TMAE [28], both as a function of wavelength.

the wire chambers (sensitive, e.g. sufficient multiplication, to single electrons, but no sparking with several hundred electrons) and to the preamplifier electronic (not too much dead time after a big pulse).

The overall resolution allowed the separation of pions and kaons up to a momentum of about 100 GeV/c, which was exactly the design goal. This lead to a good number of physics results [29–38], which would not have been possible to obtain without the RICH detector.

The SELEX Phototube RICH Detector

At Fermilab an new hyperon beam experiment, called SELEX, was proposed in 1987 [39]. The key elements to perform a successful charmed-baryon experiments are 1) a high resolution silicon vertex detector and 2) a extremely good particle identification system based on RICH. During the following years, a prototype for the SELEX RICH was constructed and tested successfully [40], based in some part on experience gained by our Russian collaborators [41,42]. The real detector was constructed in 1993-1996, ready for the SELEX data taking period from July 1996 to September 1997. First results for the final detector were reported in [43], and publications from this year contain all details and performance descriptions of this detector [44,45].

A layout of the vessel is shown in fig. 12. The radiator gas is Neon at atmospheric

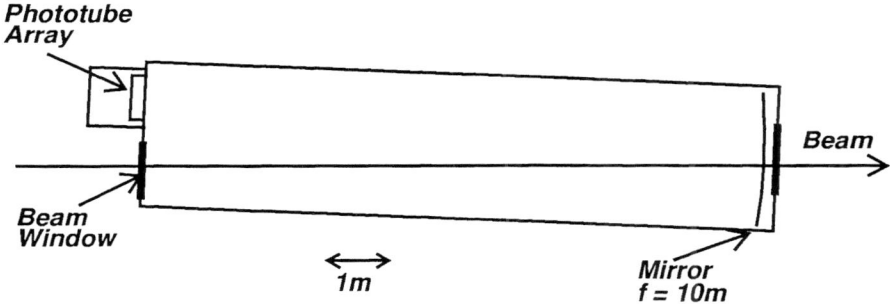

FIGURE 12. Layout of the SELEX RICH detector. The vessel has an overall length of a little over $L = 10\,\text{m}$, the mirrors have a radius of $R = 1985\,\text{m}$. The whole vessel is tilted by $2.4°$ to avoid that the particle trajectories go through the photomultipliers. Reprinted from Nuclear Instruments and Methods **A431**, J. Engelfried et al., *The SELEX Phototube RICH Detector*, pages 53-69, 1999, with permission from Elsevier Science [44].

pressure and room temperature (see fig. 7), filled into the vessel with a nice gas-system [46]: First the vessel is flushed for about 1 day with CO_2 (a cheap gas). After this the gas (mostly CO_2 and little air) is pumped in a closed system over a cold trap running at liquid Nitrogen temperature, freezing out CO_2 and the remaining water vapor. At the same time Neon gets filled into the vessel to keep the pressure constant. This part of the procedure takes about 1/2 day, and the vessel contains afterwards only Neon and about 100 ppm of Oxygen which is removed by pumping the gas over a filter of activated charcoal for a few hours, ending with an Oxygen contents of $< 10\,\text{ppm}$ in the radiator. After this all valves were closed and the vessels sits there for the whole data taking of more than 1 year at a slight ($\approx 1\,\text{psi}$)

overpressure[3].

The mirror array at the end of the vessel is made of 11 mm low expansion glass, polished to an average radius of $R = (1982 \pm 5)$ cm, coated with Aluminum and a thin overcoating of MgF_2, which gives > 85 % reflectivity at 155 nm. The quality of the mirrors was measured with the Ronchi technique [47] to assure a sufficient surface quality of the mirrors. The total mirror array covers 2 m × 1 m and consists of 16 hexagonally shaped segments. The mirrors are fixed with a 3-point mount consisting of a double-differential screw and a ball bearing to a low mass honeycomb panel. The mirrors are mounted on one sphere, and were aligned by sweeping a laser beam coming from the center of curvature over the mirrors.

The photo detector is a hexagonally closed packed 89 × 32 array of 2848 half-inch photomultipliers. A side view is shown in fig. 13. In a 3 in. thick aluminum plate

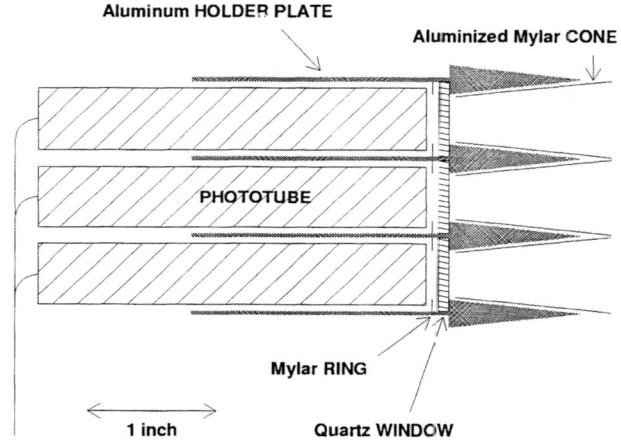

FIGURE 13. Partial cross section through the Phototube Holder Plate. For a more detailed description see text. Reprinted from Nuclear Instruments and Methods **A431**, J. Engelfried et al., *The SELEX Phototube RICH Detector*, pages 53-69, 1999, with permission from Elsevier Science [44].

holes are drilled from both sides, a 2 in. deep straight hole holds the photomultiplier, and a conical hole on the radiator side holds aluminized mylar Winston cones, which form on the radiator side hexagons, leading to a total coverage of the surface. The 2848 holes are individually sealed with small quartz windows. For the central region of the array, a mixture of Hamamatsu R760 and FEU60 tubes were used,

---

[3] Actually the closed detector still sits untouched in the PC4 pit at Fermilab. We did not open it yet.

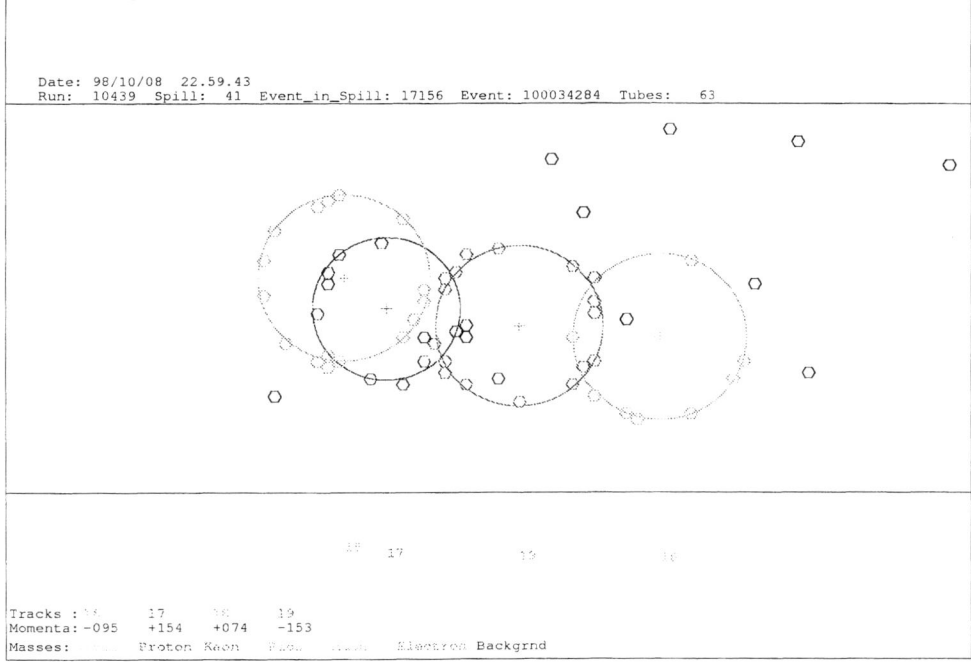

FIGURE 14. Single event display. The small hexagons represent a hit phototube, the circle shows the ring for the most probable hypothesis, and the numbers denote the track numbers, with there momenta shown at the bottom.

in the outside rows only FEU60 tubes are present. The nearly 9000 cables (signal, hv, ground) are routed to the bottom (hv) or top, where the signal cables are connected to preamp-discriminator-ecl-driver hybrid chips and finally readout via standard latch modules[4].

A single event display of the detector is shown in fig. 14, demonstrating the clear multitrack capability and the low noise of this detector. To analyze an event, the ring center is predicted via the known track parameters, and a likelihood analysis [48] for different hypothesis (the momentum is known!) is performed to identify the particle. The final performance for this detector is shown in fig. 15. The detector is nearly 100 % efficient, even below the proton threshold the efficiency is above 90 %. In the SELEX offline analysis, the RICH is one of the first cuts applied to extract physics results. SELEX presented already several results at conferences [49–54], and one paper is submitted for publication [55].

---

[4] Since the phototubes are detecting single photons, no ADCs are necessary.

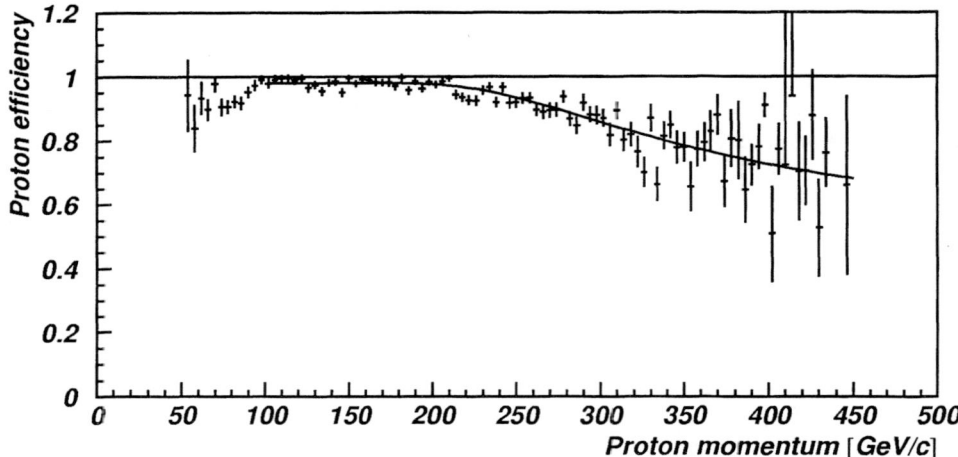

FIGURE 15. Efficiency for identifying a proton as function of its momentum. The likelihood of the track to be a proton has to be at least as big as to be a pion. Reprinted from Nuclear Instruments and Methods **A431**, J. Engelfried et al., *The SELEX Phototube RICH Detector*, pages 53-69, 1999, with permission from Elsevier Science [44].

Two RICHes for the CKM Experiment

Last year a new experiment called CKM [56] was proposed at Fermilab. The goal of the experiment is to measure the branching ratio for $K^+ \to \pi^+ \nu \bar{\nu}$ to an accuracy of 10 % (SM prediction is 10^{-10}) to measure the CKM matrix element V_{td}. To withstand the high expected physics background, the experiment will use, in addition to a conventional magnetic spectrometer, a velocity spectrometer consisting of two phototube RICH detectors, one to measure the incoming K^+, the second the outgoing π^+. The design of the detectors is based on the SELEX RICH. The HEP group in San Luis Potosí is involved in the design, construction, and testing of parts of these detectors.

ACKNOWLEDGEMENT

The author wishes to thank the organizers for the opportunity to present this course. This work was partly financed by FAI-UASLP and CONACyT. Figures 12, 13, and 15 are reprinted from Nuclear Instruments and Methods **A431**, J. Engelfried et al., The SELEX Phototube RICH Detector, pages 53-69, 1999, with permission from Elsevier Science.

REFERENCES

1. R.C. Fernow, Introduction to experimental particle physics, Cambridge University Press (1986).
2. K. Kleinknecht, Detectors for particle radiation, Cambridge University Press, 2. english edition (1998).
3. C. Caso et al., Review of Particle Physics. *Eur. Phys. J.* **C3** (1998) 1-794. http://pdg.lbl.gov
4. A.H. Walenta, J. Heintze, B. Schürlein, *Nucl. Instr. Meth.* **92** (1971) 373.
5. H.M. Fischer et al., The OPAL Jet Chamber Full Scale Prototype. *Nucl. Instr. Meth.* **A252** (1986) 331-342.
6. H. Breuker et al., Particle Identification with the OPAL Jet Chamber in the Region of the Relativistic Rise. *Nucl. Instr. Meth.* **A260** (1987) 329.
7. R.D. Heuer, A. Wagner, The OPAL Jet Chamber. *Nucl. Instr. Meth.* **A265** (1988) 11-19.
8. P. Matthew, Construction and Evaluation of a high Resolution Silicon Microstrip Tracking Detector and Utilization to determine Interaction Vertices. *Ph.D. Thesis, Carnegie-Mellon University, Pittsburg* (1997).
9. S.A. Kleinfelder et al., Lawrence Berkeley Laboratory Note.
10. W. Brückner et al., Silicon μ-strip detectors with SVX-chip readout. *Nucl. Instr. Meth.* **A348** (1994) 444-448.
11. J. Russ, et al., *IEEE Trans. Nucl. Sci.* **NS36** (1989) 471.
12. E. Crescio, The ALICE Silicon Drift Detector. *These proceedings*.
13. A. Roberts, *Nucl. Instr. and Meth.* **9** (1960) 55.
14. J. Séguinot and T. Ypsilantis, *Nucl. Instr. and Meth.* **142** (1977) 377.
15. E. Nappi, T. Ypsilantis (Eds.), Proceedings of the First Workshop on Ring Imaging Cherenkov Detectors. *Nucl. Instr. and Meth.* **A343** (1994) no. 1.
16. T. Ekelöf (Ed.), Proceedings of the Second International Workshop on Ring Imaging Cherenkov Detectors. *Nucl. Instr. and Meth.* **A371** (1996) no. 1/2.
17. A. Breskin, R. Chechik, T. Ypsilantis (Eds.), Proceedings of the Third International Workshop on Ring Imaging Cherenkov Detectors. *Nucl. Instr. and Meth.* **A433** (1999) no. 1/2.
18. P.A. Cherenkov, *Phys. Rev.* **52** (1937) 378.
19. I. Frank, I. Tamm, *C. R. Acad. Sci. URSS* **14** (1937) 109.
20. A. Bideau-Mehu et al., *J. Quant. Spectrosc. Radiat. Transfer* **25** (1981) 395.
21. H.-W. Siebert et al., *Nucl. Instr. and Meth.* **A343** (1994) 60.

22. J. Engelfried et al., A high-statistics experiment on the U(3100) and on charmed-strange baryons. *Letter of Intent CERN/SPSC/87-3, SPSC/I165* (1987).
23. A. Forino et al., Proposal for a new hyperon beam experiment at the CERN SPS using the Omega facility. *CERN/SPSC/87-43, SPSC/P233* (1987).
24. J. Engelfried, *Ph.D. Thesis, Heidelberg University* (1992), unpublished.
25. W. Beusch et al., *Nucl. Instr. and Meth.* **A323** (1992) 373.
26. U. Müller et al., *Nucl. Instr. and Meth.* **A371** (1996) 27.
27. U. Müller et al., *Nucl. Instr. and Meth.* **A433** (1999) 71.
28. R.A. Holroyd et al., *Nucl. Instr. and Meth.* **A261** (1987) 440.
29. WA94 Collaboration, S. Abatzis et al., Strange Particle Production in Sulphur-Sulphur Interactions at 200 GeV/c per Nucleon. *Nucl. Phys.* **A256** (1994) 499c-502c.
30. WA89 Collaboration, M.I. Adamovich et al., Measurement of the polarization of Λ^0, $\bar{\Lambda}^0$, Σ^+, and Ξ^- produced in a Σ^- beam of 330 GeV/c. *Z. Phys.* **A350** (1995) 379-386.
31. WA89 Collaboration, M.I. Adamovich et al., Measurement of the Ω_c^0 lifetime. *Phys. Lett.* **B358** (1995) 151-161.
32. WA89 Collaboration, M.I. Adamovich et al., Ξ^- production by Σ^-, π^- and neutrons in the hyperon beam experiment at CERN. *Z. Phys.* **C76** (1997) 35-44.
33. WA94 Collaboration, S. Abatzis et al., Charged Particle Production in S–S Collisions at 200 GeV/c per Nucleon. *Phys. Lett.* **B412** (1997) 148.
34. WA89 Collaboration, M.I. Adamovich et al., First observation of the $\Xi^- \pi^+$ decay mode of the $\Xi^0(1690)$ hyperon. *Eur. Phys. J.* **C5** (1998) 621-624.
35. WA89 Collaboration, M.I. Adamovich et al., Charge asymmetries for D, D_s and Λ_c production in Σ^-–nucleus interactions at 340 GeV/c. *Eur. Phys. J.* **C8** (1999) 593-601.
36. WA89 Collaboration, M.I. Adamovich et al., First Observation of Σ^-–e^- elastic scattering in the hyperon beam experiment WA89 at CERN. *Eur. Phys. J.* **C8** (1999) 59-66.
37. WA89 Collaboration, M.I. Adamovich et al., Production of Ξ^* resonances in Σ^- induced reactions at 345 GeV/c. Accepted in *Eur. Phys. J.* **C**.
38. WA89 Collaboration, M.I. Adamovich et al., Determination of the Total $c\bar{c}$ Production Cross Section in 340 GeV/c Σ^-–Nucleus Interactions. Submitted to *Eur. Phys. J.* **C**.
39. J. Russ et al., A proposal to construct SELEX, *Fermilab P781* (1987), unpublished. J. Russ, *Nucl. Phys.* **A585** (1995) 99.
40. M.P. Maia et al., *Nucl. Instr. and Meth.* **A326** (1993) 496.
41. V.A. Dorofeev et al., *Physics of Atomic Nuclei* **57** (1994) 227.
42. A. Kozhevnikov et al., *Nucl. Instr. and Meth.* **A433** (1999) 164.
43. J. Engelfried et al., *Nucl. Instr. and Meth.* **A409** (1998) 439.
44. J. Engelfried et al., *Nucl. Instr. and Meth.* **A431** (1999) 53-69.
45. J. Engelfried et al., *Nucl. Instr. and Meth.* **A433** (1998) 149.
46. R. Richardson and R. Schmitt, *Adv. in Cryo. Eng.* **41B** (1996) 1907.
47. L. Stutte, J. Engelfried and J. Kilmer, *Nucl. Instr. and Meth.* **A369** (1996) 69.
48. U. Müller et al., *Nucl. Instr. and Meth.* **A343** (1994) 279.
49. A. Kushnirenko, for the SELEX Collaboration, Charm Physics Results from SELEX. *Proceedings of Heavy Quark 98*.

50. SELEX Collaboration, I. Eschrich et al., Hyperon Physics Results from SELEX. *Proceedings of Heavy Quark 98.*
51. SELEX Collaboration, V. Kubarovski et al., Radiative Width of the $a2$ Meson. *Proceedings of ICHEP 98.*
52. SELEX Collaboration, J. Russ et al.: First Charm Hadroproduction Results from SELEX. *Proceedings of ICHEP 98.*
53. SELEX Collaboration, F.G. Garcia, S.Y. Jun et al., First Charm Baryon Physics from SELEX (E781). *Proceedings of DPF99.*
54. SELEX Collaboration, M. Iori et al., Charm hadroproduction results from Selex. *Proceedings of EPS-HEP99.*
55. SELEX Collaboration, S.Y. Jun et al., Observation of the Cabibbo-suppressed decay $\Xi_c^+ \to pK^-\pi^+$. Submitted to *Phys. Rev. Lett.*
56. R. Coleman et al., CKM – Charged Kaons at the Main Injector – A proposal for a Precision Measurement of the Decay $K^+ \to \pi^+ \nu \bar{\nu}$ and Other Rare K^+ Processes at Fermilab Using the Main Injector. *FERMILAB-P-0905* (1998), unpublished.

PHYSICS AT THE ELECTRON-PROTON COLLIDER HERA

A. De Roeck

CERN, 1211 Geneva 23, Switzerland

Abstract. Physics results from the electron-proton collider HERA are presented. Particular attention is given to the proton structure function measurements. Diffraction, photon structure and hadronic final states are discussed as well. The high Q^2 region is revisited two years after the reported hint of an excess of events above Standard Model expectation.

I INTRODUCTION AND KINEMATICS

A new breakthrough in Deep Inelastic lepton-hadron Scattering (DIS) has been achieved by HERA, which provides for the first time collisions of electrons and protons in a collider mode, thereby increasing $\sqrt{s}$, the centre of mass system (CMS) energy of the scattering process, by an order of magnitude compared to the traditional fixed target experiments. Collisions are produced by 27.5 GeV electrons on 820 GeV protons, yielding a CMS energy of 300 GeV, and are recorded by two experiments H1 [1] and ZEUS [2]. A third experiment, HERMES [3], makes use of the polarised electron beam and studies DIS on a (polarised) fixed target.

HERA is above all a large microscope which studies the structure of the proton with a resolution given by the four momentum transfer, Q^2, from the electron to the proton (or quark therein). The larger the Q^2, the smaller the distances one can resolve, due to the Heisenberg uncertainty relation. For its largest Q^2 values HERA can resolved distances down to about 10^{-16} cm, an order of magnitude better than fixed target experiments.

A wealth of new results has been obtained in DIS (Q^2 > few GeV^2) and photoproduction ($Q^2 \sim 0$) during the first years after the startup of HERA. Major highlights are measurements of the proton structure in terms of the structure function F_2 and the gluon distribution $xg(x)$. Further, a perhaps unexpected large fraction of the events (about 10%) has been found to contain a large rapidity gap, as illustrated in Fig. 1, and is referred to as diffraction. The figure also illustrates some of the kinematic variables used which are $Q^2 = -q^2$ (photon virtuality) and

$W^2 = (q+P)^2$ (invariant mass of the hadronic system squared), where q and P denote the four-vectors of the virtual photon and the proton respectively. Further we define the Bjorken-x variable $x = Q^2/(2P \cdot q)$ and the inelasticity $y = P \cdot q/(P \cdot P_e)$, with P_e the four-vector of the incoming electron. The Bjorken-x variable is in the Quark Parton Model (QPM) – i.e. in absence of QCD effects – a measure of the momentum fraction of the proton carried by the struck quark.

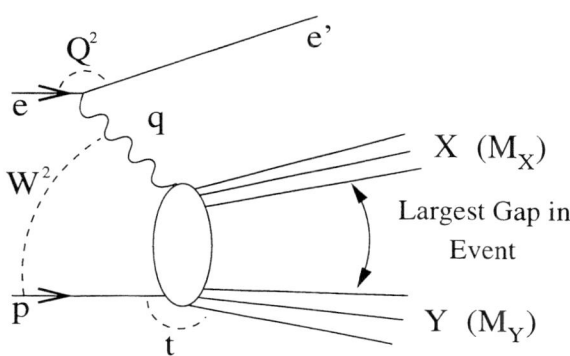

FIGURE 1. Schematic view of a DIS event with definition of some kinematic variables (see text). Some of the events were found to have a large rapidity gap (see section 5), in which case there are two distinct hadronic systems X and Y.

II PROTON STRUCTURE

It has been a vintage decade for structure function measurements from DIS processes. The kinematic region presently covered in the momentum transfer squared, Q^2, and the scaling variable Bjorken-x is shown in Fig. 2. The down right corner summarizes the reach of the heroic fixed target experiments with muon, electron and neutrino beams. The extension of this region achieved by the HERA collider experiments is clearly seen.

Structure functions measure the structure of extended objects such as the proton: they relate the scattering cross-sections to the parton distributions in the nucleon. For neutral current DIS (i.e. with γ, Z^0 exchange) we have

$$\frac{d^2\sigma_{NC}^{Born,e^{\pm}}}{dQ^2 dx} = \frac{4\pi\alpha^2}{xQ^4}[y^2 xF_1(x,Q^2) + (1-y)F_2(x,Q^2) \mp (y-\frac{y^2}{2})xF_3(x,Q^2)] \quad (1)$$

where at not too high Q^2 (i.e. $Q^2 < 3000$ GeV^2) the parity violating structure function $F_3(x,Q^2)$, and not too high y (i.e. $y < 0.6$) the longitudinal structure function $F_L = F_2 - 2xF_1$, are both small and can be taken as corrections to the

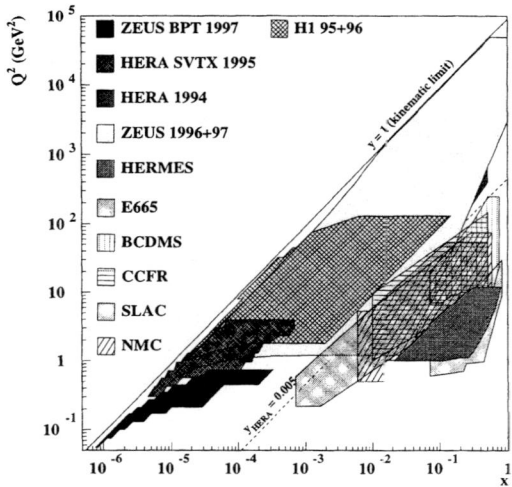

FIGURE 2. Regions in the x and Q^2 plane covered by the DIS experiments.

cross-section. More on scattering formulae and basics of DIS is given in [4]. Hence, the differential cross-section is essentially proportional to F_2, which in leading order (LO) gives $F_2(x, Q^2) = \sum_{flavours} e_i^2 x [q_i(x, Q^2) + \bar{q}_i(x, Q^2)]$ where q_i are the quark distributions in the proton, and e_i^2 the quark charges squared. Expressions for F_2 in next-to-leading order (NLO) are available (see e.g. [4,5]) and generally used to analyse the data. Quark distributions are key elements in the understanding of hard processes at hadron colliders, such as the LHC. The production rate of a given particle, e.g. the Higgs, depends directly on the parton distributions of the partons involved to produce a Higgs. The precise knowledge of parton distributions is therefore of major importance.

There are various ways to measure kinematic variables at HERA. A classical way is to use only the energy and angle of the scattered electron, θ_e, E'_e:

$$Q^2 = 4E_e E'_e \cos^2 \theta/2; \quad y_e = 1 - (E'_e/E_e)\sin^2 \theta_e/2; \quad x = Q^2/sy \qquad (2)$$

with E_e the incident electron beam energy. The angles are defined with respect to the proton beam direction. Since the HERA detectors also allow for measurements of the hadronic final state, many more formulae are available [6], which use hadrons or a combination of hadrons and the scattered electron.

The regions of interest in Fig. 2 are:

- The region $Q^2 \to 0$: i.e. the region where one can study the transition from real photoproduction to DIS.

- The region $Q^2 \to 10^5$ GeV^2: i.e. the region of Electroweak Unification, and also where new phenomena could be expected, such as signs of substructure.

- The region $x \to 10^{-4} - 10^{-6}$: i.e. the region where, as will be discussed, the parton densities become very large.

- The region $x \to 1$: i.e. the 'traditional' region which is dominated by the valence quarks.

A Very Low Q^2 Data

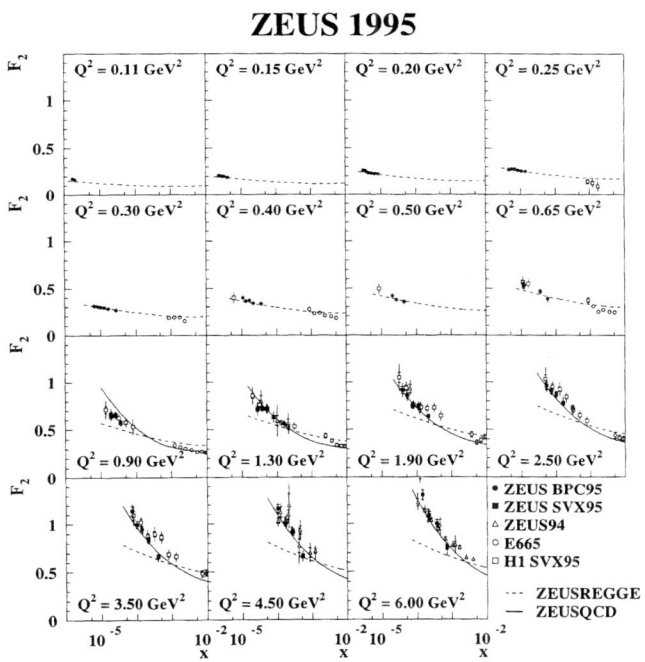

FIGURE 3. Low Q^2 data for different Q^2 bins, together with the ZEUS Regge fit and ZEUS QCD fit.

The first F_2 measurements in the newly reached low x region at HERA have led to a surprise. Contrary to the tendency anticipated from measurements at fixed target energies, the structure function was found to rise steeply with decreasing x. This dramatic effect is shown in Fig. 3, in the lowest row ($Q^2 > 3.5$ GeV^2). A lot of speculation started immediately after the first HERA results were released, mostly in terms of anticipated new QCD effects, to which we turn later. At the start only data above Q^2 of about 5 GeV^2 were available and an important question was what would happen with F_2 at lower Q^2. Would this rise persist? We knew that for real photoproduction, i.e. data with $Q^2 = 0$, the cross-section is almost flat.

Both collaborations have made an effort to probe this region of lower Q^2, which is limited by the small angle of the scattered electron (see eq. 2). By taking data with a shifted vertex and instrumenting the region around the beampipe with extra detectors, smaller Q^2 values could be reached [7,8] as shown in Fig. 3. The smallest value reached, $Q^2 = 0.045$ GeV^2 [9], is shown in Fig. 4. The tendency is clear: the rise of the F_2 at small x is reduced strongly for decreasing Q^2.

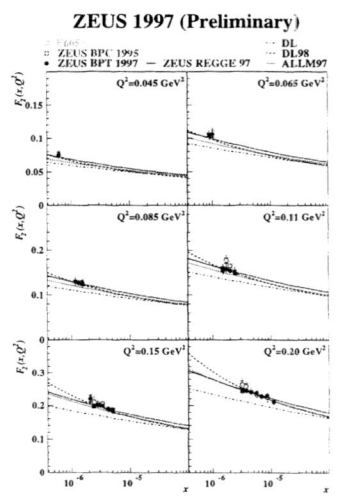

FIGURE 4. Preliminary low Q^2 data together with Regge inspired models.

A classical test of perturbative QCD (pQCD) is to check the Q^2 dependence of the structure function data at different x values. Indeed, evolution equations have been formulated which allow the value of $F_2(x, Q^2)$ to be calculated, when it is known at $F_2(x, Q_0^2)$ i.e. at a scale Q_0^2. The so called DGLAP (Dokshitzer-Gribov-Lipatov-Altarelli-Parisi) [10] evolution equations have been turned into a well oiled machinery over the last 15 years, and have been successful in describing the fixed target data [5]. Its success has been extended to the HERA data in the region of Q^2 larger than a few GeV^2. Fig. 3 shows the result of the QCD fit (labelled ZEUSQCD in the figure) in the low Q^2 region. These general QCD fits to data work well, as will be further emphasised in the next section. Note however that this success is partially a result of the allowed freedom of the choice of 'starting distribution' at the scale Q_0^2: the so called non-perturbative input. This 'non-perturbative input' can unfortunately not (yet) be accurately calculated within QCD.

Below $Q^2 = 1$ GeV^2 it becomes difficult to accommodate the QCD fit to the data, and a different description is necessary. ZEUS has shown that a Regge type of description of the type $F_2 \sim m_0^2/(m_0^2+Q^2) \cdot (A_{I\!R} W^{2(\alpha_{I\!R}-1)} + A_{I\!P} W^{2(\alpha_{I\!P}-1)})$ works well

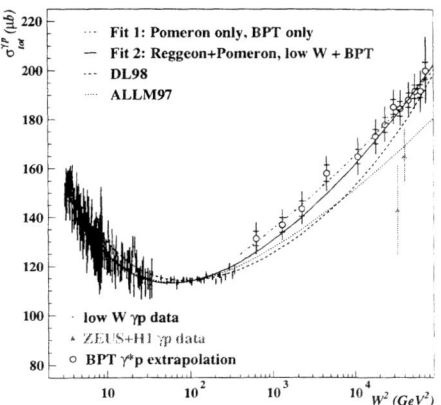

FIGURE 5. Total γp cross-section data as function of W with Regge inspired models.

in that region (with $\alpha_{I\!R}$ fixed to 0.5 and $\alpha_{I\!P}$ fitted to 1.097 ± 0.002). Many models which attempt to describe the transition region from DIS to photoproduction have have been compared to these low x, low Q^2 data, with varying success (see e.g. [5]). Much remains to be understood on the transition region from high Q^2 to low Q^2, or hard to soft processes, a topic which is intensively studied, using the recent HERA data.

ZEUS went even a step further and 'extrapolated' their low Q^2 data to $Q^2 = 0$, using the Regge predicted Q^2 dependence, given above. The result is shown in Fig. 5 as function of γp CMS energy squared W^2 which is proportional to $1/x$ at fixed Q^2. The data are however found to be systematically higher than the 'direct' γp cross-section measurements, using events where Q^2 was really zero. One should note that no theoretical uncertainty for the extrapolation has been included. The best test to verify this effect would be a more precise direct measurement of the γp cross-section, which is under way. In case the extrapolated data are confirmed, this would mean that the total cross-section in γp rises faster than in e.g. pp, and would be a very interesting reflection of the different hadronic interaction of a photon compared to a proton (see also section 4).

B Medium Q^2 Data

With medium Q^2 data we mean data with Q^2 above a few GeV^2, to be safely in the pQCD region, and Q^2 below a few 1000 GeV^2, outside the region where Z^0 exchange becomes important. With approximately 30 pb^{-1} collected per experiment, HERA has delivered very precise F_2 data in this region, where total errors

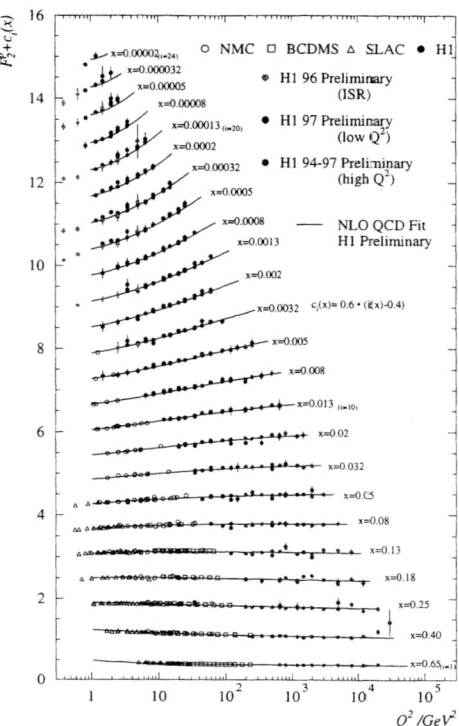

FIGURE 6. Structure function F_2 versus Q^2 with H1 DGLAP fit.

of better than 5% are achieved in a large part of it. The errors will be improved to the 2% level shortly. This is very important as it reflects directly into the precision with which we know the quark distributions in the proton.

A summary of the H1 data versus Q^2 is shown in Fig. 6. Scaling violations of F_2 are clearly observed. The F_2 data are found to exhibit a continued rise with decreasing x, as shown in Fig. 7, for all Q^2 values. This means that the number of partons increases correspondingly. However this cannot continue forever, as the proton would be completely 'filled' up with partons and saturate. Such a state is of high theoretical interest since it has a large parton density where parton annihilation diagrams should become important, and α_s is large, allowing pQCD techniques to be used to study this region.

The rise of F_2 can be quantified by a fit of the data to the form $x^{-\lambda}$ for $x < 10^{-2}$ at each Q^2 bin. The result is shown in Fig. 8; note that the value of $\langle x \rangle$ (upper scale) is different at each Q^2 (lower scale) value, due to the kinematic range of the data. The power λ rises from around 0.1 at small Q^2 to around 0.3 at large Q^2.

The analysis of the rise of the F_2 at small x, shown first in 1993, has originally led

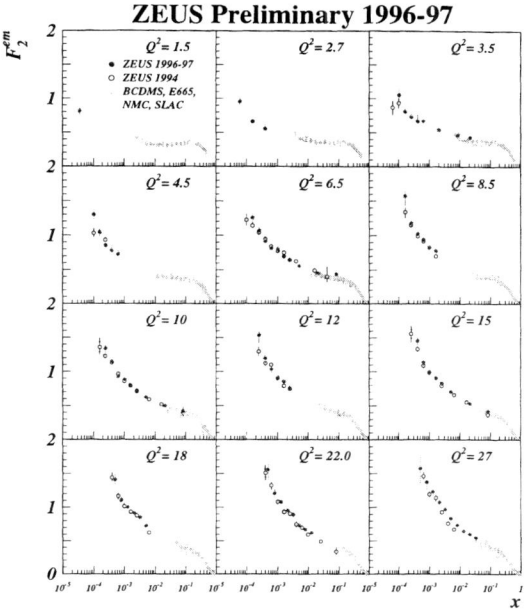

FIGURE 7. Structure function F_2 versus x at medium Q^2.

to exploring new pQCD avenues such as the ones due to the low x QCD evolution or the BFKL (Balitsky-Fadin-Kuraev-Lipatov) pomeron [11]. Meanwhile it has become clear that 'traditional' QCD evolution based on the DGLAP [10] evolution equation predictions can describe the data for Q^2 values larger than 1 GeV^2, as demonstrated by the fit (solid line) shown in Fig. 6 (and in Fig. 3), provided a suitable non-singular non-perturbative input is taken at a low starting scale Q_0^2 for the evolution ($Q_0^2 \sim 1$ GeV^2, details of the fits and data are given in [8,12,13]). However, also improved evolution equations which include $\ln 1/x$ resummation terms or attempts for a unified evolution equation can describe the data equally well [14,15]. Hence we conclude that the data at present do not require effects beyond DGLAP, but they can not exclude that specific low x effects are already at work in the HERA regime [5]. We will turn back to other possible signals for the $\ln 1/x$ terms in section 3.

The DGLAP evolution equations have been further pursued to analyze the scaling violations of F_2 and in particular to extract the gluon density at small x. The NLO order result in the $\overline{MS}$ scheme is shown in Fig. 9. The error bands include systematic and statistical uncertainties [8]. The knowledge of the gluon at low x has reached a precision of about 10-15%. A remarkable effect is seen: while the gluon grows very strongly with decreasing x for Q^2 values above a few GeV^2, it collapses completely for $Q^2 \sim 1$ GeV^2. Such a dramatic change seems to hint that

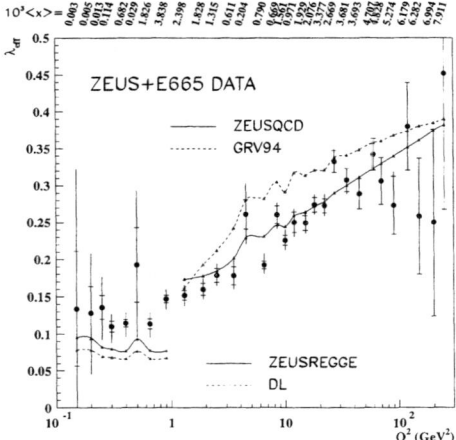

FIGURE 8. Effective λ (from $F_2 \sim x^{-\lambda}$) as function of Q^2 for ZEUS and E665 data, together with the ZEUS QCD and ZEUS Regge calculations.

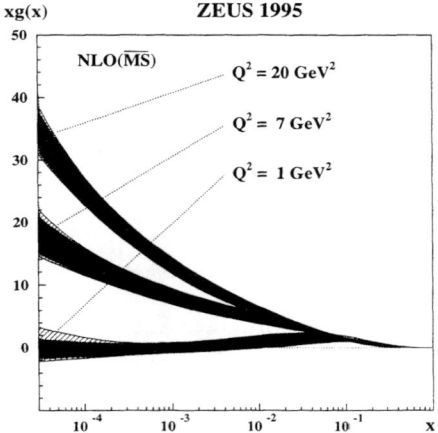

FIGURE 9. The gluon distribution versus x for 3 different Q^2 values, with error bands including statistical and systematical error sources.

either DGLAP is at the verge of breaking down at these scales, or other effects such as higher twists –generally not taken into account in such fits– are required and/or parton screening effects are becoming important.

Further attempts have been made to try to understand the F_2 data. In one such attempt the logarithmic slope $dF_2/d\ln Q^2$ is studied versus x. Fig. 10 shows the derivation of this quantity for the NMC fixed target and ZEUS data. Note that the $\langle x \rangle$ of the data depends on Q^2. While $dF_2/d\ln Q^2$ continues to rise with decreasing Q^2, down to $Q^2 = 0.5$ GeV^2, $\langle x \rangle \sim 0.001$ for the fixed target data, the ZEUS data show a turn-over around $Q^2 = 10$ GeV^2, $\langle x \rangle \sim 0.001$.

The original DGLAP based parton density distributions predictions, such as GRV94 [16] do not follow this turn over and continue to rise. The DGLAP QCD fit to these new data *does* exhibit the turn over, but at the expense of the resulting 'valence gluons' as mentioned above. Originally this plot lead to exciting discussions on whether these data could be a sign of saturation. The argument followed basically the observation of Prytz [17]: when gluons dominate the evolution at small x, then $xg(x) \sim dF_2(x/2, Q^2)/d\ln Q^2$. This would mean that at low x the gluon is consistent with the saturation condition $xg(x) \sim R_0^2 Q^2$, with R_0^2 a constant that can be related to the size in the proton over which the saturation manifests itself.

This is however a too naive interpretation of these data, and ignores the fact that at lower x the measurement is made in a different interval of Q^2, due to the available region in x and Q^2. If the rise of the slope $xg(x) \sim dF_2(x, Q^2)/d\ln Q^2$ is reduced at smaller Q^2, the combined shift in x and Q^2 can *cause* this apparent turn over with decreasing x. Whether on top of this reduction due to the restricted range of the data available, there is still a reduction due to saturation effects, is presently still debated [18].

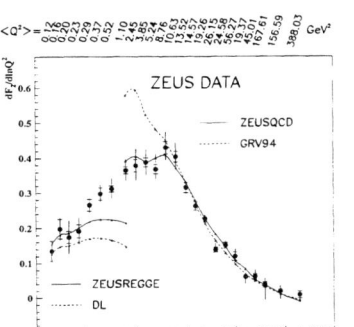
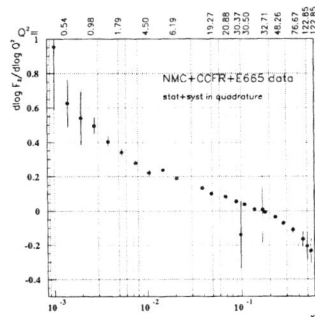

FIGURE 10. $dF_2(x, Q^2)/d\ln Q^2$ as function of x calculated by fitting F_2 data in bins of x to the form $a + \ln Q^2$: (left) ZEUS data, (right) NMC data. For details see [12]. The results are compared with models and phenomenological fits of the F_2 data.

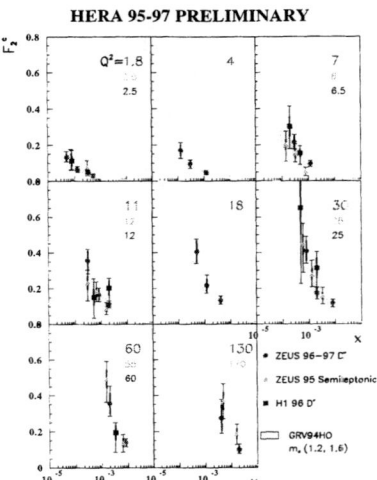

FIGURE 11. F_2^c as function of x for different Q^2 bins, compared with the prediction from GRV parton densities.

C F_2^C and F_L at Low X

Extracting the gluon via the DGLAP fits, as discussed above, is a rather indirect extraction. It would be reassuring if more direct measurements could confirm this measurement of the gluon density. A good candidate for this is the measurement of DIS events which contain charm.

Charm in DIS is mainly produced via the Boson Gluon Fusion (BGF) process, and thus is a direct and independent measurement of the gluon density. Events with charm are detected via $D^* \to D^0 \pi_{slow} \to K\pi\pi_{slow}$ or via semi-leptonic decays. The charm part of F_2 has been quantified as a structure function F_2^c defined as: $\frac{d^2\sigma_{c\bar{c}}X}{dxdQ^2} = \frac{2\pi\alpha^2}{xQ^4}[1+(1-y)]F_2^c(x,Q^2)$. The measurements of F_2^c are shown in Fig. 11. One observes a strong rise of F_2^c with decreasing x, as for F_2 and an agreement with the QCD prediction (GRV94 gluon density) based on the BGF mechanism and using the gluon from the QCD fit. Hence F_2 and F_2^c are driven by the same gluon. It would be interesting to check if this would be true also at smaller Q^2 values, but unfortunately here one hits the validity boundary of the approach.

Further checks on the gluon distribution can be made by studying dijets events and the structure function F_L which at low x are both dominated by the gluon density. Experimentally with dijets events one cannot reach very low x values, due to the requirement of jets with E_T typically above 5 GeV. The gluon extracted from dijets is found to be consistent with the one extracted via the QCD fit of F_2

F_L cannot be directly accessed from the present data: measurements at a different energy would be required to determine the cross-section at the same x, Q^2 value but for different y value(s), and thus allow to decouple F_2 and F_L in Eq. 1. Two indirect methods have been developed by H1. The first method is based on extrapolating the QCD fit, made in a region of low $y < 0.3$ where F_L essentially does not contribute, to the region of high $y \sim 0.7$, where sensitivity to F_L can be expected. The other is based on assuming a linear extrapolation of dF_2/dy. Both methods lead to similar results, and are consistent with being driven by the same gluon as extracted form the DGLAP fits of F_2.

Hence, all methods we have at hands teach us that when applying them to data, consistently the same gluon distribution emerges!

D Polarized Structure Functions at HERA

If both beams are polarised, then the polarised structure function g_1 of the proton can be measured. It tells us what part of the proton spin is carried by the quarks. Results of the EMC experiment at the end of the eighties [19] where shocking! It appeared that the quarks did not contribute at all. Where is the spin of the proton?

Encouraged by these results several experiments were followed up after this discovery by the EMC. One such experiment is HERMES which uses the natural transverse polarization of electrons in the HERA ring due to the Sokolov-Ternov effect of synchrotron radiation. With the aid of spin rotators this polarization is turned into a longitudinal one. HERMES is a spectrometer with particle identification possibilities, to detect the scattered electron and much of the hadronic final state from interactions of the polarised electrons from HERA on a polarized fixed target, at a CMS energy of 7 GeV. Apart from inclusive distributions HERMES has a strong potential to study semi-inclusive asymmetries by tagging hadrons from the final state.

Fig. 12 shows the structure function $g_1(x, Q^2)$ for the proton, calculated from the longitudinal asymmetries. In the QPM this structure function is the difference of the (charged weighted) density of quarks with spin parallel to the one with spin anti-parallel to the proton spin. The results are compared with other fixed target experiments, and in general a good agreement is found. This is a non-trivial result as the HERMES experiments uses quite different techniques (gas storage cell for the polarized target) than the other fixed target experiments, and thus constitute a powerful systematic check of the results.

The results of all new experiments are in agreement, and it confirms the earlier EMC result that the contribution of the quarks to the proton spin is small. Where is the missing spin of the proton? Analyses within the framework of perturbative QCD of polarised parton distributions can be performed, much like the one for the unpolarised case discussed before [20], and show that the contribution of the polarised gluon distribution may well be large, but its present error is almost 100%. HERMES recently used its final state particle analysis to try to measure the contribution of the gluons to the spin directly, via high p_T particles, much like the jet analysis at H1 and ZEUS. This is shown in Fig. 13, and the result indicates

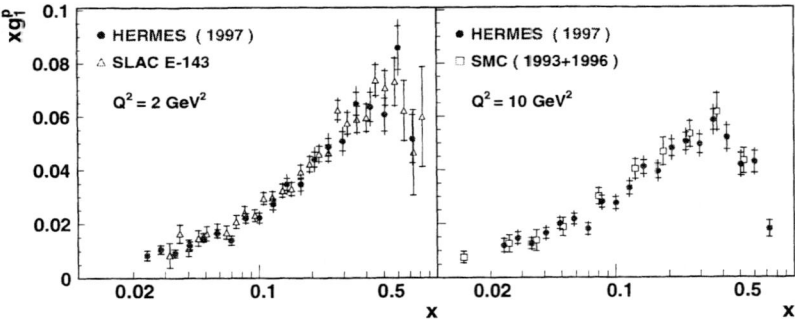

FIGURE 12. Measurements of the $g_1(x)$ polarized structure function of the proton from HERMES (full circles), compared with results from E-143 (open triangles) and SMC (open squares).

that indeed the polarised gluon contribution could well be very large! Certainly HERMES and other data in the near future will help us to unravel the proton 'spin puzzle'.

If the protons at HERA were to be polarized as well, the collider experiments could study polarized ep scattering in the same kinematic range as the presently unpolarized measurements, a very exciting project both for machine and physics [21].

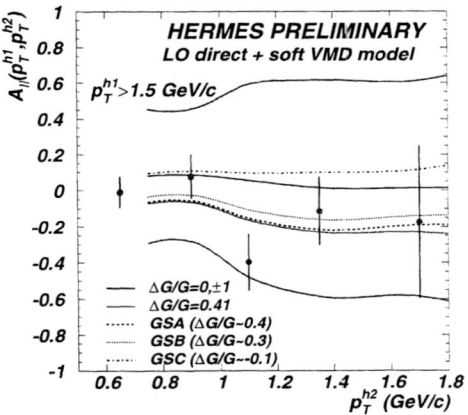

FIGURE 13. Measurements of the asymmetry for high p_T particles from HERMES

III HADRONIC FINAL STATES

The HERA detectors can measure the hadronic final state of the DIS and photoproduction events. The hadronic final state allows for measurements in the perturbative and non-perturbative sector of QCD, such as jet rate, extraction of α_s, hadronization, heavy flavour analysis, etc.

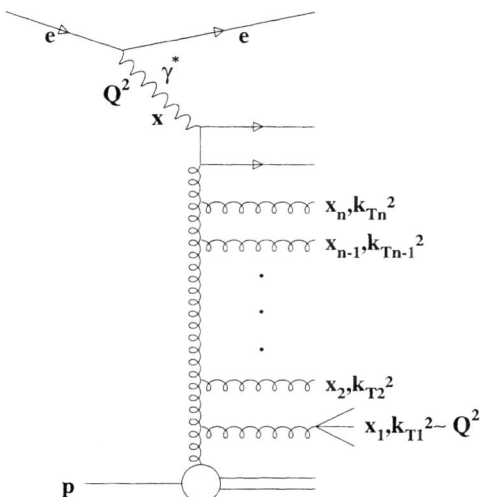

FIGURE 14. Diagram for a low x Deep Inelastic Scattering event.

Some applications of the final state information were discussed before, such as the extraction of the gluon distribution from dijet events or events with charm. Another application is the measurement of the strong coupling constant α_s. This quantity can be accessed by e.g. measuring the rate or cross-section of events with two jets in the final state, since these are α_s corrections to the QPM. Presently values of $\alpha_s = 0.120 \pm 0.007$ (ZEUS) and $\alpha_s = 0.118 \pm 0.006$ (H1) have been reported based on these observables. Furthermore the fragmentation of the quark kicked out of the proton by the virtual photon has been compared with the fragmentation of quarks produced in e^+e^- annihilation. When allowance is made for the different higher order contributions in DIS and e^+e^-, in general good 'fragmentation universality' is observed.

The final state measurements are particularly interesting for low x dynamics studies. A low x interaction can be pictured by a ladder diagram as shown in Fig. 14: An initial cascade develops from the proton, whereby at each parton branching vertex the x value gets reduced (shared between the outgoing partons), until a quark at low x is probed by the virtual photon. The multiplicity and the x distribution of these emitted partons differ significantly in different approximations of QCD dynamics at small x.

At low x, pQCD evolution is complicated by the occurrence of two large logarithms in the evolution equations, namely $\ln 1/x$ and $\ln Q^2$. A complete perturbative treatment in the low x region is not yet available, and different approximations are made resulting in different parton dynamics. At high Q^2 and high x pQCD requires the resummation of contributions of $\alpha_s \ln(Q^2/Q_0^2)$ terms, yielding the DGLAP [10] evolution equations. However at small x the contribution of large leading $\ln 1/x$ terms may become important. Resummation of these terms leads to the BFKL [11] evolution equation. Hence a pertinent and exciting question is whether these $\ln 1/x$ contributions to the parton evolution can be observed experimentally in final states (remember that F_2 did not appear to be a sensitive enough measurement for this study).

Differences between different dynamical assumptions for the parton cascade are expected to be most prominent in the phase space region towards the proton remnant direction, i.e. away from the scattered quark. In the HERA laboratory frame this corresponds to a region of small polar angles and has been generically termed "forward region". In previous analyses results have been presented on forward jet and forward inclusive charged and neutral pion production [22,23]. The data showed an excess over expectation from DGLAP (Monte Carlo model) calculations, but it appeared that this could be remedied by adding 'resolved photon' diagrams to the DIS diagram [24].

New data on forward single π^0 production, shown in Fig. 15, became available recently at larger transverse momentum, p_T. A calculation based on pQCD which uses the BFKL formalism for the perturbative part, and fragmentation functions for the hadronization, is available [25] and compares well to the data. The resolved model shows however some deficiencies. There are remaining questions on the scales to be used in all these calculations, but these forward pion data could well be the first which show evidence for BFKL effects in the hadronic final state. When confirmed, this would be a very important observation, which will guide theory on the importance of low x effects in present data.

Heavy flavours, in particular charm quark production, have been studied over the last 5 years at HERA. The production is generally well understood in terms of Boson Gluon Fusion. Recently, beauty production has been measured and the outcome was a surprise [26]: the cross-section, be it measured in a limited range of phase space, was found to be five times larger than the one predicted by the (LO) program AROMA. For $Q^2 < 1$ GeV^2, $0.1 < y < 0.8$ and $p_T^\mu > 2.0$ GeV/c, $35^0 < \theta^\mu < 130^0$, with p_T^μ, θ^μ the transverse momentum and polar angle of the muon measured in the lab frame respectively, the cross-section is (preliminary) $\sigma_{vis}(ep \to b\bar{b}X) = 0.93 \pm 0.8(stat.)^{+0.21}_{-0.12}(syst.)$ nb. AROMA predicts 0.19 nb. It turns out however that in NLO this discrepancy is reduced to a factor of about 1.7.

Finally, we mention the interesting possibility to measure instanton events in DIS at HERA. Instantons describe the quantum tunneling between different gauge rotated classical vacua in QCD and give important information on the complicated structure of the ground state in the theory of strong interactions. It has been

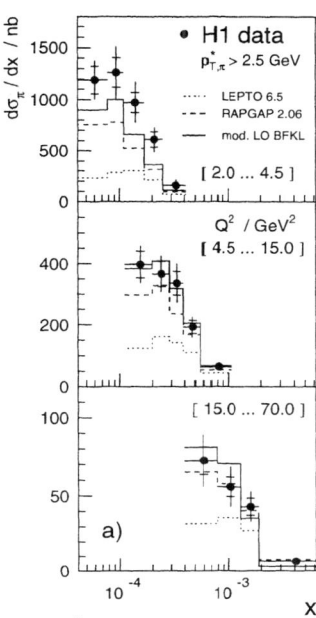

FIGURE 15. Inclusive differential π^0 cross-sections measured in the forward region for $0.1 < y < 0.6, 5^0 < \theta_\pi < 25^0$ and $E_\pi > 8.2$ GeV.

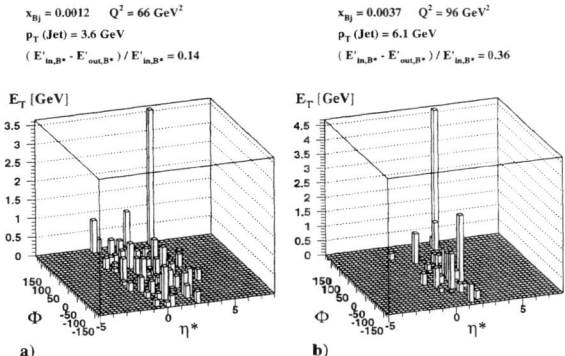

FIGURE 16. Transverse energy distribution in the $\eta - \phi$ plane for instanton induced processes (Monte Carlo).

suggested that these can have measurable effects in DIS, a nice overview of which is given in [27]. Instantons would be characterised by quark flavour democracy, a band of about 1-2 units in pseudorapidity with homogenous many-particle production and locally increased transverse energy. The H1 experiment has produced an upper limit of 300 pb in a selected kinematical region based on the analysis of the multiplicity distribution [28]. It is expected that this limit can be improved by a factor 5 to 10 soon, which would meet the expected –be it with large uncertainty– cross-section of $O(100)$ pb. Two Monte Carlo events are shown in Fig. 16, a 'gold plated' one, and a 'typical' one where the homogenous instanton particle production got distorted by the high E_T current quark jet.

IV   PHOTON STRUCTURE

The observation that the photon has a hadronic structure was first made more than 30 years ago when the cross-sections of hadronic photoproduction interactions were demonstrated to have dependences on energy and momentum transfer which were similar to those in hadron-hadron interactions. With the advent of the quark-parton model, and subsequently QCD, more quantitative predictions for this hadronic structure became available, and its gross features were identified experimentally in e^+e^- interactions [29]. Subsequently these features were also observed in hard photoproduction processes at HERA [30].

Photoproduction at HERA can be studied for (quasi) real photons, i.e. photons which have a virtuality, Q^2, close to zero by selecting events with the absence of a scattered electron in the central detector (typically $Q^2 < 1$ GeV^2, and $\langle Q^2 \rangle < 10^{-2}$ GeV^2), or by explicitly requiring the electron to be detected in a small angle tagger located close to the electron beampipe, approximately 30 m away from the central detector ($10^{-8} < Q^2 < 10^{-2}$ GeV^2).

Measurements of the photon structure functions in e^+e^- collisions, via 'DIS' $e\gamma$ scattering, are directly sensitive to the quark structure of the photon. Only through QCD evolution studies can information be extracted concerning the gluon component of this structure, but the presently available data have not been precise enough for such an analysis. Recently, photoproduction studies of jets and high p_T charged particles in photoproduction events at the ep collider HERA have shown sensitivity to the partonic content of the photon. Here the photon structure is probed by the partons of the proton, rather than by a virtual photon as in $e\gamma$ collisions. Hence these data are sensitive to both the quark and gluon content of the photon. Leading Order (LO) diagrams are shown in Fig. 17, for so called direct (Fig. 1a; the photon couples directly to the partons in the proton) and resolved (Fig. 1b; the partons of the hadronic component of the photon scatter on the partons of the proton) processes in photoproduction. In LO QCD only the latter process contains information on the partonic structure of the photon.

For dijet events one can construct an experimental variable from the transverse momentum E_T and pseudorapidity η of the jets:

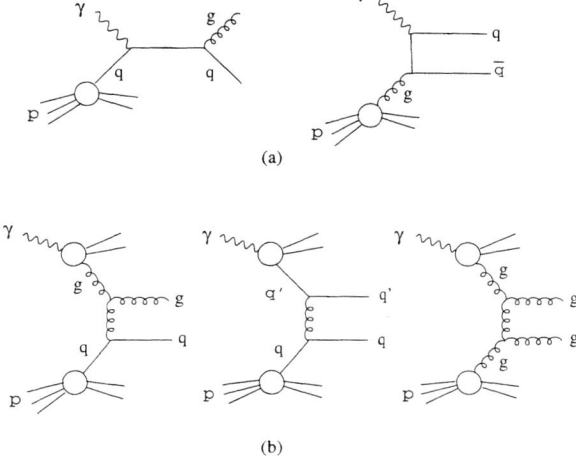

FIGURE 17. LO (a) direct and (b) resolved processes.

$$x_\gamma^{rec} = \frac{E_{T,jet1}e^{-\eta_{jet1}} + E_{T,jet2}e^{-\eta_{jet2}}}{2E_\gamma} \quad (3)$$

which is in LO closely related to x_γ, the fraction of the photon momentum carried by the parton involved in the hard scattering. Fig. 18 shows [31] the distribution of x_γ^{rec} for a sample of photoproduction events with 2 jets with an E_T larger than 14 GeV for the first jet, and larger than 11 GeV for the second jet. Comparison with the HERWIG model, tracked through the detector simulation, shows that the region of $x_\gamma^{rec} > 0.75$ is populated dominantly with events for which the photon interacts directly with the partons in the proton (hence for which $x_\gamma = 1$) and the region $x_\gamma^{rec} < 0.75$ contains dominantly events where the photon structure is resolved.

The x_γ^{rec} distribution is unfolded to the true x_γ distribution with Monte Carlo events, and from the resulting spectra the resolved quark induced and direct photon interactions are subtracted. What remains is the part which is the resolved gluon induced cross-section, which can then easily be converted in a measurement of the gluon distribution in the photon. The result is shown in Fig. 19 [32]. The average scale of this measurement, taken to be the transverse momentum squared, P_T^2, of the scattered partons leading to the jets, has been determined from Monte Carlo and amounts to 75 GeV$^2/c^2$.

It is essential to verify the procedure to extract $xg(x)$ by other methods. A new method recently exploited [33] uses instead of jets charged tracks with a high transverse momentum p_T. This method thus avoids two large systematic errors entering in the jet analysis: the energy scale uncertainty of the calorimeter and the uncertainty of the jet energy measurement due to overlap with energy deposits

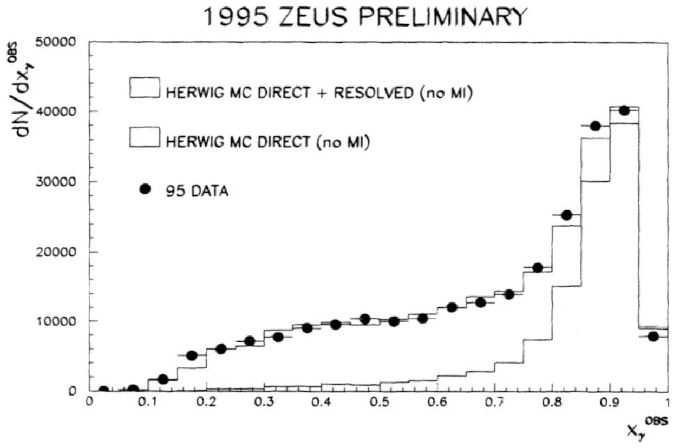

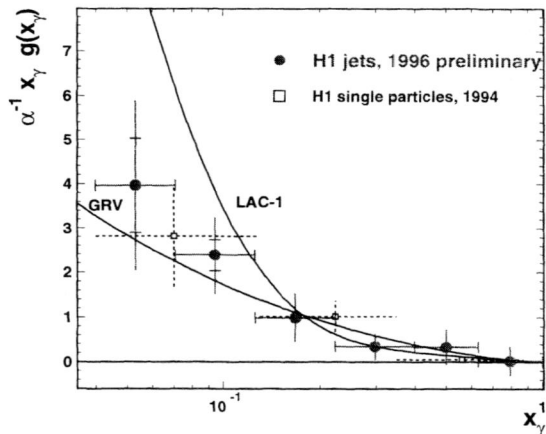

FIGURE 18. The distribution dN/dx_γ^{rec}, for a sample of dijet events compared with the HERWIG Monte Carlo program for real γp interactions, separately for the direct and total contribution.

FIGURE 19. The LO gluon distribution in the photon from charged tracks (full circles; $< P_T^2 > = 38$ GeV$^2/c^2$) and jets (open circles; $< P_T^2 > = 75$ GeV$^2/c^2$). The inner error bars are statistical, the full bars are the statistical and systematic errors added in quadrature. The curves are GRV and LAC1.

from soft multiple interactions which may occur on top of the hard scattering process. The drawback of this method is a stronger sensitivity to fragmentation uncertainties. Using events which have at least one measured track of a charged particle with $p_T > 2.6$ GeV/c the variable x_γ^{rec} was calculated:

$$x_\gamma^{rec} = \frac{\sum p_T e^{-\eta}}{E_\gamma} \qquad (4)$$

where the sum runs over all tracks with $p_T > 2$ GeV/c. The contribution of processes in which quarks are resolved in the photon (for example the middle diagram in Fig. 17(b)) were calculated and subtracted from the data as for the jet analysis. The average transverse momentum squared of the hard partonic scattering for this data sample amounts to $<P_T^2> = 38$ GeV$^2/c^2$, as derived from Monte Carlo studies, and is taken as the scale for comparisons with parton densities.

The jet and particle measurements are found to be consistent as shown in Fig. 19. The results confirm that the contribution of the gluon to the photon structure is significant. The gluon density tends to rise with decreasing x_γ. Predictions for parton distributions in the photon are generally in agreement with the data. HERA has given for the first time a direct insight in the 'gluonic' structure of the photon.

V   DIFFRACTION

The data taken in 1992 at the ep collider HERA led to the first observation of DIS events with an interval in rapidity around the proton remnant direction devoid of hadron production - a "rapidity gap" [34,35]. Using the data taken in 1993, first measurements were made of the contribution of such events to the DIS cross-section, quantified in terms of a structure function $F_2^{D(3)}$ [36,37] (see also the discussion in [4]). The picture in mind is that the proton emits a colour singlet object (lets call it – as many do– a pomeron) and this object is hit by the virtual photon. The cross-section behaviour of the data was found to be in agreement with what one would expect from diffraction, well known from hadronic interactions, and usually described in terms of pomeron exchange. This leads to the exciting possibility of measuring the partonic structure of the pomeron, an object which originates from soft hadronic scattering phenomenology, but is not yet understood in QCD.

Additional kinematical variables to describe a diffractive process are indicated in Fig. 1. They include

$$t = (P - Y)^2 \qquad x_{I\!P} = \frac{Q^2 + M_X^2}{Q^2 + W^2} \qquad \beta = \frac{Q^2}{Q^2 + M_X^2},$$

neglecting t and the proton mass in the definitions of β and $x_{I\!P}$. For pomeron exchange, $x_{I\!P}$ is the fraction of the proton momentum carried by the pomeron and β can be interpreted as the fraction of pomeron momentum carried by the struck quark. Both rapidity gap and measurement of leading protons have been used to select diffractive events.

The results are presented in terms of a structure function $F_2^{D(3)}$ which is related to the differential $ep \to e'XY$ cross-section by

$$\frac{d^3\sigma_{ep \to eXY}^D}{dx_{I\!P}\, d\beta\, dQ^2} = \frac{4\pi\alpha^2}{\beta Q^4}(1 - y + \frac{y^2}{2})F_2^{D(3)}(\beta, Q^2, x_{I\!P}). \tag{5}$$

The measurement of $x_{I\!P} \cdot F_2^{D(3)}(\beta, Q^2, x_{I\!P})$ is shown as a function of $x_{I\!P}$ in Fig. 20 for different values of β and Q^2 for the ZEUS and H1 measurements. The two measurements agree in the region of overlap both in shape and in normalization. There is however a significant difference between the H1 and ZEUS data at low Q^2. In order for this concept of pomeron structure to make sense, the data must be factorisable in the variable $x_{I\!P}$ (which describes the flux of the pomerons emitted by the proton, although one should take 'emitted' not too literally: it could well be that the photon sees a region in the proton which is simply not colour connected to the rest of the proton) and β, Q^2, which describe the structure of the pomeron. This has been indeed shown to work well [36,37].

Then one can treat these data as pomeron structure function data and analyse the structure with QCD evolution equations, which result in a set of parton distributions. The result of the H1 fit is that the exchange is dominated by gluons. The parton distributions are shown in Fig. 21. The fit which has the largest flexibility w.r.t. the gluon yields a "leading" gluon at low Q^2, in which the exchange is dominated by gluons carrying fractional momentum, $x_{g/I\!P} > 0.6$. But less leading gluon solutions are also possible. In any case the gluon is "hard": the integrated momentum fraction carried by gluons amounts to 90% (4.5 GeV^2) and decreases to 80% (75 GeV^2).

The usefulness and fundamentality of this picture and the corresponding extracted parton distributions must be demonstrated e.g. by transporting the distributions to different processes, such as diffraction in $p\bar{p}$ at the Tevatron, or by performing consistency checks with hadronic final state properties of diffractive events at HERA. The results on the latter are encouraging [38], while the debate on the former is still ongoing [39] and complicated by final state interactions.

New detectors in the beamline, about 100 m away from the central detector in the proton direction, have been used to tag leading neutrons. In the case of DIS with a leading neutron in the final state, the process can be viewed as the emition of a charged meson, e.g. a pion which is then probed by the virtual photon. Contributions due to the ρ and a_2 meson have been estimated to be an order of magnitude less in the kinematical region selected here [40]. Hence the partonic structure of this 'pion' is probed in the same way the partonic structure of the pomeron is probed in diffractive events. The results of such a measurement [41] are shown in Fig. 22. Note that an ambiguity in such an extraction is the assumed flux, which affects the normalization. Within this DIS scattering picture, HERA delivers for the first time data on the pion structure at small 'x' values.

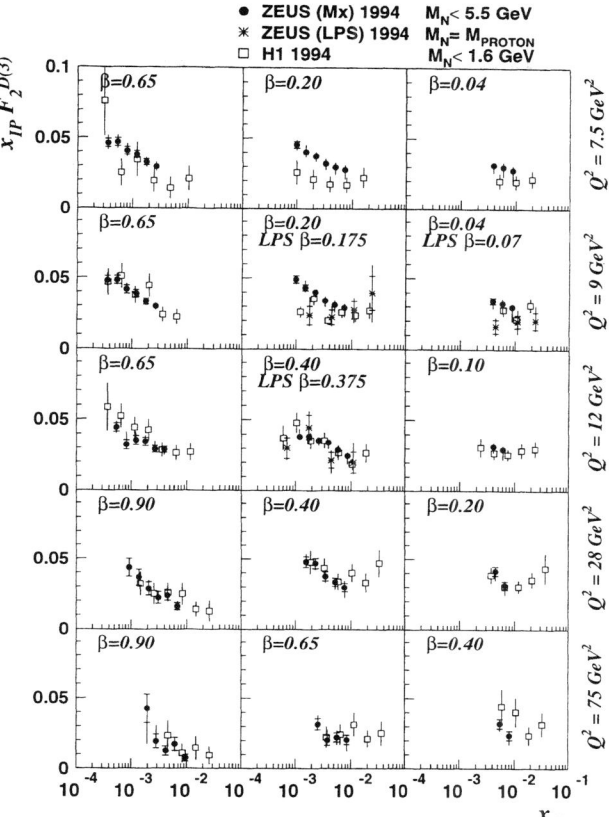

FIGURE 20. ZEUS and H1 results on the structure function $F_2^{D(3)}$ obtained with various methods.

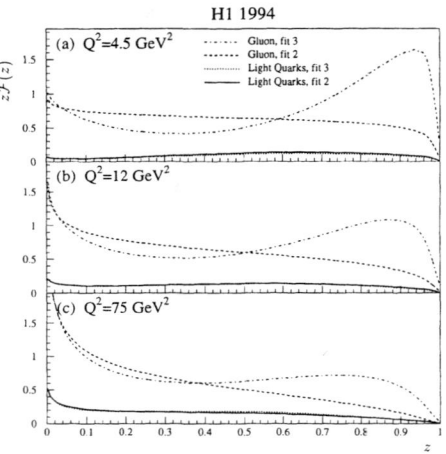

FIGURE 21. Parton distributions in the pomeron from a QCD fit.

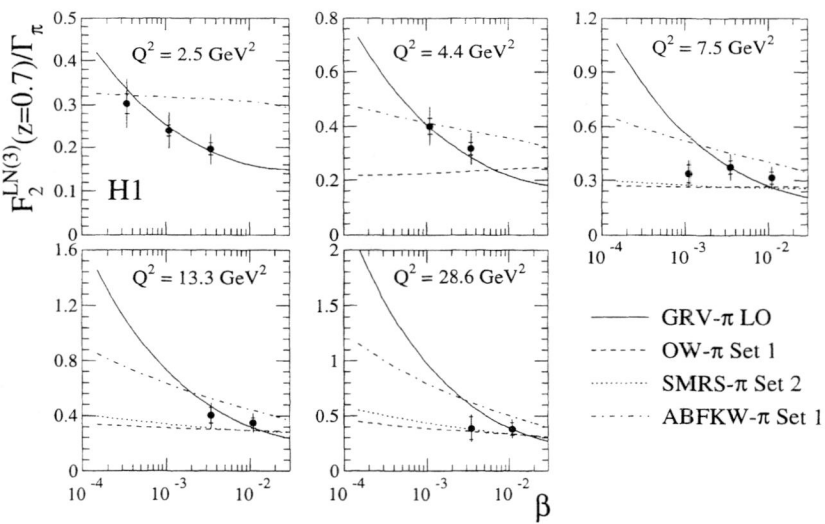

FIGURE 22. Extracted 'pion structure' as function of the fractional momentum variable β and Q^2, with various parametrizations.

VI HIGH Q^2 REGION

The HERA data now extends to values of $Q^2 \sim 40000$ GeV^2, where the exchange of Z^0 (for neutral currents NC) and W bosons (charged currents CC) becomes important. Several aspects on the physics of the high Q^2 region have been covered in [42].

For high Q^2 NC events the contribution of structure function F_3 becomes important and this also leads to generalized charges for the definition of F_2 and F_1. The data is conveniently represented in terms of a reduced cross-section, $\tilde{\sigma}_{NC}(x, Q^2) = \frac{xQ^4}{2\pi\alpha^2} \frac{1}{Y_+} \frac{d^2\sigma_{NC}}{dxdQ^2}$ with $Y_+ = 1 + (1-y)^2$. This cross-section equals F_2 in the limit of γ exchange only and for $F_L = 0$. Fig. 23 shows the reduced cross-section as function of Q^2 for H1 data. A slight excess over Standard Model expectation is seen for $x = 0.45$. This is potentially of large interest, since at the largest values of Q^2 we probe the proton with the best resolution, i.e we can resolve the smallest objects in the proton. Also this is the region where eventual highly massive new particles could be produced. We turn to this in the next section.

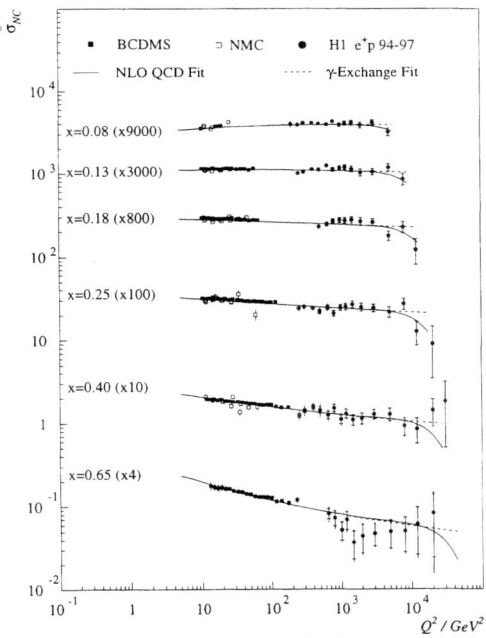

FIGURE 23. NC reduced cross-section measured at high x, compared with the Standard Model expectations (NLO QCD fit including Z^0 exchange –solid line– and without Z^0 exchange – dashed line).

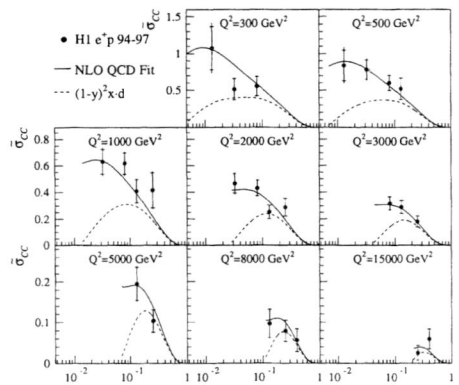

FIGURE 24. e^+p CC reduced cross-section as a function of x, with the full NLO QCD fit (solid line) and the d quark contribution only.

For charged current events we have

$$\frac{d^2\sigma(e^+p)}{dxdQ^2} = \frac{G_F^2}{2\pi}\left(\frac{M_W^2}{M_W^2+Q^2}\right)^2[(\overline{u}+\overline{c})+(1-y)^2(d+s)]$$

and

$$\frac{d^2\sigma(e^-p)}{dxdQ^2} = \frac{G_F^2}{2\pi}\left(\frac{M_W^2}{M_W^2+Q^2}\right)^2[(u+c)+(1-y)^2(\overline{d}+\overline{s})]$$

The reduced e^+p cross-section $\tilde{\sigma}_{CC}(x,Q^2) = \frac{2\pi x}{G_F}\left(\frac{M_W^2+Q^2}{M_W^2}\right)\frac{d^2\sigma_{CC}}{dxdQ^2} = x[(\overline{u}+\overline{c})+(1-y)^2(d+s)]$ is shown in Fig. 24. A DGLAP fit to the data is shown together with its d quark contribution. It demonstrates that the cross-section is dominated by d valence quarks at high x, while sea quark contributions dominate for $x < 0.1$. Since 1998 HERA runs with electrons instead of positrons, hence measurements like the one shown in Fig. 25 become accessible, namely the ratio of $e^-p/e^+p \sim (u+c)/((1-y^2)(d+s))$ which has sensitivity to the u/d ratio. This ratio has been measured in fixed target experiments but its interpretation is still controversial, and a good referee for many static nucleon models. The experimental precision is not yet good enough to pin down this ratio, but significant measurements from HERA will become available with the future ten times increase of the luminosity.

Fig. 26 shows the cross section for neutral current and charged current interactions. While the cross sections are orders of magnitude different at $Q^2 < 1000$ GeV^2, they become of similar size at large Q^2, and demonstrate nicely the Electroweak Unification of forces.

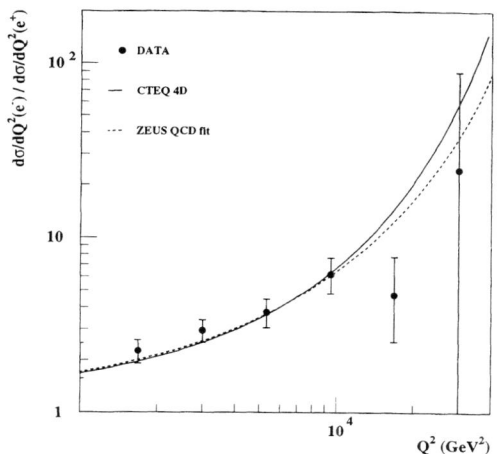

FIGURE 25. Ratio of e^-p/e^+p CC cross-section versus Q^2, with two model predictions.

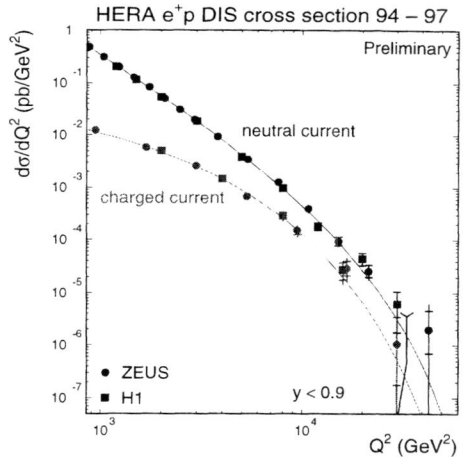

FIGURE 26. Comparison of Neutral Current and Charged Current cross sections

VII BEYOND THE STANDARD MODEL

In February 1997 both experiments announced a tantalizing excess of detected events over Standard Model (SM) expectation at high Q^2 ($Q^2 > 15000$ GeV^2) [43,44]. Although the compatibility of the excess seen by the two experiments is at the limit, and statistical fluctuations cannot be excluded yet, the community has taken up on these observations and exciting speculations on their origin have been presented. Meanwhile the luminosities have increased by a factor

2.5 to a total of 37 pb^{-1} (H1) and 47 pb^{-1} (ZEUS) respectively.

Almost identical analyses as for the 94-96 data have been performed. The experiments calculate the kinematics of the DIS scattering from different variables, which are differently affected by detector effects and radiative corrections. For neutral current (NC) events H1 uses the electron variables (angle and energy of the scattered electron) while ZEUS uses the angle of the scattered electron and current quark (measured from the hadronic final state). For the charged current (CC) events both experiments can use hadronic variables only.

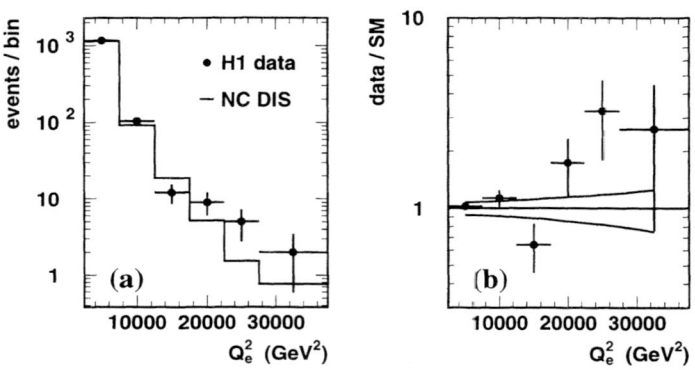

FIGURE 27. The left plot shows the Q^2 spectrum for the H1 neutral current sample. The points with error bars indicate the data and the histogram shows the NC DIS expectations. The right plot shows the Q^2 spectrum divided by the SM expectations. The smooth curves indicate the uncertainty in the expectations.

Fig. 27 shows the distribution of the H1 event sample in Q^2, with SM NC DIS expectations showing still an excess of events at large Q^2. For $Q^2 > 15000$ GeV^2, H1 observes in total 22 events where 14.1 ± 2.0 are expected. The excess is however reduced by adding the new 1997 data. For the latter 10 events are found and 8.1 expected, compatible with the SM prediction within errors. Furthermore, in the data collected up to 1996 7 events with $y > 0.4$ in the 25 GeV mass (M) window centred on 200 GeV were observed where 1 event was expected. The total sample including the 1997 data now yield 8 events with 2.9 ± 0.5 expected. Hence the formation of a resonance at a mass of around 200 GeV is not confirmed by the new data. In all the data is somewhat higher than the SM prediction, but the excess is much weaker including the 1997 data, than what was observed before. Similar conclusions are made for charged current event rates and the ZEUS data [45].

Hence the extra luminosity did not clarify the matter, but it did not confirm the substantial excess seen in the previous data. It is clear that if there is an excess, it will be much smaller than anticipated two years ago, and at least five times the current luminosity is required to settle the issue. This could be achieved soon after 2001, when HERA will be upgraded to higher luminosity, of the order of 150 pb^{-1}/year/experiment.

The data can be used for contact interactions limits and leptoquark limits. They give a measure of either the scale of new physics which can be probed (e.g. due to the increased resolution) or new massive particles being produced. H1 quotes (preliminary) a limit of 275 GeV for the leptoquark mass, assuming the coupling $\lambda = \sqrt{4\pi\alpha}$ and branching ration $\beta = 1$. ZEUS quotes contact interaction limits in the range of 1.5 to 5 TeV.

In conclusion, the data at HERA at present contains no evidence for new physics beyond the Standard Model, but a few (low statistics) deviations still need to be clarified with future high statistics data.

VIII SUMMARY

HERA is above all a machine which measures structure of matter very precisely in a new kinematic regime. Precise F_2 data are now available in a large range: $0.045 < Q^2 < 30000$ GeV^2 and $6 \cdot 10^{-7} < x < 0.65$. At low x and $Q^2 > 1$ GeV^2 F_2 shows a steep rise: $F_2 \sim x^{-\lambda}$ with $\lambda \sim 0.1 - 0.3$. This rise is still subject of debate, but fits of the NLO DGLAP QCD equation to data are found to work well for $Q^2 > 1$ GeV^2. A turn over of $dF_2/d\ln Q^2$ at small Q^2, x is seen which is in part due to the restricted range available of the data for fitting these slopes, but could contain hints to new dynamics, such as saturation effects.

The gluon distribution has been extracted in NLO in the region $3.10^{-5} < x < 10^{-1}$ from QCD fits of F_2. The extracted gluon is consistent with the one derived from charm production in DIS, 2-jet events and the longitudinal structure function F_L.

Polarized data from HERMES will help in the understanding of what constitutes the spin of the proton. The results on $g_1(x, Q^2)$ demonstrate the high quality of the data. New measurement show sensitivity to the polarised gluon, which appears to be large, as expected from pQCD fits.

The structure of real photons has been probed, in particular its gluonic content, and was found to be large. Final state analyses hint that $\ln 1/x$ effects may be visible in the data. A special class of events with a large empty rapidity gap in the final state have become a full field of analysis of their own, and have strongly advanced our knowledge on diffractive exchange. Also a first extraction of the pion structure has been made, based on data which have a leading neutron.

For the high Q^2 region, the issue remains unsettled. The new 1997 data does not confirm a significant excess, and much more luminosity will be needed. This can be achieved with the HERA luminosity upgrade.

In the years 1999-2000 HERA will collect an additional luminosity of 30-40 pb^{-1}. After 2001 HERA will operate in the luminosity upgrade mode, leading to ~ 150 pb^{-1}/year, for a total running period of 3-4 years. The next step (> 2005) could be a fully polarised HERA or eA collisions at HERA.

This need not be the end of Deep Inelastic Scattering, which is by all means the cleanest way to probe the structure of matter. When cleverly planned, future facilities can be used for lepton-hadron scattering [46]. Examples are a possible linear e^+e^- collider targeting on an existing proton machine or LHC & LEP. This would allow to extend the present kinematic range by another one or two orders of magnitude in x and Q^2. If the F_2 and gluon distribution continue to grow with decreasing x this may bring us unambiguously in a region of parton saturation. Also it allows us to look to the proton with a factor 10 improved resolution (due to the increased Q^2), and –who knows– maybe the quark starts to reveal some structure....

ACKNOWLEDGMENTS

I warmly thank the organizers for their efficient organization and hospitality in Merida.

REFERENCES

1. H1 Collab., I. Abt et al., Nucl. Instr. and Meth. **A386** (1997) 310 and **A386** (1997) 348.
2. ZEUS Collab., *The ZEUS Detector*, Status Report (1993).
3. HERMES Collab., K. Ackerstaff et al., Nucl. Instrum. Meth. **A417** (1998) 265.
4. C. Garcia-Canal, these proceedings.
5. A. M. Cooper-Sarkar, R. C. E. Devenish and A. De Roeck, Int. J. Mod. Phys. **A13** (1998) 3385.
6. U. Bassler and G. Bernardi, Nucl. Instrum. Meth. **A361** (1995) 197.
7. H1 Collab., C. Adloff et al., Nucl. Phys. **B497** (1997) 3.
8. ZEUS Collab., J. Breitweg et al., Eur. Phys. J. **C7** (1999) 609.
9. ZEUS Collab., Contributed paper to EPS99, Tampere, paper 493 (1999).
10. Yu. L. Dokshitzer, JETP **46** (1977) 641.
    V. N. Gribov and L. N. Lipatov, Sov. Journ. Nucl. Phys. **15** (1972) 78.
    G. Altarelli and G. Parisi, Nucl. Phys. **B126** (1977) 298.
11. E.A. Kuraev, L.N. Lipatov, V.S. Fadin, Sov. Phys. JETP **45** (1972) 199;
    Y.Y. Balitsky, L.N. Lipatov, Sov. J. Nucl. Phys. **28** (1978) 822.
12. ZEUS Collab., J. Breitweg et al., Phys. Lett. **B407** (1997) 432.
13. H1 Collab., S. Aid et al., Nucl. Phys. **B470** (1996) 40.
14. J. Kwiecinski, A. D. Martin and A. M. Stasto, Phys. Rev. **D56** (1997) 3991.
15. R. S. Thorne, Phys. Lett. **B392** (1997) 463, hep-ph/9701241 and hep-ph/9710541.
16. M. Glück, E. Reya and A. Vogt, Z. Phys. **C67** (1995) 433.
17. K. Prytz, Phys. Lett. **B311** (1993) 286.
18. E. Gotsman et al., Nucl. Phys. **B539** (1999) 535.
19. J. Ashman et al., Phys. Lett. **B206** (1988) 364.
20. G. Altarelli et al., Nucl. Phys. **B496** (1997) 337.

21. Workshop on Physics with Polarized Protons at HERA, DESY-proceedings-1998-01 (1998), Eds. A. De Roeck and T. Gehrmann.
22. H1 Collab., C. Adloff et al., Nucl. Phys. **B538** (1999) 3.
23. ZEUS Collab., J. Breitweg et al., Eur. Phys. J. **C6** (1999) 239.
24. H. Jung, L. Jonsson, H. Kuster, Eur. Phys J. **C9** (1999) 383.
25. J. Kwiecinski, A.D. Martin, J.J. Outhwaite, Eur. Phys. J. **C9** (1999) 622.
26. H1 Collab., Contributed paper to ICHEP98, Vancouver, paper 575 (1998).
27. T. Carli et al., DESY preprint 99-067 (1999).
28. S. Aid et al., Z. Phys. **C72** (1996) 573.
29. S. Söldner-Rembold, XVIII Symposium on Lepton-Photon Interactions, World Scientific, Eds. A. De Roeck and A Wagner, Hamburg 1997, p97.
30. M. Erdmann, The Partonic Structure of the Photon, S.T.M.P. Vol.138, Springer Verlag, Heidelberg, 1997.
31. ZEUS Collab., Contributed paper to ICHEP98, paper 810.
32. H1 Collab., Contributed paper to EPS99, Tampere, paper 157ad.
33. H1 Collab., C. Adloff et al., Eur. Phys. J. **C10** (1999) 372
34. ZEUS Collab., M. Derrick et al., Phys. Lett. **B315** (1993) 481.
35. H1 Collab., T. Ahmed et al., Nucl. Phys. **B429** (1994) 477.
36. H1 Collab., C. Adloff et al., Z. Phys. **C76** (1997) 613.
37. ZEUS Collab., J. Breitweg et al., Eur. Phys. J. **C6** (1999) 43.
38. H1 Collab., C. Adloff et al., Phys. Lett. **B428** (1998) 206.
39. P. Marage, Review talk at EPS99, Tampere, to appear in proceedings.
40. B. Kopeliovich et al., Z. Phys **C73** (1996) 125.
41. H1 Collab., C. Adloff et al., Eur. Phys. J. **C6** (1999) 587.
42. E. Elsen, these proceedings.
43. H1 Collab., C. Adloff et al, Z. Phys. **C74** (1997) 191.
44. ZEUS Collab., J. Breitweg et al, Z. Phys. **C74** (1997) 207.
45. G. Bernardi, DESY seminar, March 1998; A Quadt, DESY seminar, March 1998.
46. A. De Roeck and M. Strikman, in proc. of 6th International Workshop on Deep Inelastic Scattering and QCD (DIS 98), Brussels, Belgium, 4-8 Apr 1998.

Charm Hadroproduction

R. Vogt

Nuclear Science Division, Lawrence Berkeley National Laboratory, Berkeley, CA, 94720[1]
and
Physics Department, University of California at Davis, Davis, CA, 95616

Abstract. These lectures provide a brief overview of heavy quark production in QCD. We begin with a review of heavy quark production beyond leading order. Some puzzling aspects of charm production, not described by perturbative QCD alone, are touched upon. The intrinsic charm model is then introduced and some of its predictions are described. It is shown how the intrinsic charm probability is obtained from the charm structure function. The model is compared to hadroproduction data with a Σ^- beam and the expected results for a new experiment at CERN are described. Only charm is discussed here but other, heavier, quarks can be treated identically.

CHARM PRODUCTION IN PERTURBATIVE QCD

Since the discovery of the J/ψ opened up the field of heavy quark production, both theoretical and experimental advances have been made. Theoretical work has shown that the next-to-leading order, NLO, corrections to the leading order, LO, Born cross section can be large. This is especially true for charm quarks, the 'lightest' heavy quarks. Calculations beyond NLO have been carried out near the heavy quark production threshold where soft and virtual gluons can be resummed to all orders. Early leading logarithm, LL, resummations included only terms which, at $n^{\rm th}$ order, are proportional to $(-\alpha_s^n/n!)[\ln^{2n-1}(s_4/m_c^2)/s_4]_+$. This resummation is incomplete because it ignores the color structures in the diagrams which can be important for heavy quarks. The next-to-leading logarithmic, NLL, terms, proportional to $(-\alpha_s^n/n!)[\ln^{2n-2}(s_4/m_c^2)/s_4]_+$, have recently been worked out and included in the resummation calculations. The resummation techniques have been tested against top quark production at the Tevatron and have been shown to work well because m_t is large. Resummation of the charm cross section is more

---

[1] This work was supported in part by the Director, Office of Energy Research, Division of Nuclear Physics of the Office of High Energy and Nuclear Physics of the U. S. Department of Energy under Contract No. DE-AC03-76SF00098.

problematic because m_c is small. In this section we briefly review the state of these calculations and their applications to charm production.

At LO charm production arises from two processes, $q\bar{q}$ annihilation and gluon fusion. At NLO, in addition to virtual corrections to the Born diagrams, the real processes $q\bar{q} \to c\bar{c}g$, $gg \to c\bar{c}g$, and $q(\bar{q})g \to c\bar{c}(\bar{q})q$ must also be included. The gluon fusion contribution is dominant except when the longitudinal momentum fraction, x_F, or transverse momentum, p_T, of the charm quark is large. The double differential cross section for $c\bar{c}$ pair production by hadrons A and B is

$$E_c E_{\bar{c}} \frac{d\sigma_{AB}}{d^3 p_c d^3 p_{\bar{c}}} = \sum_{i,j} \int dx_1\, dx_2\, F_{i/A}(x_1, \mu_F) F_{j/B}(x_2, \mu_F) E_c E_{\bar{c}} \frac{d\hat{\sigma}_{ij}(s, m_c, \mu_R)}{d^3 p_c d^3 p_{\bar{c}}} \quad (1)$$

where $s = x_1 x_2 S$. Here i and j are the interacting partons and $F_{i/A}(x, \mu_F)$ are the number densities of gluons, light quarks and antiquarks evaluated at momentum fraction x and factorization scale μ_F. The inclusive charm quark production cross section is obtained by integrating over the momentum of the other, unobserved, quark. The partonic cross section, $\hat{\sigma}_{ij}$, can be expressed in terms of dimensionless functions which depend only on $\eta = s/4m_c^2 - 1$ [1],

$$\hat{\sigma}_{ij}(s, m_c, \mu_R) = \frac{\alpha_s^2(\mu_R)}{m_c^2} \sum_{k=0}^{\infty} (4\pi \alpha_s(\mu_R))^k \sum_{l=0}^{k} f_{ij}^{(k,l)}(\eta) \ln^l\left(\frac{\mu_R^2}{m_c^2}\right) \quad (2)$$

$$\sim \frac{\alpha_s^2(\mu_R)}{m_c^2} \left\{ f_{ij}^{(0,0)}(\eta) + 4\pi \alpha_s(\mu_R) \left[f_{ij}^{(1,0)}(\eta) + f_{ij}^{(1,1)}(\eta) \ln\left(\frac{\mu_R^2}{m_c^2}\right) \right] + \cdots \right\}.$$

Only $f_{gg}^{(0,0)}$ and $f_{q\bar{q}}^{(0,0)}$ contribute at leading order. The strong coupling constant is evaluated at the renormalization scale μ_R. Both μ_F and μ_R are of the order of m_c. The physical cross section should be independent of μ_R. A strong μ_R dependence suggests that the predictive power of the calculation is weak [2].

In Fig. 1 the variation of the total $c\bar{c}$ production cross section with $\mu = \mu_F = \mu_R$ is shown for pp production at 800 GeV (a) and $\pi^- p$ production at 340 GeV (b) using the GRV HO pion and nucleon distributions [3,4]. As μ/m_c increases, the cross section varies less rapidly but the rather large theoretical K factor between the LO and NLO cross sections, ≈ 2.5, implies that further higher-order corrections are important. The pp production data were used to fix m_c and μ at NLO with two parton distribution functions, MRS D$-'$ [5] and GRV HO [3]. The NLO calculations of $\sigma_{c\bar{c}}^{\rm tot}(S)$ are compared with the pp and pA data [6,7] in Fig. 1(c). The cross section per nucleon is given, assuming a linear nuclear dependence [8]. Since $m_c < Q_{0,\rm MRS}$, $\mu = 2m_c$ is used with $m_c = 1.2$ GeV for MRS D$-'$. The GRV HO distributions have a much lower Q_0, so that $\mu = m_c$ and $m_c = 1.3$ GeV is used. Both sets give an equivalent description of the data with a tendency to underestimate the total cross section. In Fig. 1(d) the π^--induced data [6,7] is compared to calculations using the same parameters. Although the calculations again somewhat underpredict the data, the same parameters seem reasonable for both pion and proton projectiles.

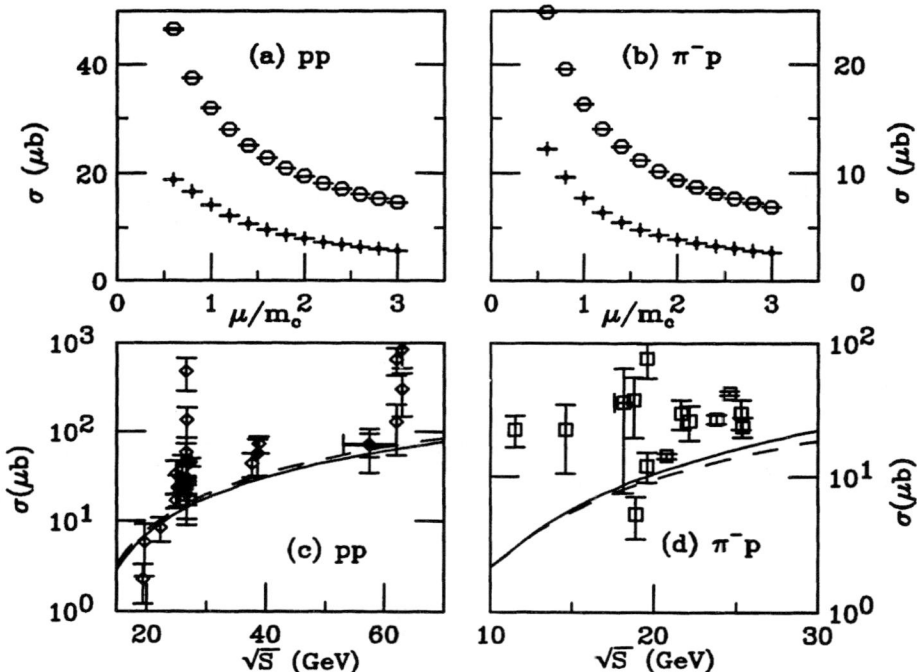

FIGURE 1. The $c\bar{c}$ production cross section, calculated with $m_c = 1.3$ GeV, $\mu = m_c$, and GRV HO parton distributions for pp production at 800 GeV (a) and in π^-p production at 340 GeV (b). The circles show the NLO calculation, the crosses, the Born result. NLO calculations of $\sigma_{c\bar{c}}^{tot}$ for (c) pp and pA and (d) π^-p and π^-A measurements [6,7] with MRS D$-'$, SMRS P2 [9], $m_c = 1.2$ GeV, $\mu = 2m_c$ (solid) and GRV HO, $m_c = 1.3$ GeV, $\mu = m_c$ (dashed). Reproduced from Ref. [10] with permission.

When only the total cross section is considered, $\mu \propto m_c$ is sufficient. However, for inclusive charm production, the transverse mass, $m_T = \sqrt{p_T^2 + m_c^2}$, is a better scale to use since $m_T^2 \propto (t - m_c^2), (u - m_c^2)$, regulating large logarithms that appear at large p_T while $\mu \propto m_c$ cannot do so. If $\mu \propto m_T$, then the K factor is constant as a function of p_T and x_F [11].

A complete calculation of higher order corrections beyond NLO is very difficult and is so far not possible. However, near threshold there can be large logarithms in the perturbative expansion, arising from an imperfect cancellation of the soft and virtual terms, which can be resummed to make more reliable theoretical predictions. An approximation of these contributions was used to resum the leading logarithmic terms to all orders in perturbation theory [12], analogous to resummation of the Drell-Yan process. The method, first applied to top production [12] and later extended to charm and bottom [13], relies on the proportionality of the higher order terms to the Born cross section. The resummed cross section is

$$E_c E_{\bar{c}} \frac{d\sigma_{AB}^{\text{res}}}{d^3 p_c d^3 p_{\bar{c}}} = \sum_{i,j} \int dx_1\, dx_2 F_{i/A}(x_1, \mu_F) F_{j/B}(x_2, \mu_F) E_c E_{\bar{c}} \frac{d\hat{\sigma}_{ij}^{\text{res}}(s, m_c, \mu_R)}{d^3 p_c d^3 p_{\bar{c}}} \quad (3)$$

where

$$\hat{\sigma}_{ij}^{\text{res}}(s, m_c^2) = -\int_{s_{\text{cut}}}^{s - 2m_c s^{1/2}} ds_4 f_{ij}^{\text{sch}}\left(\frac{s_4}{2m_c^2}\right) \frac{d\overline{\sigma}_{ij}^{(0)}(s, s_4, m_c^2)}{ds_4}. \quad (4)$$

The resummation scheme only works for processes where the Born cross section, $\overline{\sigma}_{ij}^{(0)}(s, s_4, m_c^2)$ exists. Therefore $ij = q\bar{q}$, gg only and not $q(\bar{q})g$. The imbalance of the Mandelstam variables, s, $t_1 = t - m_c^2$, and $u_1 = u - m_c^2$, caused by soft gluon emission is $s_4 = s + t_1 + u_1$. To LL, the lower bound of the integral is $s_{\text{cut}} = m_c^2(\mu_0/\mu)^{2z}$ where $z = 3/2$ in the $\overline{\text{MS}}$ scheme and $z = 1$ in the DIS scheme. The cutoff was introduced to keep the running coupling constant finite and monitor the sensitivity of the cross section to nonperturbative higher-twist effects. A large cutoff indicates an important nonperturbative contribution to the cross section. The contribution from soft gluon emission is contained in the function f_{ij}^{sch} where the superscript indicates the renormalization scheme, $\overline{\text{MS}}$ or DIS.

In Fig. 2 the resummed cross section is calculated with ratios μ_0/m_c chosen to agree with the ratios needed for convergence of the perturbative expansion for top and bottom production in the $\overline{\text{MS}}$ scheme: $\mu_0/m_c \approx 0.15$ for $q\bar{q}$ and 0.35 for gg. The GRV HO parton distributions are used with $\mu = m_c = 1.5$ GeV in the central curves. Any set of parton distributions that provides a consistent description of the pion and proton parton densities and has an initial scale such that $\mu = m_c$ should produce similar results. Since the exact NLO results are known, the perturbation theory improved cross sections, $\sigma^{\text{imp}} = \sigma^{\text{res}} + \sigma_{\text{exact}}^{\text{NLO}} - \sigma_{\text{app}}^{\text{NLO}}$ are shown to exploit the fact that $\sigma_{\text{exact}}^{\text{NLO}}$ is known and $\sigma_{\text{app}}^{\text{NLO}}$ is included in σ^{res}. The difference between σ^{res} and σ^{imp} is larger in pp production. The NLO cross section calculated with the same mass and scale factors used in σ^{res} and σ^{imp} is also given. Note that σ^{res} agrees with the data when $m_c = 1.5$ GeV, a larger mass than the $1.2 - 1.3$ GeV needed for the NLO calculations in Fig. 1 because $\sigma^{\text{res}} \sim 5\sigma_{\text{exact}}^{\text{NLO}}$. The results are shown up to $\sqrt{S} = 30$ GeV even though the perturbative expansion no longer converges and resummation fails in the gg channel. This causes the faster increase of σ^{res} and σ^{imp} with energy compared to $\sigma_{\text{exact}}^{\text{NLO}}$ for $\sqrt{S} > 20$ GeV. The dotted curves indicate the bounds of convergence of σ^{res} when m_c, μ and the parton densities are varied. The upper dotted curves, with $\mu = m_c = 1.3$ GeV, show that the resummation breaks down at even lower energies for the lighter quark mass. The lower dotted curves are calculated with $\mu/2 = m_c = 1.8$ GeV. The larger quark mass improves convergence at higher energies although the larger mass scale requires a larger cutoff. Neither of these extremes converge with μ_0/m_c ratios similar to those needed for heavier quarks.

The LL resummation calculation producing the results in Fig. 2 has been improved by the inclusion of the two loop running coupling constant and NLL terms

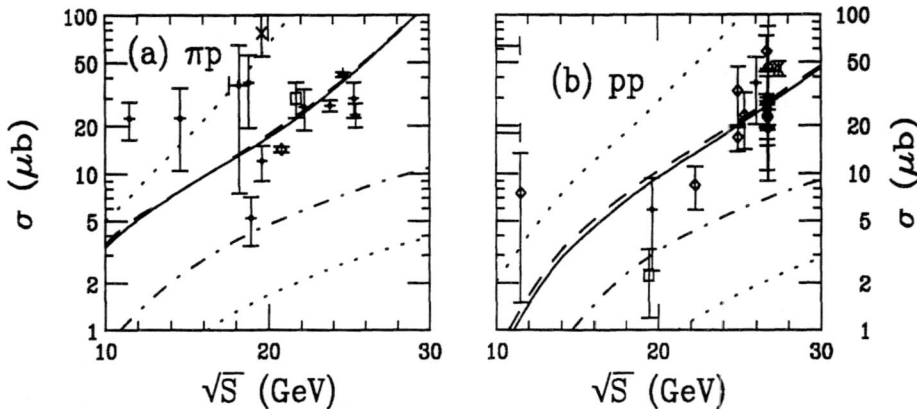

FIGURE 2. The cross sections $\sigma^{\rm res}$ (solid), $\sigma^{\rm imp}$ (dashed), and the NLO (dot-dashed) $c\bar{c}$ cross sections are shown as a function of $\sqrt{S}$ in (a) π^-p and (b) pp interactions. Both $\sigma^{\rm res}$ and $\sigma^{\rm imp}$ are calculated with $\mu_0 = 0.15 m_c$ in the $q\bar{q}$ channel and $\mu_0 = 0.35 m_c$ in the gg channel with $\mu = m_c = 1.5$ GeV and GRV HO parton distributions. Extreme values of $\sigma^{\rm res}$ are shown in the dotted lines. The upper dotted curves are calculated with $\mu = m_c = 1.3$ GeV and GRV HO while the lower dotted curves use $\mu/2 = m_c = 1.8$ GeV with MRS D$-'$ and SMRS P2. Adapted from Ref. [13] with permission of the publisher.

[14]. The function containing the terms to be resummed, $f_{ij}^{\rm sch} = \exp(E_{I,ij}^{\rm sch})$, is calculated at NLL with the exponent [15]

$$E_{I,ij}^{\rm sch} = \int_{\omega_0}^{1} \frac{d\omega'}{\omega'} \Big\{ \int_{\xi_L}^{\xi_U^{\rm sch}} \frac{d\xi}{\xi} \Big[\frac{2\gamma_{ij}}{\pi} \Big(\alpha_s(\xi) + \frac{K}{2\pi} \alpha_s^2(\xi) \Big) \Big] - \frac{3 C_F}{2\pi} \delta_{ij}^{\rm sch} \alpha_s \Big(\frac{\omega'\mu^2}{\Lambda^2} \Big)$$
$$- \Big\{ \lambda_{I,ij} \Big[\alpha_s \Big(\frac{\omega'^2 \mu^2}{\Lambda^2} \Big), \theta^* = 90° \Big] + \lambda_{I,ij}^* \Big[\alpha_s \Big(\frac{\omega'^2 \mu^2}{\Lambda^2} \Big), \theta^* = 90° \Big] \Big\} \Big\} \quad (5)$$
$$= E_{ij}^{\rm sch}(g_1) + E_{ij}^{\rm sch}(g_2) + E_{I,ij}(\lambda_{I,ij}) , \quad (6)$$

where $\xi_L = \omega'^2 \mu^2/\Lambda^2$, $\xi_U^{\rm DIS} = \omega' \mu^2/\Lambda^2$, $\xi_U^{\overline{\rm MS}} = \mu^2/\Lambda^2$, $\gamma_{q\bar{q}} = C_F = 4/3$, $\gamma_{gg} = C_A = 3$ and $\delta_{q\bar{q}}^{\rm DIS} = 1$ and 0 otherwise. The $\lambda_{I,ij}$ are eigenvalues of the anomalous dimension matrix Γ_{IJ} which appears in the renormalization group equation. The ω' integral is cut off at $\omega_0 = s_4/2m_c^2$. Because the running coupling constant diverges when $\omega'^2 \mu^2/\Lambda^2 \sim 1$, the minimum cutoff in Eq. (5) is $s_{\rm cut} \sim 2m_c^2 \Lambda/\mu$ where Λ is the QCD scale parameter in the parton densities. If $\mu = m_c$, $s_{\rm cut} \sim 2m_c \Lambda$. In general a larger value is chosen to stay away from the point of divergence. In Eq. (6), the exponent is separated into two universal scheme dependent terms independent of the color flow, the g_1 terms referring to the integral over ξ and the g_2 term proportional to $\delta_{ij}^{\rm sch}$, while the last term is independent of the scheme and arises from the NLL color dependent terms.

In the first numerical evaluation [15], the exponents were calculated at the parton-parton center of mass scattering angle $\theta^* = 90°$ which diagonalizes Γ_{IJ}, a considerable simplification. In the $q\bar{q}$ channel, only the octet component $\lambda_{2,q\bar{q}}$ contributes at 90°,

$$\lambda_{2,q\bar{q}} = \frac{\alpha_s}{\pi}\left\{-C_F(L_\beta + 1 + i\pi) + \frac{C_A}{2}\left[L_\beta - \ln\left(\frac{m^2 s}{t_1^2}\right) + i\pi\right]\right\} \quad (7)$$

where $L_\beta = (1 - 2m_c^2/s)/\beta\left[\ln\left((1-\beta)/(1+\beta)\right) + i\pi\right]$ and $\beta^2 = 1 - 4m_c^2/s$. In the gg channel, the eigenvalues are

$$\lambda_{1,gg} = -\frac{\alpha_s}{\pi}\left[C_F(L_\beta + 1) + C_A i\pi\right], \quad (8)$$

$$\lambda_{2,gg} = \lambda_{3,gg} = \frac{\alpha_s}{\pi}\left\{-C_F(L_\beta + 1) + \frac{C_A}{2}\left[L_\beta - \ln\left(\frac{m_c^2 s}{t_1^2}\right) - i\pi\right]\right\}. \quad (9)$$

The resulting exponents are shown in Fig. 3. A significant enhancement over the LL terms alone is seen when the NLL terms are included, particularly at small $s_4/2m_c^2$. The cutoff must then be adjusted to regulate the hadronic cross section. However, since the cutoff cannot be fixed by a perturbative calculation for 'light' heavy quarks, especially charm, definitive statements about a resummed NLL result are difficult to make. The NLL calculation was later extended to all angles for $q\bar{q} \to t\bar{t}$ in the DIS scheme but much larger cutoffs were needed than in the LL resummation and for NLL at $\theta^* = 90°$ [16]. Since no cutoff is needed for a fixed order expansion, the NLL terms are being added to the expansion of the partonic cross section in Eq. (2) to NNLO [1].

HIGHER TWIST CHARM

Perturbative QCD cannot fully predict the charm cross section at all orders because the charm quark is so light. The LO and NLO calculations, with appropriate K factors, can describe charm production over most of phase space but generally fail when the charm quark carries a large fraction of the initial longitudinal momentum, x_F. Typical charm fragmentation functions [17] underpredict hadroproduction at high x_F [18]. Conversely, string models tend to harden the x_F distributions too much [19,20], particularly for charm and charm-strange baryons [21]. No flavor correlations are predicted in perturbative QCD. However, asymmetries between leading [19,22,23] charm mesons, which share valence quarks with the projectile, and nonleading charm mesons, which do not, are observed. String models tend to overpredict this asymmetry [19,22,23]. Measurements of the charm structure function, $F_{2,c}(x)$ [24] at high x and Q^2 suggest that the charm distribution is harder than expected from photon-gluon fusion or QCD evolution. Anomalies are also observed in charmonium production [25,26]. Intrinsic charm [27], a higher twist effect, can dominate charm production at large x_F, explaining many of these phenomena.

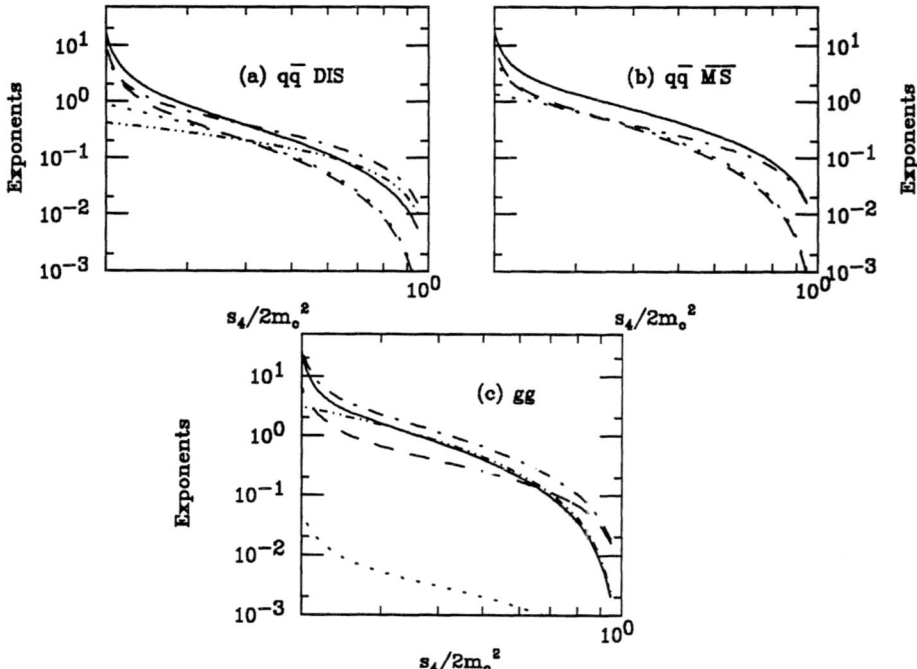

FIGURE 3. The contributions to the exponents $E^{\text{sch}}_{I,ij}$ for $m_c = 1.5$ GeV and $\sqrt{s} = 3.01$ GeV at $\theta^* = 90°$ as a function of $s_4/2m^2$. (a) The $q\bar{q}$ DIS curves are $E^{\text{DIS}}_{2,q\bar{q}}$ (solid), $E_{2,q\bar{q}}(\lambda_2)$ (dot-dashed), $E^{\text{DIS}}_{q\bar{q}}(g_1)$ (dashed), and $|E^{\text{DIS}}_{q\bar{q}}(g_2)|$ (dot-dot-dashed). The dotted curve is $E^{\text{DIS}}_{q\bar{q}}$ from Eq. (3.30) of Ref. [12]. (b) The $q\bar{q}$ $\overline{\text{MS}}$ curves are $E^{\overline{\text{MS}}}_{2,q\bar{q}}$ (solid), $E_{2,q\bar{q}}(\lambda_2)$ (dot-dashed), and $E^{\overline{\text{MS}}}_{q\bar{q}}(g_1)$ (dashed). The dotted curve is $E^{\overline{\text{MS}}}_{q\bar{q}}$ from Eq. (3.34) of Ref. [12]. (c) The gg results: $E_{gg}(g_1)$ (solid); the eigenvalue contributions, $E_{2,gg}(\lambda_2)$ (dashed) and $|E_{1,gg}(\lambda_1)|$ (dotted); and $E_{2,gg}$ (dot-dashed). $E_{1,gg}$ is indistinguishable from $E_{gg}(g_1)$. The dot-dot-dashed curve is $E^{\overline{\text{MS}}}_{gg}$ from Ref. [12].

The Intrinsic Charm Model

At fixed light-cone time, a hadron wavefunction can be expanded as a sum over the complete Fock state basis of free quark and gluon states: $|\Psi_B\rangle = \sum_m |m\rangle \, \psi_{m/H}(x_i, k_{T,i})$. The boost-invariant light-cone wavefunctions, $\psi_{m/p}(x_i, k_{T,i})$ depend on the relative momentum coordinates $x_i = k_i^+/P^+$ and $k_{T,i}$. The color-singlet states, $|m\rangle$, represent the fluctuations in the hadron wavefunction with the Fock components $|q_1q_2q_3\rangle$, $|q_1q_2q_3g\rangle$, $|q_1q_2q_3c\bar{c}\rangle$, etc. For example, intrinsic charm, IC, fluctuations [27] can be liberated by a soft interaction which breaks the coherence of the Fock state [28] provided the system is probed during the characteristic time that such fluctuations exist.

The dominant Fock state configurations are not far off shell and thus have minimal invariant mass, $M^2 = \sum_i^n \widehat{m}_i^2/x_i$ where $\widehat{m}_i^2 = m_i^2 + \langle \vec{k}_{T,i}^2 \rangle$ is the square of the average transverse mass of parton i. The general form of the Fock state wavefunction appropriate to any frame at fixed light-cone time is

$$\Psi(x_i, \vec{k}_{\perp i}) = \frac{\Gamma(x_i, \vec{k}_{\perp i})}{m_h^2 - M^2}, \qquad (10)$$

where Γ is a vertex function, expected to be a slowly-varying, decreasing function of $m_h^2 - M^2$.

The parton distributions reflect the underlying shape of the Fock state wavefunction. Assuming it is sufficient to use $\langle \vec{k}_{T,i}^2 \rangle$, the probability distribution for a general n–particle intrinsic $c\bar{c}$ Fock state is

$$\frac{dP_{\rm ic}}{dx_i \cdots dx_n} = N_n \alpha_s^4(M_{c\bar{c}}) \frac{\delta(1 - \sum_{i=1}^n x_i)}{(m_h^2 - \sum_{i=1}^n (\widehat{m}_i^2/x_i))^2}, \qquad (11)$$

where N_n normalizes the probability of the n particle Fock state. The probability for $c\bar{c}$ fluctuations to exist in a hadron is higher twist since it scales as $\alpha_s^2(m_c^2)/m_c^2$ relative to the extrinsic, EC, leading-twist production [20]. The particle distributions are then controlled by the light-cone energy denominator and phase space. Equation (11) is generalizable to an arbitrary number of light and heavy partons. Intrinsic $c\bar{c}$ Fock components with minimum invariant mass correspond to configurations with equal rapidity constituents. Thus, unlike extrinsic heavy quarks generated from a single parton, intrinsic heavy quarks carry a larger fraction of the parent momentum than the light quarks in the state [27]. In the following sections, the probability distribution is used to obtain the intrinsic charm structure function and the final-state charm hadron probability distributions.

The Charm Structure Function

The European Muon Collaboration, EMC, [24] established that photon-gluon fusion explains the bulk of the charm quark contribution to the deep inelastic structure function F_2. However, some of the data did not agree with photon-gluon fusion, the EC contribution, at all x and Q^2. A LO analysis of the data with EC and IC components showed that IC could account for the difference between photon-gluon fusion and the data at large x and Q^2. Afterwards, Hoffman and Moore [29] calculated the NLO IC corrections and, based on LO photon-gluon fusion, they determined that the probability of an IC contribution to the proton wavefunction was 0.3%. More recently, the EMC analysis was revisited with a consistent NLO treatment of both extrinsic and intrinsic charm [30].

At the values of Q^2 relevant to the EMC data, the mass of the charm quark is not negligible. Thus the charm quark cannot be treated as massless and absorbed in the parton densities. To NLO [31], $F_{2,c}$ is calculated from the inclusive virtual photon-induced reaction $\gamma^* + p \to c + X$. The partonic interaction is $\gamma^* + a_1 \to c + \bar{c} + a_2$ where a_1 and a_2 are massless partons. At LO, a_1 is a gluon but at NLO it can be any parton. The EC structure function is the sum of the longitudinal and transverse virtual photon total cross sections, $F_2 \propto (Q^2/4\pi^2\alpha)(\sigma_T + \sigma_L)$. The NLO extrinsic charm structure function is [31]

$$F_{2,c}^{\text{EC,NLO}}(x, Q^2, \mu^2, m_c^2) = \frac{Q^2 \alpha_s(\mu^2)}{4\pi^2 m_c^2} \int_{\xi_0}^1 \frac{d\xi}{\xi} \left[e_c^2 f_{g/p}(\xi, \mu^2) c_{2,g}^{(0)} \right]$$
$$+ \frac{Q^2 \alpha_s^2(\mu^2)}{\pi m_c^2} \int_{\xi_0}^1 \frac{d\xi}{\xi} \Bigg\{ e_c^2 f_{g/p}(\xi, \mu^2) \left(c_{2,g}^{(1)} + \overline{c}_{2,g}^{(1)} \ln \frac{\mu^2}{m_c^2} \right)$$
$$+ \sum_{i=q,\overline{q}} f_{i/p}(\xi, \mu^2) \left[e_c^2 \left(c_{2,i}^{(1)} + \overline{c}_{2,i}^{(1)} \ln \frac{\mu^2}{m_c^2} \right) + e_i^2 d_{2,i}^{(1)} + e_c e_i o_{2,i}^{(1)} \right] \Bigg\} . \tag{12}$$

where $\xi_0 = x(4m_c^2 + Q^2)/Q^2$. The charges are in units of e. The parton momentum distributions in the proton are denoted by $f_{i/p}(\xi, \mu^2)$ where μ, the mass factorization scale, has been set equal to the renormalization scale in the running coupling constant α_s. The scale independent parton coefficient functions, $c_{2,i}^{(l)}$, $l = 0, 1$, $\overline{c}_{2,i}^{(1)}$, $d_{2,i}^{(1)}$ and $o_{2,i}^{(1)}$ originate from the coupling of the virtual photon to the partons. The functions $c_{2,i}^{(l)}$ and $\overline{c}_{2,i}^{(1)}$ represent the virtual photon-charm quark coupling, thus appearing for both charged and neutral parton-induced reactions. The virtual photon-light quark coupling gives rise to $d_{2,i}^{(1)}$ while $o_{2,i}^{(1)}$ is an interference term proportional to $e_c e_i$.

The intrinsic charm structure function at LO in the heavy quark limit, $\widehat{m}_c \gg m_h$, $\widehat{m}_q$, is

$$\frac{dP_{\text{ic}}}{dx_i \cdots dx_n} = N_n \alpha_s^4(M_{c\overline{c}}) \frac{x_c x_{\overline{c}}}{(x_c + x_{\overline{c}})^2} \delta(1 - \sum_{i=1}^n x_i)$$
$$F_{2,c}^{\text{IC,LO}}(x) = \frac{8}{9} x c(x) = \frac{8}{9} x \int dx_1 \cdots dx_{\overline{c}} \frac{dP_{\text{ic}}}{dx_i \cdots dx_{\overline{c}} dx_c} . \tag{13}$$

If $n = 5$, the minimal intrinsic charm Fock state of the proton, $|uudc\overline{c}\rangle$, and $P_{\text{ic}} = 1\%$ [27], $N_5 = 36$. In a complete analysis of the EMC charm data, the massless result is clearly inapplicable.

Hoffmann and Moore [29] incorporated mass effects into the analysis by generalizing the operator-product expansion analysis to include both the charm quark and target masses. The final LO result, c.f. Eq. (18) in [29], is then

$$F_{2,c}^{\text{IC,LO}}(x, Q^2, \mu^2, m_c^2) = \frac{8x^2}{9(1 + 4\rho x^2)^{3/2}} \left[\frac{(1 + 4\lambda)}{\xi} c(\xi, \gamma) + 3\hat{g}(\xi, \gamma) \right] . \tag{14}$$

See Ref. [29] for definitions of the variables. The NLO IC component of the structure function is given by

$$F_{2,c}^{\text{IC,NLO}}(x, Q^2, \mu^2, m_c^2) = \frac{8}{9} \xi \int_{\xi/\gamma}^1 \frac{dz}{z} c(\xi/z, \gamma) \sigma_2^{(1)}(z, \lambda) . \tag{15}$$

	$\bar{\nu} = 53$ GeV		$\bar{\nu} = 95$ GeV		$\bar{\nu} = 168$ GeV	
PDF	α	β	α	β	α	β
CTEQ 3M [33]	1.0 ± 0.6	0.4 ± 0.6	1.2 ± 0.1	0.4 ± 0.3	1.3 ± 0.1	0.9 ± 0.5
MRS G [34]	1.0 ± 0.7	0.3 ± 0.6	1.4 ± 0.2	0.3 ± 0.3	1.5 ± 0.1	0.8 ± 0.5
GRV 94 HO [35]	1.2 ± 0.8	0.3 ± 0.6	1.4 ± 0.2	0.3 ± 0.3	1.5 ± 0.1	0.9 ± 0.5

TABLE 1. Results of the least squares fit of EC and IC contributions to the EMC data according to Eq. (16). Adapted from Ref. [30] with permission from Elsevier Science.

When the lowest order cross section is normalized so that $\sigma_2^{(0)}(z,\lambda) = \delta(1-z)$, the NLO QCD corrections to the IC contribution are given by Eq. (51) in [29]. The IC results at NLO are the sum of the kinematically corrected LO result, Eq. (14), and the full NLO correction.

The EMC data in bins of x and Q^2 were retrieved from the Durham-RAL HEP Database [32]. The EMC data was fit by the sum of the EC and IC components [30] with the normalization of both components left as free parameters

$$F_{2,c}(x, Q^2, \mu^2, m_c^2) = \alpha F_{2,c}^{\text{EC, NLO}}(x, Q^2, \mu^2, m_c^2) + \beta F_{2,c}^{\text{IC, NLO}}(x, Q^2, \mu^2, m_c^2) , \quad (16)$$

with the scale $\mu^2 = Q^2 + M_{c\bar{c}}^2$. The EC parameter α, typically larger than unity, can be considered an estimate of the NNLO contribution. Since a 1% normalization of the IC component is already assumed, the fitted β is the fraction of this normalization. The results are presented in Table 1. The error bars in the table correspond to a 95% confidence level for the central fit parameters. Several different parton distribution functions were used which produced essentially the same results. The final results for the combined model of Eq. (16) are shown in Fig. 4 for the CTEQ 3M [33], MRS G [34] and GRV 94 HO [35] parton densities. The table shows that given the quality of the data, no statement can be made about the intrinsic charm content for $\bar{\nu} = 53$ and 95 GeV. However, for $\bar{\nu} = 168$ GeV an intrinsic charm contribution of $(0.86 \pm 0.60)\%$ is indicated.

Leading Charm in the Projectile Fragmentation Region

The role of IC in hadroproduction is now discussed. The total x_F dependence is the sum of leading twist fusion and intrinsic charm. A linear nuclear target, A, dependence is assumed for leading-twist fusion while an A^α dependence is used for the intrinsic charm component [25] where $\alpha = 0.77$ for mesons and 0.71 for baryons

$$\frac{d\sigma_{hA}}{dx_F} = A\frac{d\sigma_{\text{lt}}}{dx_F} + A^\alpha \frac{d\sigma_{\text{ic}}}{dx_F} , \quad (17)$$

This A dependence is included in the calculations as appropriate.

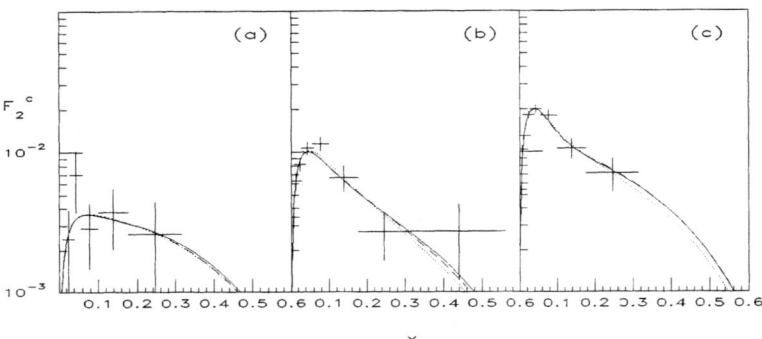

FIGURE 4. (a) The EMC data for the structure function $F_{2,c}(x, Q^2, \mu^2, m_c^2)$ at $\bar{\nu} = 53$ GeV plotted as a function of x together with the fitted results from Eq. (16) with the parameters α and β taken from Table 1. The CTEQ 3M (solid), MRS G (dashed) and GRV 94 HO (dotted) parton densities are shown. (b) Same as (a) for $\bar{\nu} = 95$ GeV. (c) Same as (a) for $\bar{\nu} = 168$ GeV. Reproduced from Ref. [30] with permission from Elsevier Science.

The calculations are at LO in α_s with a constant factor $K \sim 2-3$ included since the NLO x_F distribution is larger than the LO distribution by an approximately constant factor [11]. Neither LO nor NLO production results in the flavor correlations [7] observed in leading charm production.

The charm hadron x_F distribution at leading twist has the factorized form [18]

$$\frac{d\sigma_{lt}}{dx_F} = \frac{\sqrt{s}}{2} x_1 x_2 \int H_{AB}(x_1, x_2) \frac{1}{E_3} \frac{D_{H/c}(z_5)}{z_5} dz_5 \, dy_4 \, dp_T^2 \, , \quad (18)$$

where 1 and 2 are the initial partons, 3 and 4 are the produced charm quarks with $m_c = 1.5$ GeV, and 5 is the final-state charm hadron. The fragmentation functions, $D_{H/c}(z)$, describe the hadronization of the charm quark where $z = p_H/p_c$ is the fraction of the charm quark momentum carried by the charm hadron, assumed to be collinear to the charm quark. A delta function, $\delta(z-1)$, is consistent with low p_T hadroproduction data [36,37]. The convolution of the $q\bar{q}$ annihilation and gluon fusion cross sections [2] with the parton densities is given in $H_{AB}(x_1, x_2)$ [18]. Parton distributions of baryons other than the proton, such as the hyperon, are not available. However, using baryon number and momentum sum rules, a set of parton distributions for e.g. the Σ^- can be inferred from the proton distributions as $u_v^p = d_v^{\Sigma^-}$, $d_v^p = s_v^{\Sigma^-}$, $s^p = u^{\Sigma^-} \cdots$. The gluon distributions are assumed to be the same in the Σ^- and the proton.

The frame-independent probability distribution of an n–particle $c\bar{c}$ Fock state is given by Eq. (11) with $\widehat{m}_q = 0.45$ GeV, $\widehat{m}_s = 0.71$ GeV, and $\widehat{m}_c = 1.8$ GeV [18,38]. The intrinsic charm production cross section for a single charm hadron from the n-particle state can be related to P_{ic}^n and the inelastic hN cross section by

$$\frac{d\sigma_{ic}^{n}}{dx_F} = \frac{dP_{ic}^{n}}{dx_F}\sigma_{hN}^{in}\frac{\mu^2}{4\widehat{m}_c^2} \ . \tag{19}$$

It is assumed that P_{ic} for the minimal IC Fock state is 0.31% regardless of the projectile identity. The factor of $\mu^2/4\widehat{m}_c^2$ arises from the soft interaction which breaks the coherence of the Fock state. With $\mu^2 \sim 0.1$ GeV^2, the x_F-integrated cross sections are $\sigma_{ic}^{4}(\pi N) \approx 0.5$ μb and $\sigma_{ic}^{5}(pN) \approx 0.7$ μb at 200 GeV. The pN and πN inelastic cross sections have been measured while hyperon-nucleon cross sections, such as $\Sigma^- p$, have not. For $\sigma_{\Sigma^- N}^{in}$, σ_{pp}^{in} was scaled to $\sigma_{\Lambda p}^{in}$ at the highest measured Λp energy and the energy dependence of σ_{pp}^{in} was used thereafter.

There are two ways of producing charm hadrons from IC states. The first is by uncorrelated fragmentation. Assuming that the c quark fragments into a D meson, the D distribution is

$$\frac{dP_{ic}^{nF}}{dx_D} = \int dz \prod_{i=1}^{n} dx_i \frac{dP_{ic}^{n}}{dx_1 \ldots dx_n} D_{D/c}(z)\delta(x_D - zx_c) \ , \tag{20}$$

with with $D_{H/c}(z) = \delta(z-1)$. If the projectile contains the corresponding valence quarks, the charm quark can also hadronize by coalescence with the valence spectators. The coalescence mechanism thus introduces flavor correlations between the projectile and the final-state hadrons. The coalescence distributions do not include any binding energy or mass effects so that

$$\frac{dP_{ic}^{nC}}{dx_H} = \int \prod_{i=1}^{n} dx_i \frac{dP_{ic}^{n}}{dx_1 \ldots dx_n}\delta(x_H - x_{H_1} - \cdots - x_{H_{n_V}}) \ . \tag{21}$$

The coalescence function is simply a delta function combining the momentum fractions of the quarks in the Fock state configuration that make up the valence quarks of the final-state hadron.

Not all charm hadrons can be produced from the minimal intrinsic charm Fock state configuration, $|n_V c\bar{c}\rangle$ where n_V represents the valence quarks of the projectile hadron. However, coalescence can also occur within higher fluctuations of the intrinsic charm Fock state. For example, in the proton, the $D^+(c\bar{d})$ and $\Xi_c^0(dsc)$ can be produced by coalescence from $|uudc\bar{c}d\bar{d}\rangle$ and $|uudc\bar{c}s\bar{s}\rangle$ configurations. These higher Fock state probabilities are [39,40]: $P_{icu} = P_{icd} \approx 70.4\% \ P_{ic}$ and $P_{ics} \approx 28.5\% \ P_{ic}$.

Note that a significant leading effect is present only in the minimal configuration, i.e. there is no difference between D^+ and D^- mesons produced from $|n_V c\bar{c}d\bar{d}\rangle$ states. As more partons are included in the Fock state, the coalescence distributions soften and approach the fragmentation distributions, eventually producing charm hadrons with less momentum than uncorrelated fragmentation from the minimal $c\bar{c}$ state if a sufficient number of $q\bar{q}$ pairs are included. There is then no longer any advantage to introducing more light quark pairs into the configuration.

As a concrete example of how the total probability distributions of charm hadron production from the intrinsic charm model is calculated, D^+ and D^- production with a Σ^- beam is described in detail. The probability distributions for all the final-state charm hadrons from Σ^-, p and π^- projectiles can be found in Ref. [41]. In the $|ddsc\bar{c}\rangle$ configuration, the D^- is produced by coalescence and by uncorrelated fragmentation while the D^+ is only produced by uncorrelated fragmentation so that

$$\frac{dP_{D^-}^5}{dx_F} = \frac{1}{2}\left(\frac{1}{10}\frac{dP_{ic}^{5F}}{dx_F} + \frac{1}{2}\frac{dP_{ic}^{5C}}{dx_F}\right) \tag{22}$$

$$\frac{dP_{D^+}^5}{dx_F} = \frac{1}{10}\frac{dP_{ic}^{5F}}{dx_F} \tag{23}$$

where F refers to uncorrelated fragmentation and C to coalescence into the specific final-state with the associated probability distribution. When $|ddsc\bar{c}q\bar{q}\rangle$ configurations, where $q\bar{q} = u\bar{u}$, $d\bar{d}$ and $s\bar{s}$, are included, the total intrinsic charm probability distribution for these mesons is:

$$\frac{dP_{D^-}}{dx_F} = \frac{1}{2}\left(\frac{1}{10}\frac{dP_{ic}^{5F}}{dx_F} + \frac{1}{2}\frac{dP_{ic}^{5C}}{dx_F}\right) + \frac{1}{2}\left(\frac{1}{10}\frac{dP_{icu}^{7F}}{dx_F} + \frac{2}{5}\frac{dP_{icu}^{7C}}{dx_F}\right) \tag{24}$$

$$+ \frac{1}{2}\left(\frac{1}{10}\frac{dP_{icd}^{7F}}{dx_F} + \frac{3}{5}\frac{dP_{icd}^{7C}}{dx_F}\right) + \frac{1}{2}\left(\frac{1}{10}\frac{dP_{ics}^{7F}}{dx_F} + \frac{2}{5}\frac{dP_{ics}^{7C}}{dx_F}\right) \tag{25}$$

$$\frac{dP_{D^+}}{dx_F} = \frac{1}{10}\frac{dP_{ic}^{5F}}{dx_F} + \frac{1}{10}\frac{dP_{icu}^{7F}}{dx_F} + \frac{1}{2}\left(\frac{1}{10}\frac{dP_{icd}^{7F}}{dx_F} + \frac{1}{8}\frac{dP_{icd}^{7C}}{dx_F}\right) + \frac{1}{10}\frac{dP_{ics}^{7F}}{dx_F}. \tag{26}$$

The model calculations are compared to the WA89 data [21] on carbon and copper targets in Fig. 5. The agreement with the data is quite reasonable given both the low statistics of the data and the normalization to the first data point rather than fitting the normalization to the data. The average x_F of the coalescence distribution is more than a factor of two larger than that of a Σ_c^0 production by independent fragmentation of a c quark, producing a shoulder in the x_F distribution. Because the Σ_c^0 is produced by coalescence with $\approx 30\%$ less average momentum from the 7-particle Fock states, the intermediate x_F region is partially filled in by the higher Fock state contributions, resulting in a smoother x_F distribution. Similar results are found for the other charm hadrons. The A dependence of the reduces the IC contribution in larger targets.

Another way to quantify leading charm production is through the asymmetry between leading and nonleading charm, defined as

$$A(x_F) = \frac{d\sigma_L/dx_F - d\sigma_{NL}/dx_F}{d\sigma_L/dx_F + d\sigma_{NL}/dx_F} \tag{27}$$

where L represents the leading and NL the nonleading charm hadron. High statistics data has previously been available only from π^- beams where a significant enhancement of D^- over D^+ production was seen at $x_F > 0.3$ [19,22,23], in qualitative agreement with intrinsic charm, Ref. [20]. The model [39] also correctly predicted the symmetric production of D_s^- and D_s^+ mesons and Λ_c^+ and $\overline{\Lambda}_c^+$ baryons by π^- beams [37,42,43].

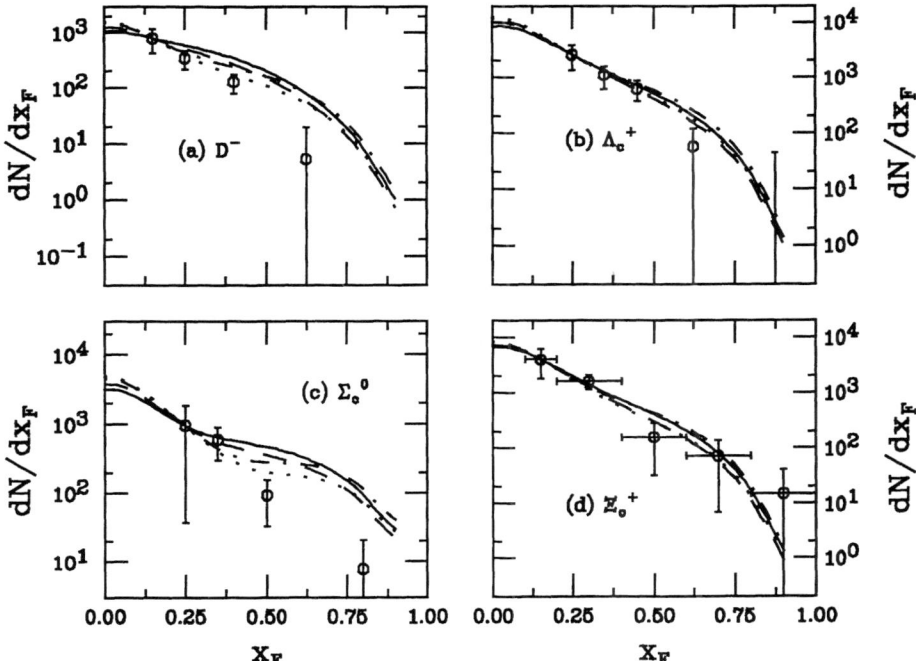

FIGURE 5. Model predictions are compared to the $\Sigma^- A$ data of Ref. [21] for (a) D^-, (b) Λ_c^+, (c) Σ_c^0 and (d) Ξ_c^+. The solid and dashed curves represent the full model, with the intrinsic charm probability distributions given in Ref. [41] for carbon and copper targets respectively. The dot-dashed and dotted curves contrast the results for carbon and copper targets respectively with the simplified model which considers only fragmentation from the minimal Fock state and coalescence only from the state with the minimum number of partons necessary to produce it. Reproduced from Ref. [41] with permission from Elsevier Science.

The WA89 collaboration also presented the D^-/D^+, D_s^-/D_s^+ and $\Lambda_c^+/\overline{\Lambda}_c^+$ asymmetries from their Σ^- data [44]. In Fig. 6 the calculations are compared to this data as well as a prediction of the D^-/Ξ_c^0 asymmetry, both of which are produced from the minimal Fock configuration. The full model, Eqs. (24) and (26), gives a larger asymmetry between D^- and D^+ at low x_F than found previously [20,39] because D^- production at intermediate x_F is enhanced by coalescence production from the 7-particle configurations. The results are in qualitative agreement with the data, shown in Fig. 6(a). The measured D_s^-/D_s^+ and $\Lambda_c^+/\overline{\Lambda}_c^+$ asymmetries are larger than predicted at intermediate x_F. This could result from the crude assumption that all final-state hadrons are produced by independent fragmentation with the same probability. The asymmetry between D^- and Ξ_c^0 is interesting because the $|ddsc\bar{c}\rangle$ state of the Σ^- can be thought of as a virtual $D^-\Xi_c^0$ fluctuation.

Charm hadron production with 650 GeV Σ^- beams [45] on copper targets is

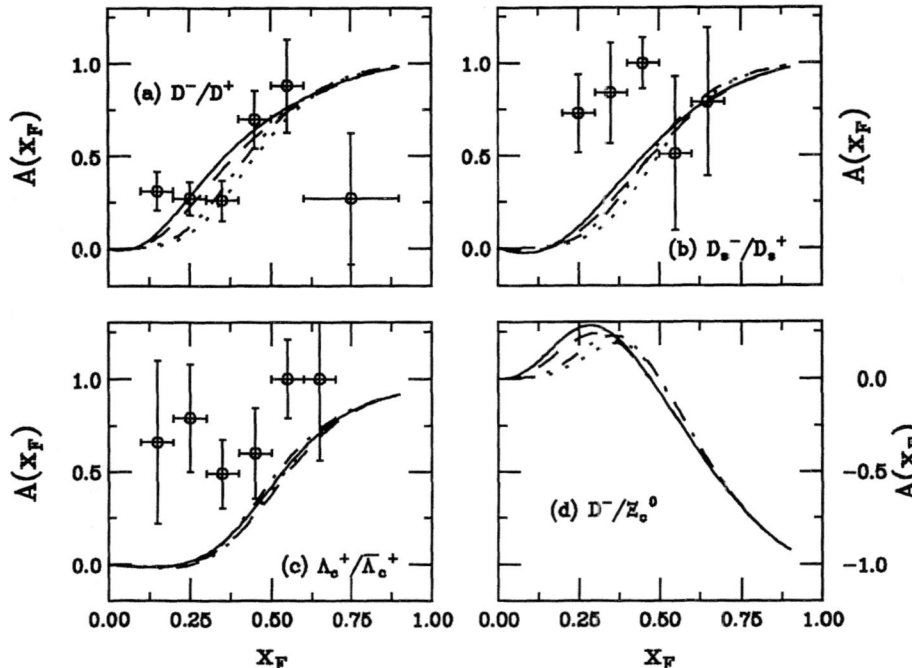

FIGURE 6. Model predictions are compared to the $\Sigma^- A$ data of Ref. [44] for the following asymmetries: (a) D^-/D^+, (b) D_s^-/D_s^+ and (c) $\Lambda_c^+/\overline{\Lambda}_c^+$, as well as a prediction of the (d) D^-/Ξ_c^0 asymmetry. The solid and dashed curves represent the full model, with the intrinsic charm probability distributions given in Ref. [41] for carbon and copper targets respectively. The dot-dashed and dotted curves for carbon and copper targets respectively contrast the results with the simplified model which considers only fragmentation from the minimal Fock state and coalescence only from the state with the minimum number of partons necessary to produce it. Reproduced from Ref. [41] with permission from Elsevier Science.

shown in Fig. 7. The Ξ_c^0 is clearly the hardest distribution, followed by the Σ_c^0. The Ξ_c^0 leads the Σ_c^0 because the more massive valence s quark carries more of the Σ^- velocity than the d valence quarks. The Ξ_c^+ leads the Λ_c^+ in the 7-particle $u\overline{u}$ state for the same reason. The D^- and D_s^-, also produced from the 5-particle state have the hardest meson distributions but lag the baryons. The D^+ and D_s^+ have the softest distributions with the D^+ slightly harder because the quarks in the $d\overline{d}$ configuration get slightly more velocity than the $s\overline{s}$ configuration with the more massive strange quarks. The asymmetries are somewhat reduced at higher energies, due to the larger leading-twist cross section.

A new letter of intent at the CERN SPS has proposed measuring charm in heavy ion collisions. Aside from nucleus-nucleus collisions, data is also needed from pA and pp interactions to provide a baseline for the heavy-ion charm measurement.

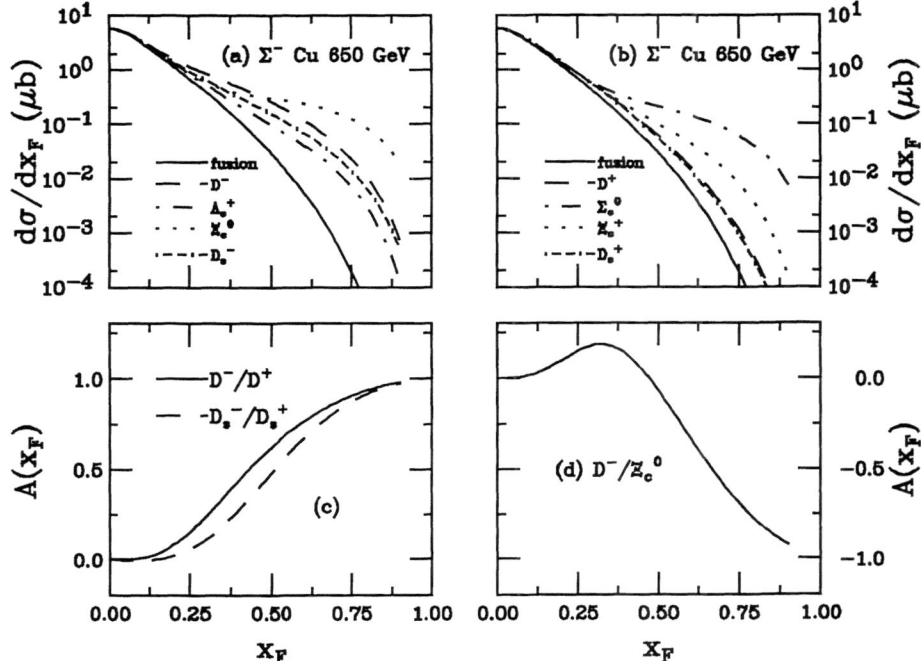

FIGURE 7. Model predictions for charm hadron production are given for Σ^-Cu interactions at 650 GeV. The individual x_F distributions are given in (a) and (b). All cross sections are compared to the leading twist fusion calculation in the solid curve. In (a) the hadron distributions are D^- (dashed), Λ_c^+ (dot-dashed), Ξ_c^0 (dotted) and D_s^- (dot dashed dashed). In (b) the hadron distributions are D^+ (dashed), Σ_c^0 (dot-dashed), Ξ_c^+ (dotted) and D_s^+ (dot dashed dashed). Predictions of the asymmetries are given in (c) for D^-/D^+ (solid) and D_s^-/D_s^+ (dashed) while the prediction for the D^-/Ξ_c^0 asymmetry is given in (d). Reproduced from Ref. [41] with permission from Elsevier Science.

It is also possible to take data with secondary $\pi^\pm$ beams. Since $x_{F_{max}} = 0.6$, it represents an opportunity to study asymmetries at high statistics with both π and p beams. The D^-/D^+ asymmetry has been well measured in πA interactions but the A dependence has not been studied in detail. The D production data in proton-induced interactions has been poor but is improving. This new experiment could significantly advance the state of the art in both areas.

Figure 8 shows a prediction of the x_F distributions of D^- and D^+ production for $\pi^\pm$ beams on carbon and tungsten targets at 200 GeV. In π^--induced interactions, the D^- is leading while in π^+-induced interactions, the D^+ is the leading particle. In a symmetric nucleus like carbon the π^- and π^+ charm production cross sections are nearly identical. When a neutron-rich target such as tungsten is used, the π^- cross section is larger. The A dependence of the asymmetry can be measured in the new experiment and quantitative differences between π^- and π^+ beams may be observed with high statistics.

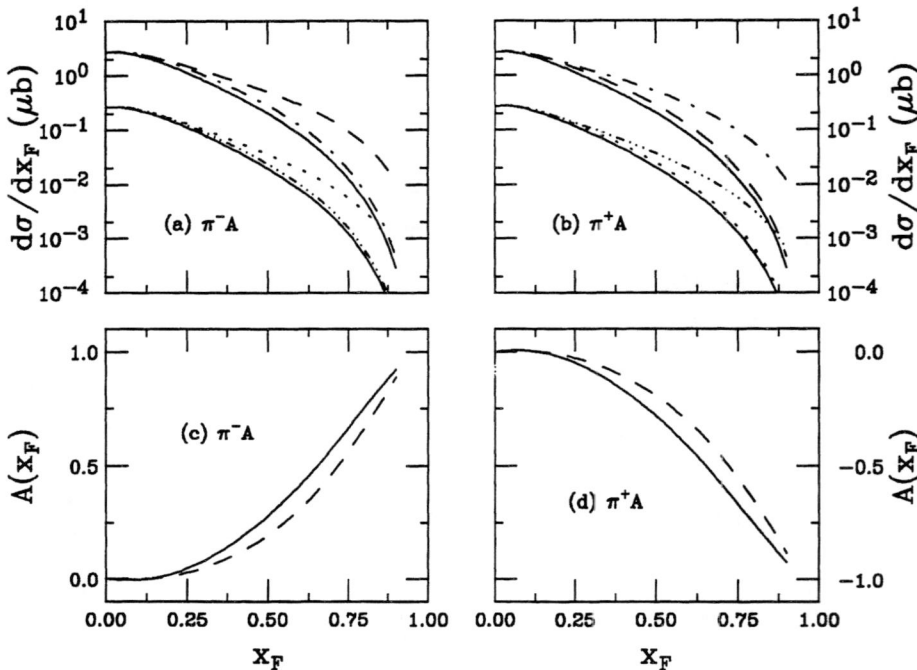

FIGURE 8. Predictions for D^- and D^+ production in $\pi^- p$ and $\pi^+ p$ interactions at 200 GeV. The curves in (a) and (b) illustrate the dependence of leading charm on the projectile and target. The fusion curves include no IC with carbon (upper solid) and tungsten (lower solid) targets while the other curves assume $P_{ic}^4 = 0.31\%$. They are D^- (dashed) and D^+ (dot dashed) on a carbon target while the dotted and dot-dot-dot-dashed curves are the D^- and D^+ calculations on a tungsten target. The D^-/D^+ asymmetries are shown for carbon (solid) and tungsten (dashed) with π^- (c) and π^+ (d) projectiles.

The energy dependence of the asymmetries can be observed in pA studies. Since Λ_c and D^- can both be produced by coalescence from $|uudc\bar{c}\rangle$, an interesting asymmetry, Λ_c/D^-, between leading charm baryons and leading charm mesons can be measured as well as the standard D^-/D^+ asymmetry. The IC cross section is only weakly dependent on energy so that the smaller charm cross section at 200 GeV would result in a larger asymmetry at moderate x_F than at 450 GeV. At larger x_F the asymmetry goes to unity at both energies because the fusion cross section is very small. The absolute cross sections and asymmetries are shown in Fig. 9. High statistics charm data with a proton beam at both energies would significantly advance the physics of leading charm production.

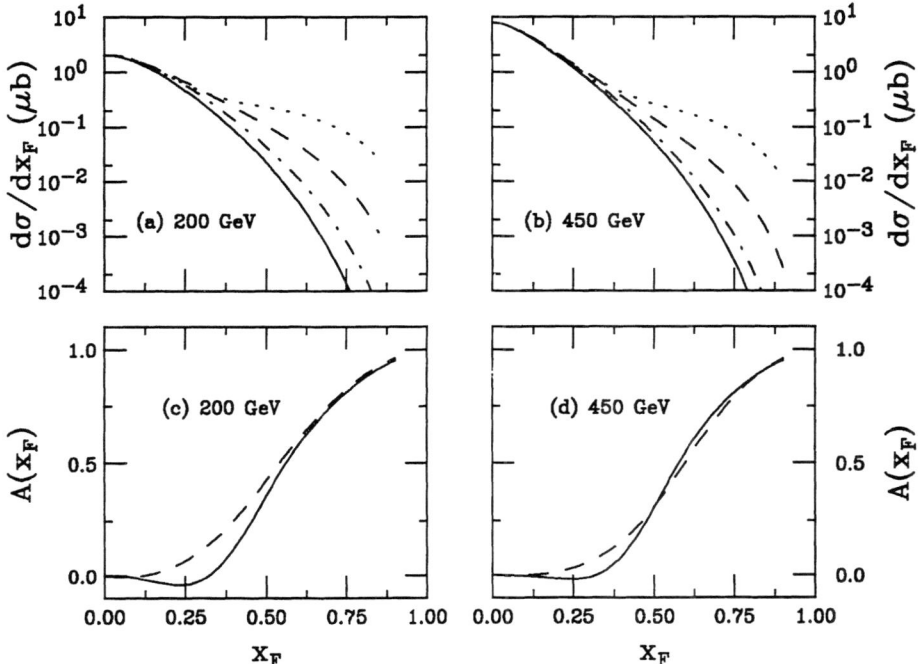

FIGURE 9. Predictions of the energy dependence of charm hadron production by a proton beam on lead targets. The curves in (a) and (b) illustrate the dependence of leading charm on the projectile energy. The fusion curve (solid) includes no IC while the other curves assume $P_{ic} = 0.31\%$. They are D^- (dashed), D^+ (dot dashed) and Λ_c (dotted). The D^-/D^+ (solid) and Λ_c/D^- (dashed) asymmetries are shown at 200 GeV (c) and 450 GeV (d).

SUMMARY

Recent theoretical advances in heavy quark production have been introduced. Although the discussion has centered on charm production, applications of higher order calculations to top and bottom production have resulted in much greater success, see *e.g.* Refs. [14,15].

At high x_F, where perturbative charm production does not agree well with data, intrinsic charm may dominate. Several examples have been shown, in particular for leading charm production. Predictions have been made for a new experiment which can more fully probe the large x_F region with proton beams.

ACKNOWLEDGMENTS

I thank S.J. Brodsky, T. Gutierrez, B.W. Harris, E. Laenen, N. Kidonakis, S. Moch, and J. Smith for enjoyable collaborations. I thank the organizers for their hospitality.

REFERENCES

1. Kidonakis, N., Laenen, E., Moch, S., and Vogt, R., in preparation.
2. Ellis, R.K., in *Physics at the 100 GeV Scale*, Proceedings of the 17th SLAC Summer Institute, Stanford, California, 1989, edited by E.C. Brennan (SLAC Report No. 361), p. 45.
3. Glück, M., Reya, E., and Vogt, A., *Z. Phys.* **C53**, 127 (1992).
4. Glück, M., Reya, E., and Vogt, A., *Z. Phys.* **C53**, 651 (1992).
5. Martin, A.D., Roberts, R.G., and Stirling, W.J., *Phys. Lett.* **B306**, 145 (1993).
6. See Tavernier, S.P.K., *Rep. Prog. Phys.* **50**, 1439 (1987) for references to data prior to 1988.
7. See Frixione, S., Mangano, M.L., Nason, P., and Ridolfi, G., *Nucl. Phys.* **B431**, 453 (1994) for data since 1988.
8. Alves, G.A. *et al.*, (E769 Collab.), *Phys. Rev. Lett.* **70**, 722 (1993).
9. Sutton, P.J., Martin, A.D., Roberts, R.G., and Stirling, W.J., *Phys. Rev.* **D45**, 2349 (1992).
10. Vogt, R., in *cbt Workshop*, Proceedings of LISHEP 95 Session C: Heavy Flavour Physics, Rio de Janeiro, Brazil, 1995, edited by F. Caruso *et al.* (Editions Frontieres), p. 51.
11. Vogt, R., *Z. Phys.* **C71**, 475 (1996).
12. Laenen, E., Smith, J., and van Neerven, W.L., *Nucl. Phys.* **B369**, 543 (1992).
13. Smith, J., and Vogt, R., *Z. Phys.* **C75**, 271 (1997).
14. Kidonakis, N., and Sterman, G., *Phys. Lett.* **B387**, 867 (1996).
15. Kidonakis, N., Smith, J., and Vogt, R., *Phys. Rev.* **D56**, 1553 (1997).
16. Kidonakis, N., and Vogt, R., *Phys. Rev.* **D59**, 074014 (1999).
17. Peterson, C., Schlatter, D., Schmitt, I., and Zerwas, P., *Phys. Rev.* **D27**, 105 (1983).
18. Vogt, R., Brodsky, S.J., and Hoyer, P., *Nucl. Phys.* **B383**, 643 (1992).
19. Adamovich, M.I. *et al.*, (WA82 Collab.), *Phys. Lett.* **B305**, 402 (1993).
20. Vogt, R., and Brodsky, S.J., *Nucl. Phys.* **B438**, 261 (1995).
21. Werding, R., (WA89 Collab.), in Proceedings of *ICHEP94*, 27th International Conference on High Energy Physics, Glasgow, Scotland (1994).
22. Alves, G.A., *et al.*, (E769 Collab.), *Phys. Rev. Lett.* **72**, 812 (1994).
23. Aitala, E.M., *et al.*, (E791 Collab.), *Phys. Lett.* **B371**, 157 (1996).
24. Aubert, J.J., *et al.*, (EM Collab.), *Nucl. Phys.* **B213**, 31 (1983); *Phys. Lett.* **94B**, 96 (1980); *Phys. Lett.* **110B**, 73 (1982); Strovink, M., in proceedings of the *1981 International Symposium on Lepton and Photon Interactions at High Energies* (Bonn) edited by W. Pfeil, p. 594.
25. Badier, J., *et al.*, (NA3 Collab.), *Z. Phys.* **C20**, 101 (1983).
26. Badier, J., *et al.*, (NA3 Collab.), *Phys. Lett.* **114B**, 457 (1982); **158B**, 85 (1985).
27. Brodsky, S.J., Hoyer, P., Peterson, C., and Sakai, N., *Phys. Lett.* **B93**, 451 (1980); Brodsky, S.J., Peterson, C., and Sakai, N., *Phys. Rev.* **D23**, 2745 (1981).

28. Brodsky, S.J., Hoyer, P., Mueller, A.H., and Tang, W.-K., *Nucl. Phys.* **B369**, 519 (1992).
29. Hoffmann, E., and Moore, R., *Z. Phys.* **C20**, 71 (1983).
30. Harris, B.W., Smith, J., and Vogt, R., *Nucl. Phys.* **B461**, 181 (1996).
31. Laenen, E., Riemersma, S., Smith, J., and van Neerven, W.L., *Nucl. Phys.* **B392**, 162 (1993).
32. http://cpt1.dur.ac.uk:80/HEPDATA.
33. Lai, H.L., Botts, J., Huston, J., Morfín, J.G., Owens, J.F., Qiu, J.W., Tung, W.K., and Weerts, H., *Phys. Rev.* **D51**, 4763 (1995).
34. Martin, A.D., Roberts, R.G., and Stirling, W.J., *Phys. Lett.* **B354**, 155 (1995).
35. Glück, M., Reya, E., and Vogt, A., *Z. Phys.* **C67**, 433 (1995).
36. Aguilar-Benitez, M., et al., (LEBC-EHS Collab.), *Phys. Lett.* **161B**, 400 (1985); *Z. Phys.* **C31**, 491 (1986).
37. Barlag, S., et al., (ACCMOR Collab.), *Phys. Lett.* **B247**, 133 (1990).
38. Vogt, R., Brodsky, S.J., and Hoyer, P., *Nucl. Phys.* **B360**, 67 (1991).
39. Vogt, R., and Brodsky, S.J., *Nucl. Phys.* **B478**, 311 (1996).
40. Vogt, R., and Brodsky, S.J., *Phys. Lett.* **B349**, 569 (1995).
41. Gutierrez, T., and Vogt, R., *Nucl. Phys.* **B539**, 189 (1999).
42. Aitala, E.M., et al., (E791 Collab.), *Phys. Lett.* **B411**,230 (1997).
43. Adamovich, M.I., et al., (BEATRICE Collab.), *Nucl. Phys.* **B495**, 3 (1997).
44. Adamovich, M.I., et al., (WA89 Collab.), *Eur. Phys. J.* **C8**, 593 (1999).
45. Ramberg, E. in *Hyperons, Charm and Beauty Hadrons*, Proceedings of the 2nd International Conference on Hyperons, Charm and Beauty Hadrons, Montreal, Canada, 1996, edited by C.S. Kalman et al., *Nucl. Phys.* **B** (Proc. Suppl.) **55A**, 173 (1997).

Recent results on Charm Physics from Fermilab[1]

J. C. Anjos[2] and E. Cuautle[3]

Centro Brasileiro de Pesquisas Físicas, CBPF
Rua Dr. Xavier Sigaud 150, 22290-180 Rio de Janeiro Brazil

Abstract. New high statistics, high resolution fixed target experiments producing 10^5 - 10^6 fully reconstructed charm particles are allowing a detailed study of the charm sector. Recent results on charm quark production from Fermilab fixed target experiments E-791, SELEX and FOCUS are presented.

INTRODUCTION

This review contains recent results from three Fermilab fixed target experiments dedicated to study charm physics: E791 and E871 (SELEX) of charm hadroproduction and E831 (FOCUS) of charm photoproduction.
The experiment E791 used a 500 GeV/c π^- beam incident on platinum and diamond target foils and took data in 1991-1992. A loose transverse energy trigger was used to record 2^{10} interactions. Silicon microstrip detectors (6 in the beam and 17 downstream of the target) provided precision track and vertex reconstruction. The precise location of production and decay vertices of long lived charm particles allowed to reconstruct over 2^5 charm particles.
SELEX used a 600 GeV hyperon beam of negative polarity to make a mixed beam of Σ^- and π^- in roughly equal numbers. The positive beam was composed of 92 % of protons, and 8 % of π^+. Interactions occurred in a segmented target of 5 foils, 2 Cu and 3 C. The experiment was designed to study charm production in the forward hemisphere, with good mass and decay vertex resolution for charm momenta in the range 100-500 GeV/c. A major innovation was the use of online selection criteria to identify events that had evidence for a secondary vertex. Data taking finished in 1997.
The FOCUS experiment used a photon beam with $< E_\gamma > = 170$ GeV on a Beryllium Oxide segmented target. Sucessor of Fermilab E687, FOCUS had upgrades

---

[1] This work is partially supported by CLAF/CNPq Brazil and CONACyT México
[2] janjos@lafex.cbpf.br
[3] ecuautle@lafex.cbpf.br

in vertexing, Cerenkov identification, electromagnetic calorimetry and muon identification. The target segmentation and the use of additional silicon microstrip detectors after each pair of target segments were major improvements. The experiment took data in 1996-1997 and collected over 7 billion triggers of charm candidates. FOCUS has the largest charm sample available in the world, with over 10^6 fully reconstructed charmed particles.

The data available from these three experiments will allow to make high precision charm physics, to study rare decays and to search for new physics.

CHARM PRODUCTION AND CROSS SECTIONS FOR D MESONS

Charm production is a combination of short and long range processes. Perturbative Quantum Chromodynamics (pQCD) can be used to calculate the parton cross section, the short range process. The long range process of parton hadronization has to be modeled from experimental data. The two processes occur at different time scales and should not affect each other, leading to factorization properties. The general expression for production of charm in hadroproductions is [1]

$$\sigma(P_A, P_B) = \sum_{i,j} \int dx_A dx_B f_i^A(x_A, \mu) f_j^B(x_B, \mu) \hat{\sigma}_{i,j}(\alpha_s(\mu), x_A P_A, x_B P_B) \quad (1)$$

where $f_i^A(x_A, \mu)$ is the probability of finding parton type i inside the hadron A with momentum fraction x_A, and μ is the scale at which the process occurs. The other hadron taking part in the interaction has $f_j^B(x_B, \mu)$. The elementary parton cross section $\hat{\sigma}_{i,j}(\alpha_s(\mu), x_A P_A, x_B P_B)$ term is calculated by QCD, according to parton models. Although QCD is well defined theory, solutions to most problems are quite difficult. So, to calculate the parton cross section we usually do a perturbative series expansion in terms of the strong interaction coupling constant, α_s, and calculate at leading order (LO), next-to-leading order (NLO) and so on. Nowadays most calculations include NLO. Some Feynman diagrams for hadroproduction of charm at LO and NLO are shown in Fig.1 (more details can be found in reference [1]).
Usually the differential cross sections is measured as a function of the scaled longitudinal momentum (Feynman x_F) $d\sigma/dx_F$, and as a function of the transverse momentum squared $d\sigma/dp_t^2$. A phenomenological parametrization of the double differential cross section often used to describe the experimental data is given by,

$$\frac{d\sigma}{dx_F dp_t^2} = A (1 - x_F)^n \, exp(-bp_t^2) \quad (2)$$

where A, n and b are parameters used to fit the data. In fact kinematic considerations provide predictions for the power n at high x_F and QCD calculations give average p_t values comparable to the charm quark mass.

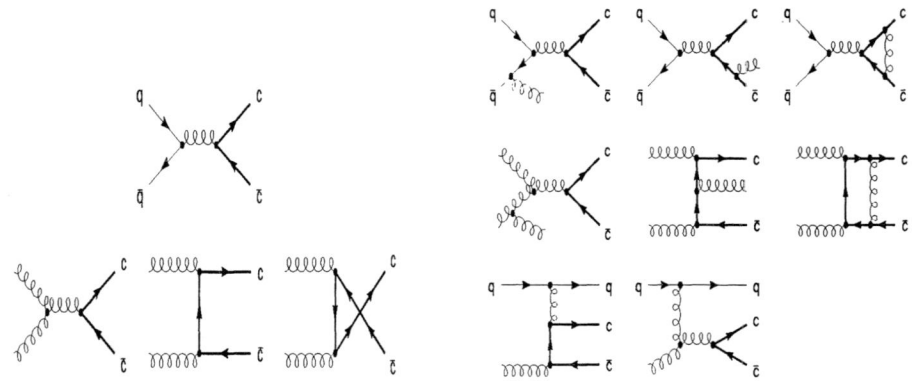

FIGURE 1. Some of Feynman diagrams for charm hadroproduction: leading order (left) and Next-to-Leading Order (right side)

E791 has measured the total forward cross section and the differential cross sections as functions of x_F and p_t^2 from a sample of 88990±460 D^0 mesons. Fig. 2 shows the differential distributions and its comparison to theoretical predictions from QCD calculations at NLO [2] and to the Monte Carlo event generator, Pythia/Jetset [3] for $c\bar{c}$ and $D\bar{D}$ production. Hadron distributions are softer than $c\bar{c}$ distributions due to fragmentation. With suitable choice for the intrinsic transverse momentum k_t of the incoming partons ($k_t = 1$ GeV/c), the Peterson fragmentation function parameter ($\varepsilon = 0.01$), and the charm quark mass $m_c = 1.5$ GeV/c^2, NLO D meson calculation provides a good match to the p_t^2 distribution and fair match to the x_F distribution. The hadronization scheme implemented in Pythia/Jetset can be adjusted to fit the data. The large number of D^0 events make it possible to clearly observe a turnover point greater than zero ($x_c = 0.0131 \pm 0.0038$) in the x_F distribution. The positive value provide evidence that the gluon distribution in the pion is harder than the gluon distribution in the nucleon [4]. The total forward neutral D meson cross section measured by E791 is $\sigma(x_F > 0) = 15.4 \pm^{1.8}_{2.3}$ μb/nucleon.

SELEX has also preliminary data on D^0 cross section. Fig. 2 (right) shows the x_F distribution for D^0 + c.c. from E791 [4] and the preliminary data from SELEX. Up to $x_F = 0.5$ data overlap and agree well. However SELEX has no strong evidence for rise at large x_F as seen in the E791 data. The continuous line is a fit to the data points with the phenomenological parametrization in Eq. 2.

Correlations in the production of Charm pairs

Observation of both of the charm particles in a hadron-produced event can give additional information on the production process and allow to test QCD predic-

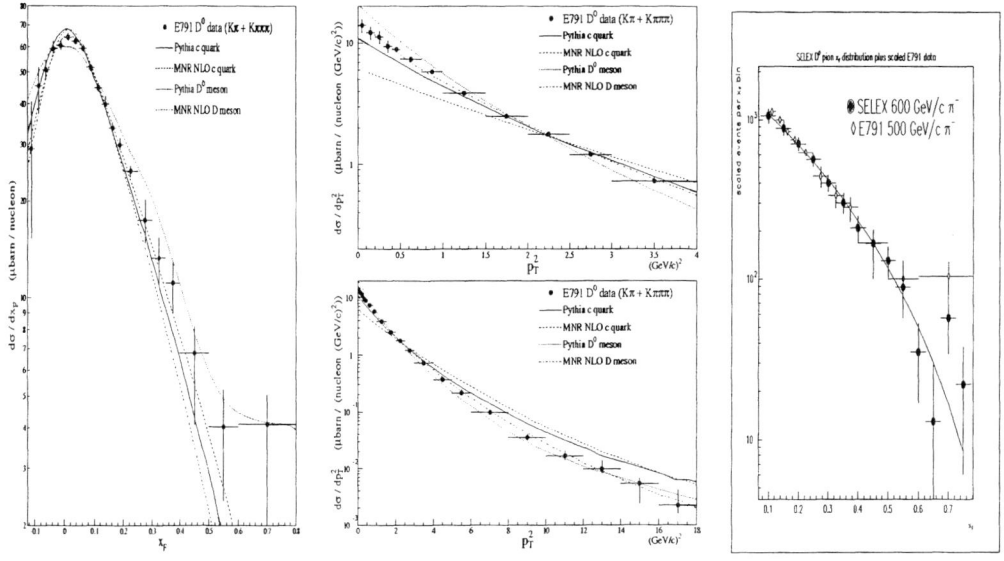

FIGURE 2. D^0 production: from E791 compared with theoretical models as a function of x_F (left), p_t^2 (central column) and compared with preliminary results from SELEX (right).

tions since both longitudinal and transverse momenta of the charm particles and angular correlations are explicitly measured.

In the simplest parton model the charm and anticharm particles are expected to be produced in opposite directions in the transverse plane. However if one assumes that the incoming partons have an intrinsic transverse momentum, k_t, this will affect the transverse momentum of the heavy quark pair, its azymuthal correlations and the transverse distribution of a single quark. The partons entering the hard interaction are indeed supposed to have a non vanishing primordial k_t, seen as a nonperturbative Fermi motion of partons inside the incoming hadrons. Typical values of k_t should thus be 300-400 MeV. However it has been noted [5] that much higher values of k_t are required, at or above 1 GeV, to reproduce charm data. This could be an indications of the importance of next to leading and higher order effects, by which emitted gluons would further modify the nearly back-to-back [6] production of the final charm hadrons.

E791 has measured correlations between D and $\bar{D}$ mesons from 791 ± 44 fully reconstructed charm meson pairs [7]. The main variables used to describe charm pair correlations are the beam direction, p_t, and either the rapidity y or the Feynman scaling variable x_F, and also the azymuthal distribution ϕ. The same way the difference and sum between these variables for two charm mesons $(D, \bar{D})$ Δx_F,

Σx_F and so on.

The measured distributions are compared to predictions of the fully differential NLO calculation for $c\bar{c}$ production [2], as well as to predictions from the Pythia/Jetset Monte Carlo event generator for $c\bar{c}$ and $D\bar{D}$ production. Default parameter have been used for the theoretical models.

For the single charm distributions shown in Fig.3, we observe that for the longitudinal momentum distributions x_F and y the experimental results and theoretical predictions do not agree. In this comparison the experimental distributions are most similar to the NLO and Pythia/Jetset $c\bar{c}$ distributions, but are narrower than all three theoretical predictions. The experimental p_t^2 distribution agrees quite well with all three theoretical distributions. As expected, both the theoretical and experimental ϕ distributions are consistent with being flat.

The experimental and theoretical longitudinal distributions for pairs Δx_F, Σx_F, Δy and Σy are shown in Fig. 4. As with the single-charm distributions, the experimental results are much closer to the two $c\bar{c}$ predictions than to the Pythia/Jetset $D\bar{D}$ predictions, but narrower than all three theoretical predictions. For the transverse distributions for charm pairs any observed discrepancy between data and theory must derive from the theory modeling the correlations between the transverse momentum of the two D mesons $p_{t,D}$ and $p_{t,\bar{D}}$ because the single charm p_t^2 and ϕ experimental distributions agree well with the theory. The $\Delta\phi$ distribution shows clear evidence of correlations (more details can be found in reference [7]).

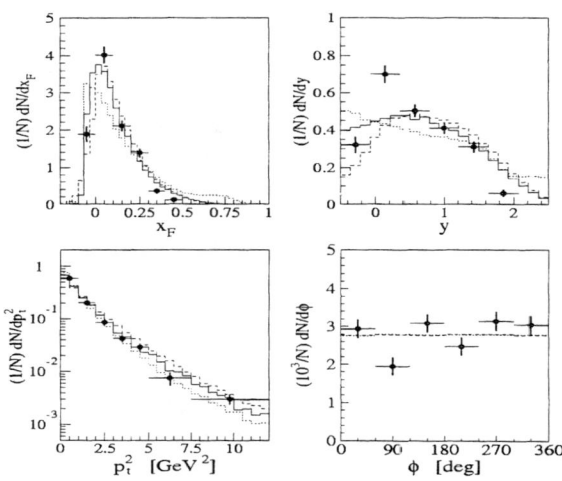

FIGURE 3. Single-charm distributions for x_F, y, p_t^2 and ϕ: weighted data (•) NLO QCD prediction (——); PYTHIA/JETSET charm quark prediction (- - - -); and PYTHIA/JETSET D meson prediction (·········). All distributions are obtained by summing the charm and anticharm distributions from charm-pair events.

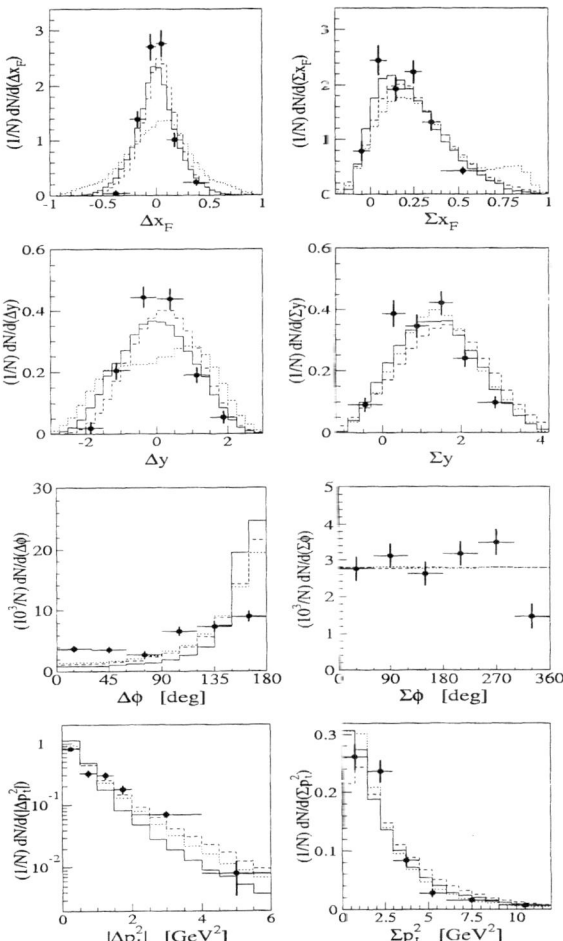

FIGURE 4. Charm-pair Δx_F, Σx_F, Δy, Σy, $\Delta\phi$, $\Sigma\phi$, Δp_t^2 and Σp_t^2 distributions; weighted data (•); NLO QCD prediction (———); PYTHIA/JETSET charm quark prediction (- - - -); and PYTHIA/JETSET D meson prediction (·········).

HADRONIZATION AND PARTICLE - ANTIPARTICLE ASYMMETRIES

The production of a charm hadron can be subdivided in two steps: the production of a $c\bar{c}$ pair followed by the hadronization of these quarks. In perturbative QCD, that describes the $c\bar{c}$ production, the x_F spectra of produced charm/anticharm quarks are identical to leading order and the effects of higher orders are very small in this respect. Therefore any asymmetry between charm and anticharm hadrons is a simple measure of nonperturbative effects coming from the hadronization process. Particle - antiparticle asymmetries can be quantified by means of the asymmetry parameter

$$A = \frac{N - \bar{N}}{N + \bar{N}} \qquad (3)$$

where N ($\bar{N}$) is the number of produced particles (antiparticles). This parameter is usually measured as a function of x_F and p_t^2.

Several experiments have reported an enhancement in the production rate of charm particles having valence quarks in common with the incident particles, relative to charge conjugate particles which have fewer or no common valence quarks. This effect is known as leading particle effect. Measurements of the asymmetry parameter A can put in evidence leading particle effects, as well as other effects like associated production of a meson and a baryon.

From the theoretical point of view, models which can account for the presence of leading particle effects use some kind of non-perturbative mechanism for hadronization, in addition to the perturbative production of charm quarks. Two examples are:

String fragmentation [8]: in this case the parton of the hard interaction and the beam remnants are connected by a string which reflect the confining color field. Successive breaking of the color flux tube stretched between a cluster, when it is kinematic possible, will create light quark-antiquark and hadrons will be produced.

Intrinsic charm model [9]: a virtual $c\bar{c}$ pair pops from the sea of the beam particle. The $c\bar{c}$ pair coalesce with the neighbor valence quarks due to their similar rapidity. This mechanism favors the production of charm particles with valence quarks in common with the beam particle at high x_F and low p_t region. A similar argument can be drawn with respect to the target.

New results on the asymmetry parameter A and evidence for leading particle effects in both meson and baryon production are available from experiments E791, SELEX and FOCUS.

E791 has measured recently the asymmetries of $D_s^\pm$ mesons [12]. Fig. 5 shows the $D_s^\pm$ asymmetry as a function of x_F and p_t^2, compared with previous $D^\pm$ results from the same experiment [13].

Preliminary results of $D^0(\bar{D}^0)$ as well as $D^\pm$ asymmetries as a function of x_F presented by SELEX [10] are shown in Fig. 6 (left), for different incident beam particles (π^-, Σ^-, p). The π^- data is compared to $D^\pm$ asymmetries from E791. The asymmetry at $x_F > 0.4$ does not rise steeply with x_F as previously reported by E791, Fig. 6 (right).

For baryons there are also preliminary results for the asymmetry parameter. Fig.7 (left) shows the E791 results for the Λ_c^+ asymmetries as function of x_F and

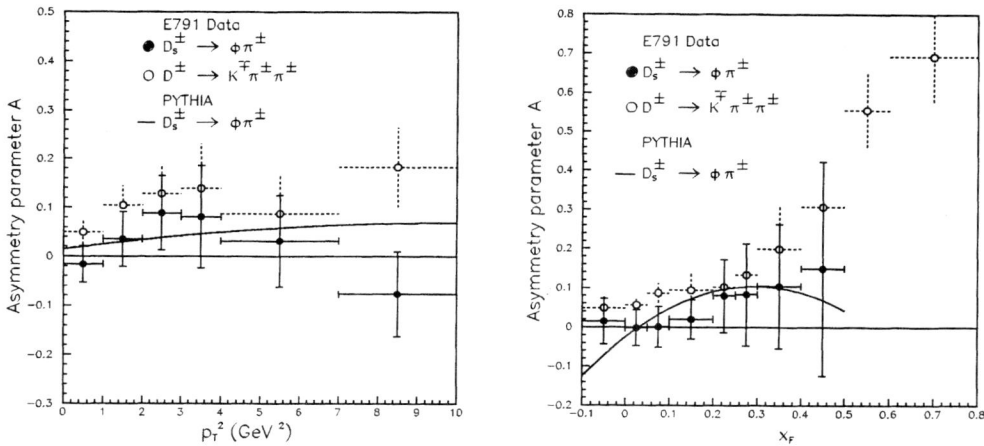

FIGURE 5. Asymmetries of D_s as a function of x_F and P_t^2, compared with $D^{\pm}$.

p_t^2, compared with predictions from Pythia/Jetset (full lines) [11]. The results show a uniform positive asymmetry of $12.7 \pm 3.4\%$ over the studied kinematical range but do not exclude a rise in the $x_F < 0$ region as predicted by Pythia/Jetset. For $x_F > 0$ the observed asymmetry does not agree with Pythia/Jetset predictions.
SELEX has also measured the Λ_C^+ asymmetry as a function of x_F for different incident beam particles (π^-, Σ^-, p) Fig. 7 (right) [10]. The asymmetry is clearly larger for the baryon beams than for the π beam. For the protons the only region in which there is $\bar{\Lambda}_c$ production is at very small x_F. Their preliminary results for the π^- beam are compatible with those from E791, Fig. 7 (left).
FOCUS has also preliminary results on baryon asymmetries. In Fig.8 we see the Λ_c asymmetry as functions of p_l, p_t^2 and x_F obtained from a sample of about 16,000 $\Lambda_c's$, compared with Pythia/Jetset Predictions. As FOCUS has a photon beam, no leading particle effect is expected in the $x_F > 0$ region. In this case the positive asymmetry observed in all the x_F range can be an indication of charm baryon and charm meson associated production, favouring a positive asymmetry.
The high statistic Λ_c sample from FOCUS allowed to obtain about 600 $\Sigma_c \to \Lambda_c \pi$. It is interesting to compare the asymmetry for charm particles with different light quark content. We present in Fig.9 (right) very preliminary results from FOCUS comparing the $\Sigma_c^{++}(uuc)$ and $\Sigma_c^0(ddc)$ to the $\Lambda_c^+(udc)$ total asymmetry.
In Fig. 9 (left) we show the comparison between the Λ^0 and the Λ_c asymmetries as a function of x_F from E791. Their similarity suggests that the ud diquark shared by the produced Λ^0 (Λ_c^+) and nucleons in the target should play an important role in the measured asymmetry in the $x_F < 0$ region. However, one expects that Λ_c asymmetry grows more slowly than the Λ^0 asymmetry due to its higher mass.

Leading particle effects were also seen by E791 in hyperon production. Preliminary results on Λ, Ξ and Ω asymmetries as a function of x_F and p_t^2 are shown in Fig.10 in comparison with predictions from Pythia/Jetset. The range of x_F covered allowed the first simultaneous study of the asymmetry in both the negative and positive x_F regions. We can clearly see leading particle effects associated with the target or with the beam particles which qualitatively agree with expectations from recombination models (see table 1) [14]. It is interesting to observe, as expected, the crossover of the Ξ asymmetry with respect to the Λ asymmetry at $x_F \simeq 0$. The positive asymmetry measured in regions $x_F > 0$ for the $\Lambda(udc)$ and for the $\Omega(sss)$ suggest the associated production of a hyperon and a kaon due to the higher

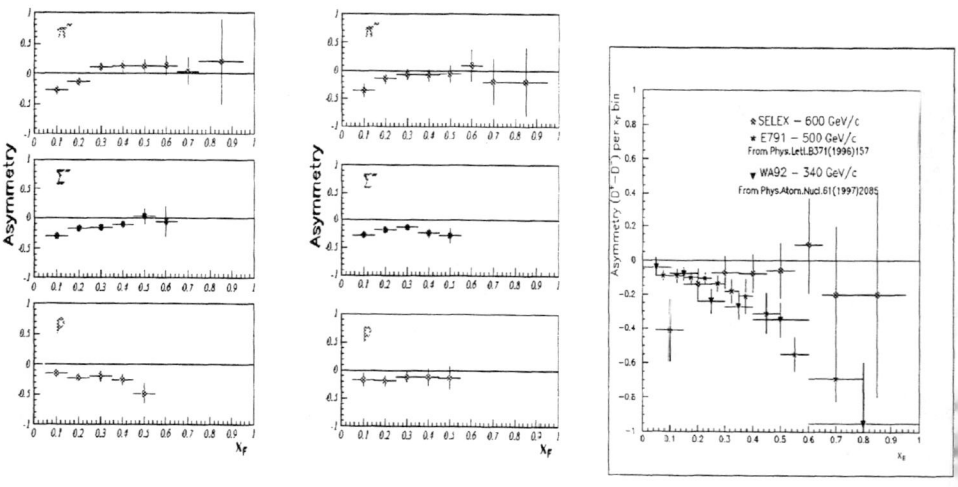

FIGURE 6. Asymmetries of D mesons from SELEX (preliminary results) and E791. The first and second columns are $D^0(\bar{D}^0)$ and $D^\pm$ asymmetries from SELEX. The right figure shows the comparison of $D^\pm$ asymmetries from SELEX and E791.

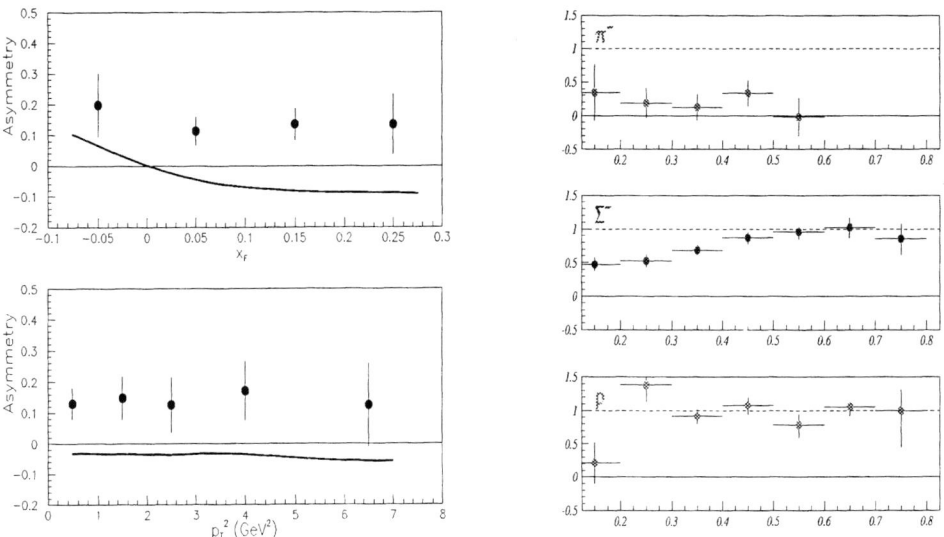

FIGURE 7. Λ_c asymmetries as function of p_t^2 and x_F from E791 (left). Λ_c asymmetries (preliminary results) as function of x_F for different beams (π^-, Σ^-, p) from SELEX (right).

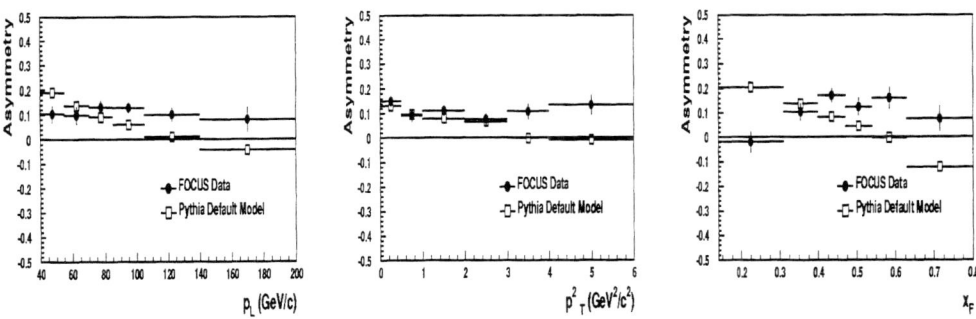

FIGURE 8. Λ_c^+ asymmetry from FOCUS as a function of p_l, p_t^2 and x_F (preliminary results), compared with Pythia/Jetset predictions

energy threshold imposed by baryon number conservation for the production of an anti-hyperon. Pythia/Jetset does not reproduce the data.

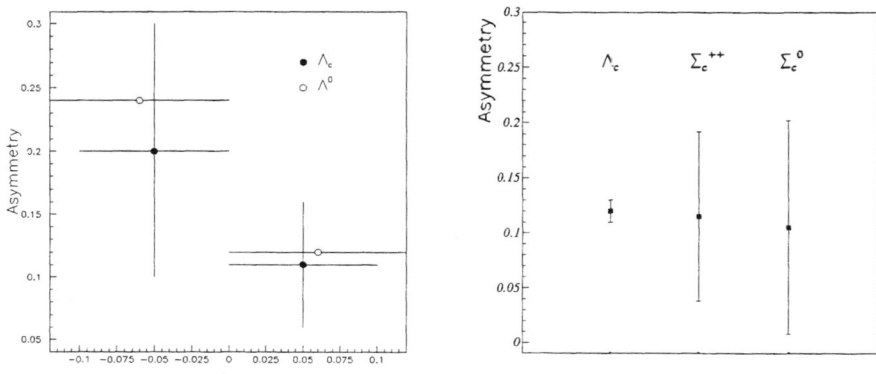

FIGURE 9. Comparison between Λ^0 and Λ_c asymmetries from E791 (left). Λ_c, Σ^{++} and Σ_c^0 total asimmetries (preliminary) from FOCUS (right).

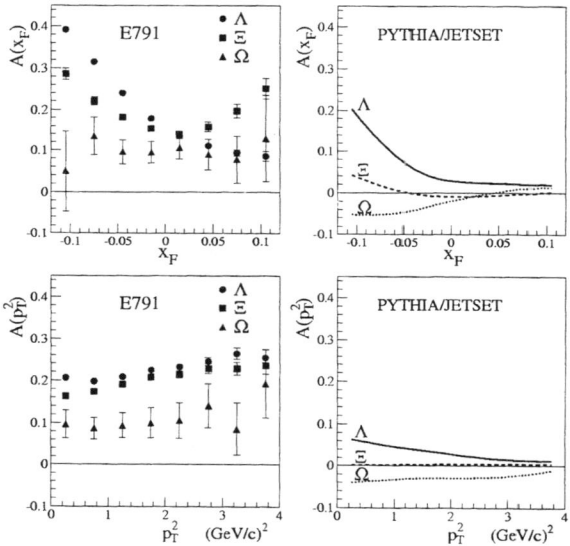

FIGURE 10. Hyperon asymmetries as a function of x_F (top) and p_T^2 (bottom) from E791 (preliminary results). The asymmetry for x_F (p_T^2) is integrated over all the p_T^2 (x_F) range of the data set. The right column show the predictions of Pythia/Jetset both as a function of x_F and p_T^2 (preliminary results).

TABLE 1. Hyperon asymmetries predictions

	$x_f < 0$ target(uud or ddu)	$x_f > 0$ beam π^- ($\bar{u}d$)
$\Lambda^0(uds)$	double leading	leading
$\overline{\Lambda}^0(\bar{u}d\bar{s})$	non leading	leading
$\Xi^-(dss)$	leading	leading
$\overline{\Xi}^+(\bar{d}\bar{s}\bar{s})$	non leading	non leading
$\Omega^-(sss)$	non leading	non leading
$\overline{\Omega}^+(\bar{s}\bar{s}\bar{s})$	non leading	non leading
Recomb. Models[a]:	$A_\Lambda > A_\Xi > A_\Omega$	$A_\Xi > A_\Lambda \sim A_\Omega$

[a] Presented in Second Latin American Symposium [14]

DOUBLE CABIBBO SUPPRESSED DECAYS

The Cabibbo suppressed charm decays can provide useful insights into the weak interaction mechanism for nonleptonic decays. The $D^+ \to K^+\pi^-\pi^+$ signal obtained from 100 % of FOCUS data set consist of ~ 300 events and is at least a factor of five larger than two previous observations by E687 and E791 (E791 observed 59 ± 13 events [15]). The preliminary branching ratio relative to $K^-\pi^+\pi^-$ is (0.72 ± 0.09) %, completely consistent with the world average of (0.68 ± 0.15)% and the E791 values of $(0.77 \pm 0.17 \pm 0.08)$% We note that this is $\sim 3\tan^4(\theta_c)$, which is roughly the ratio of the D^+/D^0 lifetime, indicating that the destructive Pauli interference present in the Cabibbo Favored D^+ decay is absent in the doubly Cabibbo suppressed (DCS) mode.

$D^+ \to K^-K^+K^+$ is an interesting DCS decay, which cannot even occur through a spectator diagram. FOCUS has the first observation of this mode, and reports a preliminary result for the Branching Ratio relative to $K^-\pi^+\pi^+$ of (0.14 ± 0.02) %. Several groups have reported observations of a $D^+ \to \phi K^+$ signal, however FOCUS did not find evidence for such decay [16].

SELEX announced the first observation of a Cabibbo suppressed decay of a charm baryon through the decay $\Xi_c^+ \to pK^-\pi^+$ [17]. Fig. 11 shows the signal of 157 ± 22 events reported by SELEX and simple spectator diagrams with external W emission for Ξ_c^+ decaying into a Cabibbo allowed and into a single Cabibbo suppressed (SCS) mode. The other Cabibbo allowed Ξ^- mode interchanges s and d quarks lines and produces a $d\bar{d}$ pair from the vacuum instead of a $d\bar{u}$ pair. FOCUS has also observed the same SCS decay, reporting a signal of 86 ± 21 events from about 70 % of their data.

E791 has published [18] results on the singly Cabibbo suppressed decay, $D^0 \to K^-K^+\pi^-\pi^+$. A coherent amplitude analysis of the resonant substructure was used to extract decay fractions. Significant phase angles among different modes indicate very strong interference. The measured branching fractions relative to $D^0 \to K^-\pi^+\pi^-\pi^+$ are presented in the table 2.

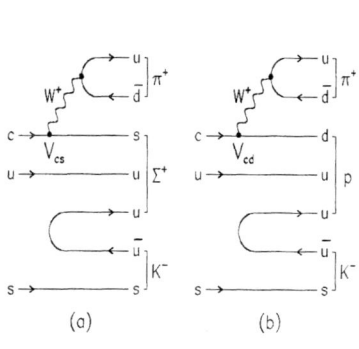 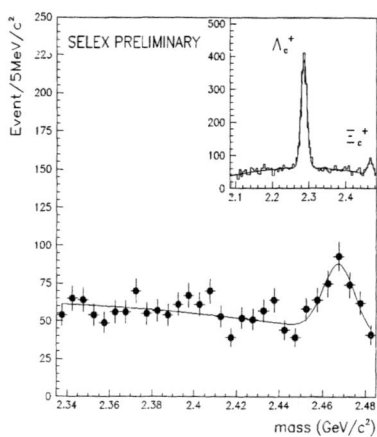

FIGURE 11. Simple spectator diagrams (a) and (b) and preliminary signal of $\Xi_c^+ \to pK^-\pi^+$ from SELEX.

TABLE 2. Branching fractions relative to $D^0 \to K^-\pi^+\pi^-\pi^+$ from E791.

Mode	Branching Fraction
$\phi\rho^0$	$(2.0 \pm 0.9 \pm 0.8)$ %
$\phi\pi^+\pi^-$	$(0.9 \pm 0.4 \pm 0.5)$ %
$\overline{K^{*0}}K^{*0}$	< 2.0 % (90 % CL)
$\overline{K^{*0}}K^+\pi^- + \overline{K^{*0}}K^-\pi^+$	< 2.0 % (90 % CL)

Hadronic charm decays, Dalitz plot Analysis

With the advent of high statistics experiments, charm meson decay have become a new way to study light meson spectroscopy. The amplitude analysis performed on Dalitz plots gives insight into the decay dynamics, providing direct information about intermediate resonances and relative decay fractions, and allowing to study final state interactions coming from the interference of the amplitudes describing competing resonant channels.

E791 has preliminary results on the decay of D^+ and D_s^+ mesons in three pions. A clear signal with 1240 ± 51 D^+ and 858 ± 49 D_s^+ was obtained after applying selection criteria aimed at identifying a clearly separated 3π vertex and after carefully estimating the backgrounds coming from possible reflections and three pion combinations. The branching ratios were normalized to $D^+ \to K^-\pi^+\pi^+$ ($34,790 \pm 232$ events) and to $D_s^+ \to \phi\pi^+$ (1038 ± 44 events) respectively. Efficien-

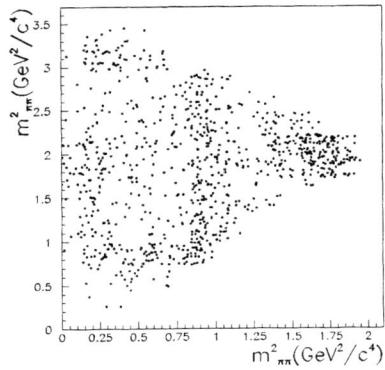 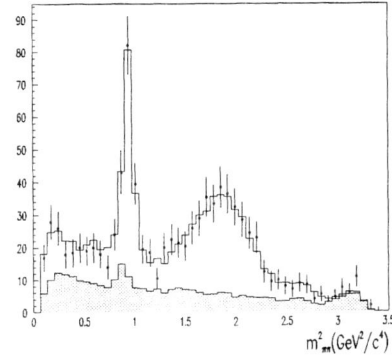

FIGURE 12. The $D_s^+ \to \pi^+\pi^-\pi^+$ Dalitz plot and its projections (preliminary results) from E791.

cies were obtained from a full Monte Carlo simulation. The branching ratio of the $D^+ \to \pi^+\pi^-\pi^+$ relative to $D^+ \to K^-\pi^+\pi^+$ obtained was $0.0329 \pm 0.0015^{+0.0016}_{-0.0026}$. Similarly the branching ratio of the $D_s^+ \to \pi^+\pi^-\pi^-$ relative to $D_s^+ \to \phi\pi^+$ was $0.247 \pm 0.028^{+0.019}_{-0.012}$.

D_s^+ Dalitz plot results from E791

Among the advantages of using charm meson decays to study light I=J=0 states is the fact that, unlike hadron-hadron scattering, in the decays of D mesons the initial state is always $J^P = 0^-$, limiting the number of possible final states. The decay $D_s^+ \to \pi^-\pi^+\pi^+$ is Cabibbo-favored without a strange meson in the final state. It can proceed via spectator amplitudes producing intermediate resonant states with hidden strangeness like the $f_0(980)$ or it can proceed via W-annihilation amplitudes producing intermediate resonant states with no strangeness. The decays like $D_s^+ \to \rho^0 \pi^+$ and the non-resonant $D_s^+ \to \pi^-\pi^+\pi^+$ would proceed via W-annihilation mechanism. It would also be responsible for the decay $D_s^+ \to f_0(1370)\pi^+$, if the $f_0(1370)$ resonance is at least partially a $u\bar{u} + d\bar{d}$ state as predicted by the simple quark model. The scale of the W-annihilation compared to the W-radiation amplitude would be indicated by the relative contribution of these channels to the $\pi^-\pi^+\pi^+$ final state [19].

The Dalitz plot of $D_s^+ \to \pi^-\pi^+\pi^+$ and the $\pi\pi$ mass projections [20] are shown in Fig. 12.

The $f_0(980)\pi^\pm$ mode is the dominant one, accounting for nearly half of the $D_s^+ \to \pi^-\pi^+\pi^+$ decay width. The $f_0(980)\pi^\pm$ is often supposed to have a large $s\bar{s}$ component, indicating a large spectator amplitude in this decay. Significant

TABLE 3. Preliminary results of f_0 mass and width of $D_s^+ \to f_0(980)\pi^+$ and $D_s^+ \to f_0(1370)\pi^+$ from E791.

	$f_0(980)$ Mass (MeV/c^2)	width (MeV/c^2)	$f_0(1370)$ Mass (MeV/c^2)	width (MeV/c^2)
E791	978 ± 4	44 ± 5	1440 ± 19	165 ± 29
PDG	978 ± 10	$40 - 100$	$1200 - 1500$	$200 - 500$

contributions of $f_0(1370)\pi^+$ and $f_2(1270)\pi^+$ components were also found. The contribution of $\rho^0(770)\pi^+$ and $\rho^0(1450)\pi^+$ components correspond to about 10 % of the $\pi^-\pi^+\pi^+$ width. This could indicate either contribution from the annihilation diagram or from inelastic final state interactions. No significant non-resonant component was found.

Preliminary f_0 masses and widths [20] from E791 and PDG are presented in the table 3.

FOCUS also has preliminary results of this decay mode. Their preliminary Dalitz plot shows the $D_s^+ \to \pi^-\pi^+\pi^+$ based on a very clean signal of ~ 1300 events reconstructed from 100 % of their data. The preliminary results indicate a negligible contribution from the ρ suggesting negligible Weak Annihilation contribution [21].

D^+ Dalitz plot results and evidence for a light scalar resonance

The proyections of Dalitz plot for single Cabibbo-suppressed decay $D^+ \to \pi^-\pi^+\pi^+$ from E791 data is shown in Fig.13. A coherent amplitude analysis was used to determine the structure of its density distribution. The fit including a non resonant amplitude and amplitudes for D^+ decaying to a π^+ and any of the five established $\pi^+\pi^-$ resonances $\rho^0(770)$, $f_0(980)$, $f_2(1270)$, $f_0(1370)$, and $\rho^0(1450)$ is shown in Fig. 13 (left). This fit is poor in the low $\pi^+\pi^-$ region and has several unsatisfactory features [22]: the NR channel dominates, different from the D_s^+ decay, and the $\rho^0(1450)\pi^+$ is more significant than the $\rho^0(770)\pi^+$ state.

It was found that allowing an additional scalar state, with mass and width unconstrained improves the fit substantially. The mass of the resonance found by this fit is 486^{+28}_{-26} MeV/c^2 and the width 351^{+51}_{-43} MeV/c^2. Referring to this $\pi^+\pi^-$ resonance as the $\sigma(500)$, it was found that $D^+ \to \sigma(500)\pi^+$ accounts for about half of the total decay rate, non-resonant decay was very small and the $\rho^0(1450)\pi^+$ fraction was much less than $\rho^0(770)\pi^+$. Preliminary results of the fit with this state are shown in Fig. 13 (right side).

Theoretically, light scalar and isoscalar resonances are predicted in models for spontaneous breaking of chiral symmetry, like the σ linear model [23]. These scalar particles have important consequences for the quark model, for understanding low energy $\pi\pi$ interactions and also for understanding the $\Delta I = 1/2$ rule.

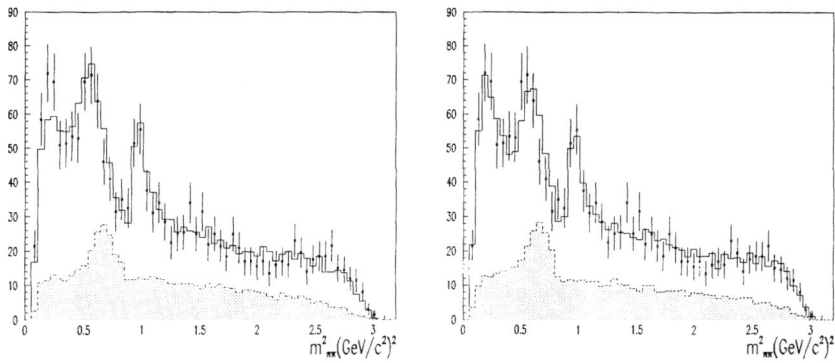

FIGURE 13. $D^+\pi^+\pi^-\pi^+$ Dalitz plot projections. The first plot is without a $\sigma\pi$ state and the last plot include a $\sigma\pi$ state in the fit (preliminary results). The shaded area is the background distribution.

Multidimensional analysis of $\Lambda_c^+ \to pK^-\pi^+$ from E791

E791 has reported recently the first amplitude analysis of the decay of a charm baryon [24,16]. The study of charm baryon decays can give information regarding the relative importance of spectator and exchange amplitudes. Exchange amplitudes are small in charm meson decays because of helicity suppression. However in charm baryon decays this effect should not inhibit exchange amplitudes due to be three body nature of the interaction. The spectator and W-exchange diagrams can contribute to $pK^{*0}(890)$, $\Lambda(1520)\pi^+$ or $pK^-\pi^+$ modes. However for the $\Delta^{++}(1232)K^-$ the W-exchange is the only diagram possible.

The charm baryon can be produced polarized and its decay products carry spin. These extra quantum numbers require five kinematic variables for a complete description of the decay, and instead of the conventional two dimensional Dalitz plot analysis, a five-dimensional amplitude analysis is required.

A sample of 946 ± 38 $\Lambda_c^+ \to pK^-\pi^+$ reconstructed decays was used by E791 to determine relative strengths and phases of resonances in the final state as well as the Λ_c production polarization. The fit projections and the polarization as function of p_t are shown in Fig.14. The resonant fractions for $\Lambda_c^+ \to pK^-\pi^+$ are shown in table 4.

The $\Delta^{++}(1232)K^-$ and the $\Lambda(1520)\pi^+$ decay modes are seen as statistically significant for the first time. The observation of a substantial $\Delta^{++}(1232)K^-$ component provides strong evidence for the W-exchange amplitude in charm baryon decays. It was also observed an increasingly negative polarization for the Λ_c as a function of p_t.

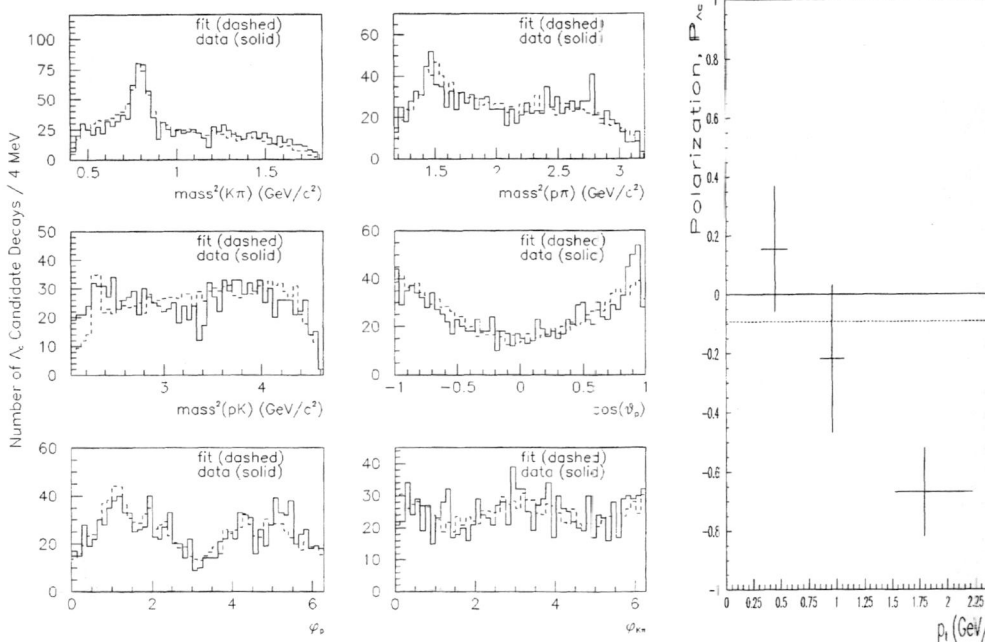

FIGURE 14. Λ_c^+ multidimensional analysis from E791: left plots shows projections of two body masses and angles. Solid line histograms are data in the range $2265 < M_{pK^-\pi^+} < 2315$ MeV/c^2. Dashed lines represent the fit in the same range. The right figure shows polarization $P_{\Lambda_c^+}$ vs. p_t.

TABLE 4. Resonant fractions for Λ_c^+ decay modes.

Decay Mode	% of $pK^-\pi^+$
$pK^{*0}(890)$	$19.5 \pm 2.6 \pm 1.8$
$\Delta^{++}(1232)K^-$	$18.0 \pm 2.9 \pm 2.9$
$\Lambda(1520)\pi^+$	$7.7 \pm 1.8 \pm 1.1$
non-resonant	$54.8 \pm 5.5 \pm 3.5$

RARE AND FORBIDDEN DECAYS

In the charm sector the rare and forbidden dilepton decay modes can be classified mainly into three categories (example of Feynman diagrams are shown in Fig. 15):
1) Flavor Changing Neutral Current decays (FCNC) such as $D^0 \to l^+l^-$ and $D^+ \to h^+l^+l^-$.
2) Lepton Family Number Violating decays (LFNV) such as $D^+ \to h^+l_1^+l_2^-$ and

$D^0 \to l_1^+ l_2^-$ where the leptons are from different generations.
3) Lepton number Violating decays (LNV) such as $D^+ \to h^- l^+ l^+$ where the leptons are of the same generation but have the same sign.
Where h stands for π, K and l for e, μ.

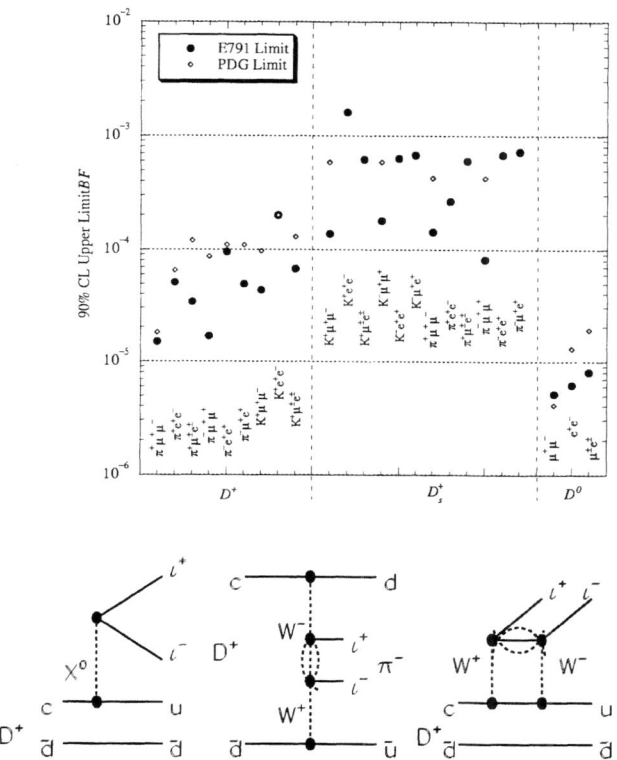

FIGURE 15. Rare decays limits from E791 (top) and some Feynman diagrams for FCNC, LFNV and LNV decay modes (bottom).

The first decay modes (FCNC) are rare, that means a process suppressed via the GIM mechanism which proceeds via an internal quarks loop in the Standard Model [25]. The FCNC decay mode $D^0 \to l^+ l^-$ can proceed via a W box diagram and the theoretical estimates [26] for the branching fraction are of the order of $\sim 10^{-19}$. The predictions for the other FCNC decay modes, $D^+ \to h^+ l^+ l^-$, are considerably larger, of the order of $\sim 10^{-9}$. These decay modes can proceed via penguin diagrams [25] and from long distance effects.

The decays modes LFNV and LNV are strictly forbidden in the Standard Model as they do not conserve lepton number. However, some theoretical extensions of the Standard Model predict lepton number violation [27], and then the observation of a signal in these modes would be evidence for new physics beyond the SM.

E791 has recently published [28] (see figure 15) a set of new limits in rare or forbidden decay modes that improve the PDG98 numbers by a factor of 10. They searched for 24 different rare and forbidden decay modes and have found no evidence for them. They therefore presented upper limits on their branching fractions. Fourteen of their limits represent a significant improvement over previous results and eight are presented for the first time.

For this study E791 used a *blind* analysis technique. The mass region where the signal is expected is *masked* throughout the analysis. Selection criteria are optimised by studying signal events generated by Monte Carlo simulation and background events obtained from data in mass windows above and below the *signal region*. The criteria were chosen to maximize the ratio $N_S/\sqrt{(N_B)}$, where N_S and N_B are the numbers of signal and background events, respectively. Only after this procedure were events within the signal window unmasked. This blind technique is used so that the presence or absence of signal does not bias the choice of the selection criteria.

FOCUS is also looking for rare and forbidden decays using the same technique. Some examples of rare decays are presented in table 5 where we compare the expected sensitivity from FOCUS to E791 results and PDG values.

TABLE 5. Rare decays limits from FOCUS (preliminary), E791 and PDG98.

Decay mode	FOCUS expected sensitivity (45 % of the data)	E791[a] 90 % C.L. limit	PDG 98 90 % C.L limit
$D^+ \to K^+ \mu^+ \mu^-$	8.1×10^{-6}	4.4×10^{-5}	9.7×10^{-5}
$D^+ \to K^- \mu^+ \mu^+$	12.1×10^{-6}		1.2×10^{-4}
$D^+ \to \pi^+ \mu^+ \mu^-$	7.8×10^{-6}	1.5×10^{-5}	1.8×10^{-5}
$D^+ \to \pi^- \mu^+ \mu^+$	7.1×10^{-6}	1.7×10^{-5}	8.7×10^{-5}
$D^+ \to \mu^- \mu^+ \mu^+$	4.4×10^{-6}		

[a] Published [28]

SEMILEPTONIC DECAYS, FORM-FACTORS

The weak decays of hadrons containing heavy quarks are influenced by strong interaction effects. Semileptonic charm decays such as $D^+ \to \overline{K}^{*0} e^+ \nu_e$, $D_s^+ \to \phi l^+ \nu_l$ are an especially clean way to study these effects because the leptonic and hadronic currents completely factorize in the decay amplitude, A, as we can see in the Eq. 4, where G_F is the Fermi coupling constant for the weak interaction and V_{cs} is the CKM matrix element. L^μ (Eq. 5) and H_μ (Eq. 6) represent the leptonic and hadronic currents [29].

$$A(D^+ \to \overline{K}^{*0} e^+ \nu_e) = \frac{G_F}{\sqrt{2}} V_{cs} L^\mu H_\mu \tag{4}$$

$$L^\mu = \bar{u}_e \gamma^\mu (1 - \gamma_5) v_\nu \tag{5}$$

$$H_\mu = (m_D + m_{K^*}) A_1(q^2) \epsilon_{mu} - \frac{A_2(q^2)}{m_D + m_{K^*}} (\epsilon \cdot p_D)(p_D + p_K)_\mu - \frac{A_3(q^2)}{m_D + m_{K^*}} (\epsilon \cdot p_D)(p_D - p_K)_\mu - i \frac{2V(q^2)}{m_D + m_{K^*}} \varepsilon_{\mu\nu\rho\sigma} \epsilon^\nu p_D^\rho p_K^\sigma \tag{6}$$

With a vector meson in the final state, there are four form factors, $V(q^2)$, $A_1(q^2)$, $A_2(q^2)$ and $A_3(q^2)$, which are functions of the Lorentz-invariant momentum transfer squared q^2, the square of the invariant mass of the virtual W [29]. The differential decay rate for $D^+ \to \overline{K}^{*0} \mu^+ \nu_\mu$ with $\overline{K}^{*0} \to K^- \pi^+$ is a quadratic homogeneous function of the four form factors. Unfortunately, the limited size of current data samples does not allow precise measurement of the q^2-dependence of the form factors; we thus assume the dependence to be given by the nearest-pole dominance model: $F(q^2) = F(0)/(1 - q^2/m_{pole}^2)$ where $m_{pole} = m_V = 2.1$ GeV/c^2 for the vector form factor V (which correspond to $J^P = 1^+$ state, D_{s1}^*), and $m_{pole} = m_A = 2.5$ GeV/c^2 for the three axial-vector form factors A [30] (corresponding to 1^- state, D_s^*).

The third form factor $A_3(q^2)$, which is unobservable in the limit of vanishing lepton mass, probes the spin-0 component of the off-shell W. Additional spin-flip amplitudes, suppressed by an overall factor of m_ℓ^2/q^2 when compared with spin no-flip amplitudes, contribute to the differential decay rate. Because $A_1(q^2)$ appears among the coefficients of every term in the differential decay rate, we can factor out $A_1(0)$ and measure the ratios:
$r_V = V(0)/A_1(0)$, $r_2 = A_2(0)/A_1(0)$ and $r_3 = A_3(0)/A_1(0)$. The values of these ratios can be extracted without any assumption about the total decay rate or the weak mixing matrix element V_{cs}.

We report E791 measurements of the form factor ratios r_v and r_2 for the muon channel [31] and combined with electron channel [32] (see Fig. 16). This is the first set of measurements in both muon and electron channels from a single experiment. We also report the first measurement of $r_3 = A_3(0)/A_1(0)$, which is unobservable in the limit of vanishing charged lepton mass.

The measurements of the form factor ratios for $D^+ \to \overline{K}^{*0} \mu^+ \nu_\mu$ presented here and for the similar decay channel $D^+ \to \overline{K}^{*0} e^+ \nu_e$ [33] follow the same analysis procedure except for the charged lepton identification. Both results in the electron and muon channels are consistent within errors, supporting the assumption that

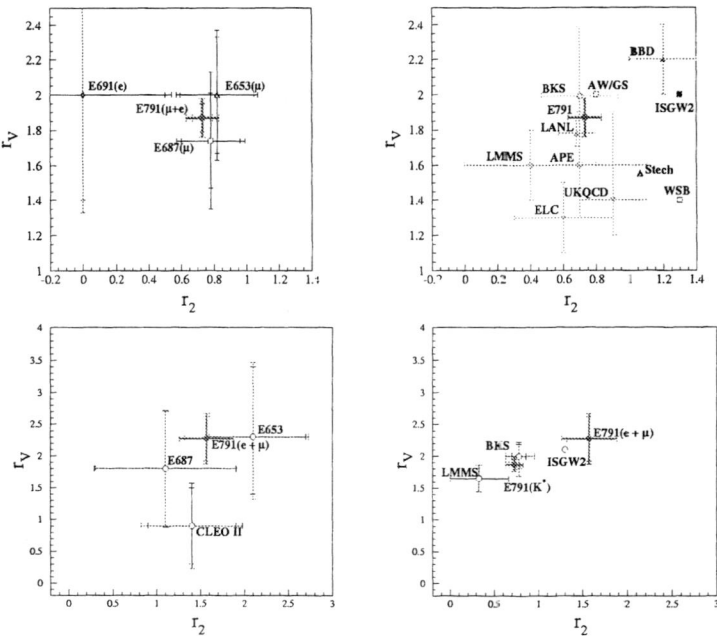

FIGURE 16. Top-left figure present the ratios r_2 and r_V for $D^+ \to \overline{K}^{*0}l\nu_l$, and top-right compare them to theoretical models. Bottom-left show ratios r_2 and r_V for $D_s^+ \to \overline{\phi}l\nu_l$, and bottom-right is a comparison with theoretical models.

strong interaction effects incorporated in the values of form factor ratios do not depend on the particular W^+ leptonic decay.

The combined results of electronic and muonic decay modes produce $r_V = 1.87 \pm 0.08 \pm 0.07$ and $r_2 = 0.73 \pm 0.06 \pm 0.08$. The combination of all systematic errors is ultimately close to that which one would obtain assuming all the errors are uncorrelated. The third form factor ratio $r_3 = 0.04 \pm 0.33 \pm 0.29$ was not measured in the electronic mode.

Table 6 compares the values of the form factor ratios r_V and r_2 measured by E791 in the electron, muon and combined modes with previous experimental results. The size of the data sample and the decay channel are listed for each case. All experimental results are consistent within errors. The comparison between the E791 combined values of the form factor ratios r_V and r_2 and other experimental results is also shown in Fig. 16 together with theoretical predictions.

The FOCUS experiments anticipate measuring r_2 and r_V to better precision than previous experiments [34]

TABLE 6. Comparison of E791 results with previous results.

Exp.	Events	$r_V = V(0)/A_1(0)$	$r_2 = A_2(0)/A_1(0)$
E791	6000 $(e+\mu)$	$1.87 \pm 0.08 \pm 0.07$	$0.73 \pm 0.06 \pm 0.08$
E791	3000 (μ)	$1.84 \pm 0.11 \pm 0.09$	$0.75 \pm 0.08 \pm 0.09$
E791	3000 (e)	$1.90 \pm 0.11 \pm 0.09$	$0.71 \pm 0.08 \pm 0.09$
E687 [35]	900 (μ)	$1.74 \pm 0.27 \pm 0.28$	$0.78 \pm 0.18 \pm 0.10$
E653 [36]	300 (μ)	$2.00^{+0.34}_{-0.32} \pm 0.16$	$0.82^{+0.22}_{-0.23} \pm 0.11$
E691 [37]	200 (e)	$2.0 \pm 0.6 \pm 0.3$	$0.0 \pm 0.5 \pm 0.2$

CHARM LIFETIMES

The study of the charm particle lifetimes is motivated by two main goals: to extract partial decay rates and to study decay dynamics. The total decay width can be expressed as a sum of the three possible classes of decays, so the lifetime of a particle can be written as

$$\tau = \frac{\hbar}{\Gamma_{tot}} = \frac{\hbar}{\Gamma_{leptonic} + \Gamma_{SL} + \Gamma_{nonleptonic}} \quad (7)$$

The leptonic partial width is normally very small ($\Gamma_{leptonic} \sim 10^{-3} - 10^{-4}$) due to helicity suppression. The observed $\Gamma_{SL}(D^+)$ and $\Gamma_{SL}(D^0)$ are equal to within 10%, as expected from isospin invariance. So, the large difference observed in the D^+ and D^0 lifetimes, $\tau(D^+)/\tau(D^0) = 2.55 \pm 0.04$, is due to a large difference in the hadronic decay rates ($\Gamma_{nonleptonic}$) for the D^+ and the D^0. If there were no other diagram but the spectator and no QCD effects we would have the same value for τ. Thus, the different lifetimes are an indication that we need to take into account contributions from diagrams where the W interact with two valence quarks, such as W-annihilation (WA) and W-exchange (WX), and any interference between them, as shows the Fig. 17.

A systematic approach now exists for the treatment of inclusive decays, based on QCD and consists of an Operator Product Expansion (OPE) in the Heavy Quark mass. In this approach the interaction is factorized into three parts: weak interaction between quarks, perturbative QCD corrections and non-perturbative QCD effects. The decay rate is given by

$$\Gamma_{HQ} = \frac{G_F^2 m_Q^2}{192\pi^3} \Sigma f_i |V_{Qq_i}|^2 [A_1 + \frac{A_2}{\Delta^2} + \frac{A_3}{\Delta^3} + \ldots] \quad (8)$$

where Δ is taken as the heavy quark mass and f_i is a phase space factor. $A_1 = 1$ contains the spectator diagram contribution; A_2 is the spin interaction of the heavy quark with light quark degrees of freedom inside the hadron and produces differences between the baryon and meson lifetimes; A_3 includes the non-spectator W-annihilation, W-exchange and Pauli Interference (PI) of the decay and the spectator

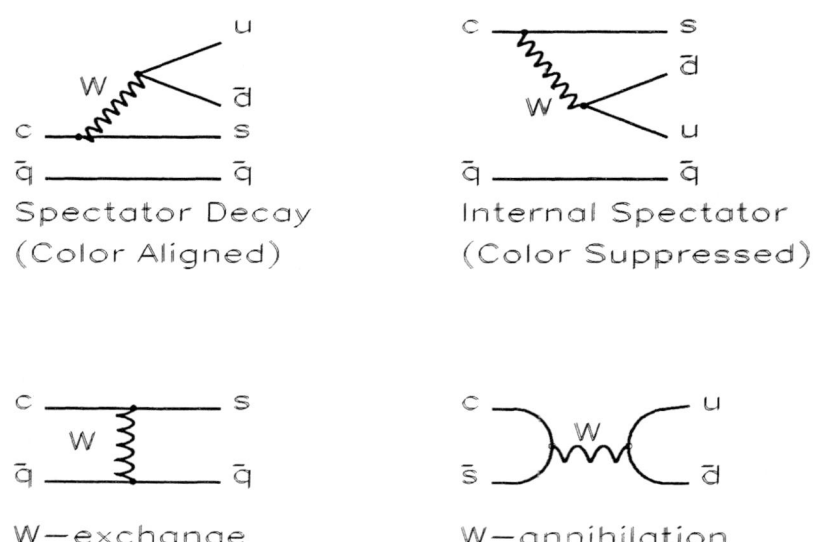

FIGURE 17. Hadronic decays diagrams for charm meson decays.

quarks contributions.

New results on charm lifetime measurements are shown in table 7. The most relevant information is the D_s^+ lifetime from E791 [38,39] and FOCUS [40] which is now conclusively above the D^0 lifetime, with a ratio

$$R_\tau = \tau(D_s^+)/\tau(D^0) = 1.22 \pm 0.02 \tag{9}$$

It is important to note that R_τ is now ten standard deviations away from unity, indicating that although not dominant, the WA/WX contributions are significant. In fact the OPE model predicts $R_\tau = 1.00 - 1.07$ without WA/WX contributions, allowing a variation of $\pm 20\%$ if the WA/WX operators are included, in agreement with the new result.

In the baryon sector SELEX and FOCUS have preliminary measurements of the Λ_c^+ lifetime, as shown in table.7 and Fig. 18. Using 100% of their data in $\Lambda_c^+ \to pK^-\pi^+$, SELEX find $\tau = 177 \pm 10(stat) fs$ [41], in a $2-3\sigma$ disagreement with PDG98 and FOCUS.

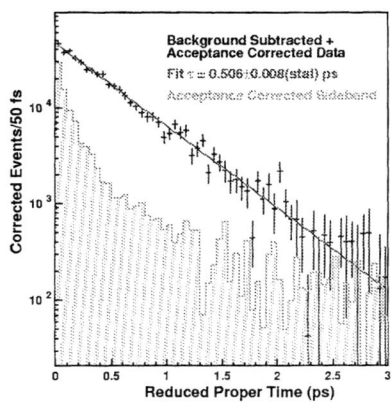
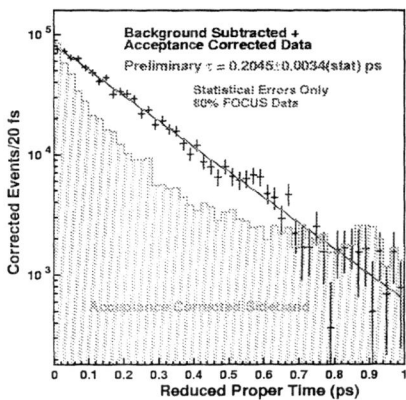

FIGURE 18. D_s^+ (left) and Λ_c^+ (right) lifetimes from FOCUS (preliminary results)

TABLE 7. Summary of new charm lifetime measurements.

Experiment	$\tau(D^+)$ fs	$\tau(D^0)$ fs	$\tau(D_s^+)$ fs	$\tau(\Lambda_c^+)$ fs
PDG'98	1057 ± 15	415 ± 4	467 ± 17	206 ± 12
E791[a]	1065 ± 48	$413 \pm 3 \pm 4$	$518 \pm 14 \pm 7$	
CLEO	$1033.6 \pm 22.1^{+9.9}_{-12.7}$	$408 \pm 4.1^{+3.5}_{-3.4}$	$486.3 \pm 15.0^{+4.9}_{-5.1}$	
FOCUS[b]			506 ± 8	204.5 ± 3.4
SELEX[b]				177 ± 10
World Average	1052 ± 12	412.8 ± 2.7	499.9 ± 6.1	201.9 ± 3.1

[a] $\tau(D^+)$ using only the $\phi\pi^+$ mode.
[b] Preliminary results with no systematic uncertainty quoted

Lifetime differences and $D^0\overline{D^0}$ mixing

E791 has published searches for a lifetime difference between the $CP - even$ and $CP - odd$ eigenstates of the D^0 [39]. To do so they compared the lifetimes of the decays $D^0 \rightarrow K^-K^+$ ($CP = +1$) and $D^0 \rightarrow K^-\pi^+$, (CP mixed) [42], shown in Fig. 19.

Defining

$$\frac{\Gamma(K^-K^+) - \Gamma(K^-\pi^+)}{\Gamma(K^-\pi^+)} = y_{CP} \qquad (10)$$

The time integrated ratio of mixed to non mixed decay rates in charm meson is given by

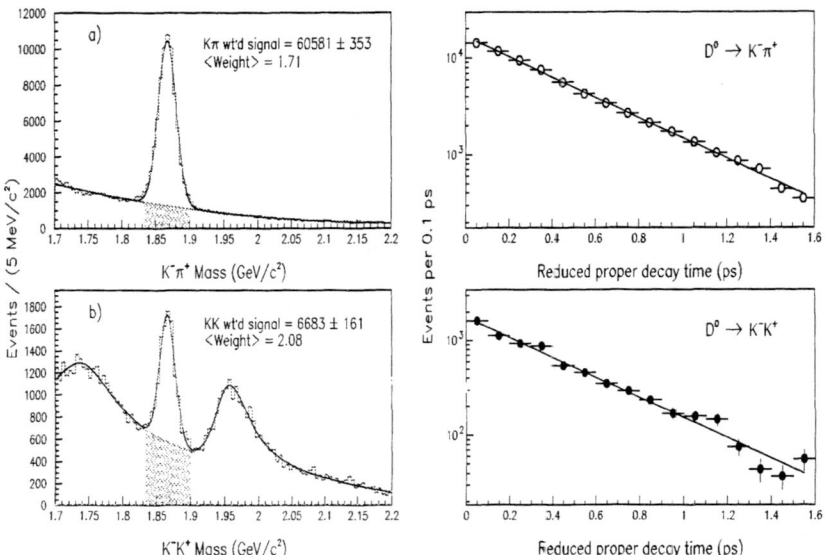

FIGURE 19. Invariant mass for $D^0 \to K^-\pi^+$ (a) and $D^0 \to K^-K^+$ (b) decays (left). Exponential fits for number of decays as a function of reduced proper decay time (right), (E791 results).

$$R_{mix} = \frac{\Gamma(D^0 \to \overline{D^0} \to \bar{f})}{D^0 \to f} = \frac{x^2 + y^2}{2} \quad (11)$$

where

$$x = \frac{\Delta m}{\overline{\Gamma}}, \quad y = \frac{\Delta \Gamma}{2\overline{\Gamma}}$$

with

$$\Delta m = m_1 - m_2, \quad \Delta \Gamma = \Gamma_1 - \Gamma_2, \quad \overline{\Gamma} = (\Gamma_1 + \Gamma_2)/2$$

where Γ_1 is for CP even states and Γ_2 for CP-odd states.
Γ_1 applies to $D^0 \to K^-K^+$ and Γ applies to $D^0 \to K^-\pi^+$ if CP is conserved.
Mixing can appear if there is either a difference in the masses of the CP eigenstates Δm or if there is a difference in the decay rates $\Delta \Gamma$.

E791 observed no difference in lifetimes and quoted $\tau(K\pi) = 0.413\pm0.003\pm0.004$ ps $\tau(KK) = 0.410\pm0.011\pm0.006$ ps, $y_{CP} = 0.008\pm0.029\pm0.010$ or $-0.04 < y_{CP} < 0.06$ (90% CL) or $\Delta\Gamma = 2(\Gamma_{KK} - \Gamma_{K\pi}) = (0.04 \pm 0.14 \pm 0.05)\ ps^{-1}$

ACKNOWLEDGEMENTS

One of us, J. C. Anjos would like to thank the conference organizers for their invitation to attend the Symposium. E. Cuautle would like to thank the Centro Brasileiro de Pesquisas Físicas (CBPF) for its kind hospitality during his postdoctoral stay. The authors would like to thank CLAF/CNPq (Brazil) and CONACyT (México) for financial support of this work.

REFERENCES

1. Nason, P., Dason, S., Ellis, R. K., *Nucl. Phys.* **303 B**, 607 (1998).
2. Mangano, M., Nason, P., and Ridolfi, G., *Nucl. Phys.* **B 373**, 295 (1992).
   Frixione, S., Mangano, M., Nason, P., and Ridolfi, G., *Nucl. Phys.* **B 431**, 453 (1994).
3. Sjöstrang, T., Pythia 7.5 and Jetset 7.4 Manual, CERN-TH 71112/93, 1995.
4. Aitala, E. M., et al., *Fermilab E791 Collaboration, Phys. Lett.* **462 B**,225 (1999).
5. Aguilar-Benitez, M., et al., *CERN LEBC-EHS Collaboration, Phys. Lett.* **164B**, 404 (1985);
   Aguilar-Benitez, M., et al., *CERN LEBC-EHS Collaboration, Z. Phys.* **C40**, 321 (1988); Aoki, S., et al., *Phys. Lett.* **209 B**,113 (1988); Kodama, K., et al., *Phys. Lett.* **263 B**, 579 (1991); Berlag, S. et al., *ACCMOR Collaboration, Phys. Lett* **257 B**, 519 (1991).
6. Appel, J.A., *Annu. Rev. Nucl. Part.Sci.* **42**, 367 (1992).
7. Aitala, E. M., et al., *Fermilab E791 Collaboration, Eur. Phys. J.* **C4**, 1 (1999)
8. Sjöstrand, T., *International Journal of Mod. Phy.* **A 3**, 751 (1988).
9. Brodsky, S.J., et al.,*Phys. Lett.* **93 B**, 451(1980).
10. Iori M., et al., *Fermilab SELEX Collaboration, Charm hadroproduction results from SELEX*, hep-ex/99100039.
    García, F.G., *Fermilab SELEX Collaboration, Charm hadroproduction from SELEX, Wine and Cheese Seminar, Fermilab Aug. 99*
11. Anjos, J.C. *Particle-antiparticle asymmetries in the production of baryons in 500 GeV/c π^-- Nucleon interactions*, (to appear, Proc.) Hyperon Physics Symposium (HYPERON 99), Fermilab, Sept. 99. Preprint hep-exp/9912039.
12. Aitala, E. M., et al., *Fermilab E791 Collaboration, Phys. Lett.* **411 B**, 230 (1997)
13. Aitala, E. M., et al., *Fermilab E791 Collaboration, Phys. Lett.* **371 B**,157 (1996).
14. Anjos, J.C., et al., *Asymmetry studies in $\Lambda^0/\bar{\Lambda}^0$, Ξ^-/Ξ^+ and Ω^-/Ω^+ production, First Tropical Workshop and Second Latin American Symposium*, San Juan, Puerto Rico, 1998. Ed. J.F. Nieves. AIP, Conference Proceedings 444, p. 540.
15. Aitala, E. M., et al., *Fermilab E791 Collaboration, Phys. Lett.* **404 B**, 187 (1997).
16. Brian, M., *Experimental Results on hadronic c decays*, (to appear, Proc.) *Heavy Flavours 8*, Southampton, UK. 1999.
17. Jun, S. Y., et al., *Fermilab SELEX Collaboration, Observation of the Cabibbo-suppressed decay $\Xi_c^+ \to pK^-\pi^+$*, hep-ex/9907062.
18. Aitala, E. M., et al.,*E791 Collaboration, Phys. Lett.* **423B**, 185 (1998).

19. Reis, A., *Light Scalars Through Charm Decays*, (to appear, Proc.) *Physics and detectors for da Φne (DAFNE99)*, Frascati, Nov. 1999.
20. Aitala, E. M., et al., *Fermilab E791 Collaboration*, preprint Fermilab-Pub-99/323-E
21. Moroni, L. *First results from FOCUS*, (to appear, Proc.) *International Europhysics Conference on High Energy Physic 99*, Finland, July 1999
22. Aitala, E. M. et al., *Fermilab E791 Collaboration*, preprint Fermilab-Pub-99/322-E
23. Nambu, Y. and Jona-Lasinio G., *Phys. Rev.* **122**, 345 (1961).
24. Aitala, E. M., et al., *Fermilab E791 Collaboration, Phys. Lett.* **471 B**, 449 (2000).
25. Schwartz, A. J., *Mod. Phy. Let.* **A8**, 967 (1993).
    Schwartz, A. J., *Recent Charm Results from Fermilab Experiment E791*, preprint hep-ex/9908054.
26. Hewett, J. L., *Heavy Flavor Physics* preprint hep-ph/9505246.
27. Pakvasa, S., *Flavor Changing Neutral Currents in Charm sector* (hep-ph/9705397).
28. Aitala, E. M., et al., *Fermilab E791 Collaboration, Phys. Lett.* **462 B**, 401 (1999).
29. Körner, J. G. and Schuler, G. A.,*Phys. Lett.* **B 226**, 185 (1989).
30. Particle Data Group, Review of Particle Physics, *Phys. Rev.* **D 50**, 1568 (1994).
31. Aitala, E. M., et al., *Fermilab Collaboration, Phys. Lett.* **450 B**, 294 (1999).
32. Aitala, E. M., et al., *Fermilab Collaboration, Phys. Lett.* **440 B** , 435 (1998).
33. Aitala, E. M., et al., *Fermilab E791 Collaboration, Phys. Rev. Lett.* **80**, 1293 (1998).
34. Brian O' Reilly, *Charmed Baryon and Semileptonic Physic at FOCUS*, (to appear, Proc.) *III International Conference on Hyperons Charm and Beauty hadrons*, Genova Italy, July 1998.
35. Frabetti, P.L., et al., *Fermilab E687 Collaboration, Phys. Lett.* **307 B**, 262 (1993).
36. Kodama, K. et al., *E653 Collaboration, Phys. Lett.* **274 B**, 246 (1992)
37. Anjos, J.C., et al., *Fermilab E691 Collaboration, Phys. Rev. Lett.* **65**, 2630 (1990).
38. Aitala, E. M., et al., *Fermilab E791 Collaboration, Phys. Lett.* **445 B**, 449 (1999).
39. Aitala, E. M., et al., *Fermilab E791 Collaboration, Phys. Rev. Lett.* **83**, 32 (1999).
40. Cheung, H.W.K., *FOCUS Collaboration, Preliminary D_s^+ lifetime from FOCUS*, APS Centennial Meeting Atlanta March 1999. Cheung, H.W.K., *Review of Charm Lifetimes*, (to appear, Proc.) *Heavy Flavours 8*, Southampton, UK. 1999.
41. Kushnirenko, A. Y., *SELEX Collaboration, Charm Physics Results from SELEX*, 4th Workshop on Heavy Quarks at Fixed Target, Batavia Illinois, USA, October 1998. Eds. H. W. K. Cheung and J. N. Butler. AIP, Conference Proceedings 459 p. 168.
42. Bianco S., *Charm Overview*, (to appear, Proc.) *Physics in Collision (PIC99)*, Michigan, June 1999 (hep-ex/9911034).
    Appel, J. A., *Charm Results on CP violation and Mixing*, (to appear, Proc.) *Physics and detectors for da Φne (DAFNE99)*, Frascati, Nov. 1999.
    Sheldon, P.D., *Charm Mixing and Rare Decays*, (to appear, Proc.) *Heavy Flavours 8*, Southampton, UK. 1999.

Deep Inelastic Scattering, Diffraction and all that[1]

C.A. García Canal*, R. Sassot[†]

Laboratorio de Física Teórica
Departamento de Física, Universidad Nacional de La Plata
C.C. 67 - 1900 La Plata, Argentina
[†] *Departamento de Física, Universidad de Buenos Aires*
Ciudad Universitaria, Pab.1 (1428) Buenos Aires, Argentina

Abstract. These lectures include an introduction to the partonic description of the proton, the photon and the 'colour singlet', as seen in inclusive and semi-inclusive DIS, in e^+e^- collisions, and in diffractive processes, respectively. Their formal treatment using structure, fragmentation, and fracture functions is outlined giving an insight into the perturbative QCD framework for these functions. Examples and comparisons with experimental data from LEP, HERA, and Tevatron are also covered.

INTRODUCTION

The discovery of asymptotic freedom, one of the most significant properties of strong interactions embodied by Quantum Chromodynamics (QCD) opened, more than 25 years ago, a new chapter in our understanding of the structure of matter which has has been actively followed by theoreticians and experimentalists ever since. The short distance structure of hadrons, together with the production of jets in hadronic collisions are paradigmatic among the strong interaction phenomena successfully accounted by QCD and even though the standing of the theory is today well established, further theoretical refinements and the corresponding experimental validation renew constantly the original enthusiasm of the high energy physics community.

These lectures intend to provide an overview of the more recent topics of high energy collisions related in a way or another to perturbative QCD. First we briefly remind the essentials of QCD, including the main features of partons (the quarks and gluons), which are the true protagonists in the story. Then, we refer to what is known about the partonic structure of three of the main benchmarks of QCD, the proton, the photon, and the singlet colour or 'pomeron'. Finally we will try to

---

[1] Partially supported by CONICET and Agencia Nacional de Promoción Científica, Argentina.

draw the connections between their corresponding structures which in some way relate the physics made in the three main HEP laboratories.

As usual, many interesting and highly active topics have been excluded from the lectures in favor of a more detailed analysis of the covered points. These include, for example, those related to the spin structure of the proton, which have driven an ongoing series of polarized experiments and a great deal of theoretical discussions [1]; heavy flavours, which involve very subtle theoretical approaches, and perturbative QCD beyond NLO, which is relevant for the most recent high precision experiments [2].

QCD

The strong interactions among quarks and gluons are described by Quantum Chromodynamics (QCD), the non-abelian gauge theory based on the gauge group $SU(3)_C$. Each quark flavour corresponds to a colour triplet in the fundamental representation of $SU(3)$ and the gauge fields needed to maintain the gauge symmetry, the gluons, are in the adjoint representation of dimension 8. Gauge invariance ensures that gluons are massless. The QCD Lagrangian may be written as

$$\mathcal{L}_{QCD} = -\frac{1}{4} F^a_{\mu\nu} F_a^{\mu\nu} + \bar{\psi}_i \left(i \gamma^\mu D_\mu - m\right) \psi_i \tag{1}$$

where

$$F^a_{\mu\nu} = \partial_\mu G^a_\nu - \partial_\nu G^a_\mu + g f^{abc} G_{b\mu} G_{c\nu} \tag{2}$$

stands for the gluon field tensor, ψ_i are the quark fields and the covariant derivative is defined by

$$D_\mu = \partial_\mu - i g T_a G^a_\mu$$

The strong coupling is represented by g and indices are summed over $a = 1, ..., 8$ and over $i = 1, 2, 3$. Finally, $T_a = \lambda_a/2$ and f_{abc} are the $SU(3)$ generators and structure constants, respectively, which are related by $[T_a, T_b] = i f_{abc} T^c$.

Like in Quantum Electrodynamics (QED), the procedure employed to deal consistently with the divergences that occur in the computation of strong interactions beyond the tree level, shows that the actual strength of the QCD coupling depends on the energy scale of the process. But in opposition to QED, this renormalized strong coupling is small at high energy (momentum), going to zero logarithmically, i.e. QCD has the property of *asymptotic freedom*. Consequently, in this regime perturbation theory is valid and tests against experimental data can be performed in terms of hadrons. Figure 1 summarize the basic QCD perturbative processes appearing in different circumstances.

Experiments with $e^- e^+$ colliders provide clean results for QCD tests. Recently, a huge amount of experimental data came from the HERA electron-proton collider [3] and also from the Tevatron at Fermilab [4]. In both cases, there is a hadronic

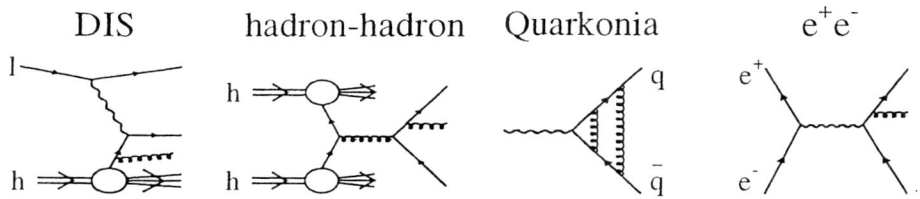

FIGURE 1. Basic processes in perturbative QCD.

remnant that make the analysis a little more involved. All this experimental evidence support the existence of quarks being colour triplets of spin 1/2 and of gluons being vector octets. Moreover, the presence of the QCD coupling has manifested itself in different measurements, as well as the above mentioned property of asymptotic freedom. This information comes mainly from the study of the so called two- and three-jets events [5].

When a given process needs a higher order in perturbation theory to be known, it is necessary to compute not only the renormalized strong coupling constant but also the appropriate corrections to the relevant cross-sections. As is usual in Quantum Field Theory, a regularization-renormalization procedure is in order, just to absorb divergences into the definition of physical quantities. This prescription requires the introduction of a new scale μ, fixing the renormalization point, and all renormalized quantities begin to depend on it. Nevertheless, different prescriptions must end with the same predictions for observables.

In order to illustrate how the general procedure works, ending with the Renormalization Group equations that guarantee that physical observables do not depend on the scale μ, let us show what happens with Green functions. Just to remember the procedure, let us begin with a single particle irreducible Green function Γ. In general, to control divergences, one has to introduce an ultra-violet cut-off Λ, or the equivalent dimensional regularization parameter, in the loop momentum integral defining the Γ. In a renormalizable theory, as QCD is, a renormalized Green function is defined as

$$\Gamma_R(p_i, g, \mu) = Z_\Gamma(g_0, \Lambda/\mu)\, \Gamma_U(p_i, g_0, \Lambda)$$

where p_i stands for the external particle momenta, g_0 and g are the bare and the renormalized couplings, respectively. This Γ_R is then finite in the limit $\Lambda \to \infty$ but it depends on the scale at which the value of the renormalized quantities are fixed, the prescription parameter μ. The function Z_Γ is a product of renormalization factors. Due to the fact that the unrenormalized Γ_U is obviously independent of μ, one has to demand

$$\frac{d\Gamma_U}{d\mu} = 0$$

and consequently, the Renormalization Group Equation (RGE)

$$\left(\mu\frac{\partial}{\partial\mu} + \beta\frac{\partial}{\partial g} + \gamma\right)\Gamma_R(p_i, g, \mu) = 0 \qquad (3)$$

has to be verified. Here γ is the *anomalous dimension*, depending on the particular Green function under consideration, and the *beta-function* is universal

$$\gamma = \frac{\mu}{Z_\Gamma}\frac{\partial Z_\Gamma}{\partial \mu} \qquad \beta(g) = \mu\frac{\partial g}{\partial \mu} \qquad (4)$$

If there is only one large momentum scale Q, or Q^2 as it is standard to quote, as it is the case here, one can express all p_i in terms of a fixed fraction x_i of Q. Then, defining the so called evolution variable

$$t = \frac{1}{2}\ln\left(\frac{Q^2}{\mu^2}\right) \qquad (5)$$

it is possible to introduce the momentum dependent, or *running* coupling through the integral

$$t = \int_{g(0)}^{g(t)} \frac{dg'}{\beta(g')}$$

and the general solution of the RGE reads

$$\Gamma(t, g(0), x_i) = \Gamma(0, g(t), x_i)\, exp\left[\int_{g(0)}^{g(t)} dg'\, \frac{\gamma(g')}{\beta(g')}\right]$$

This solution explicitly shows that the Q-scale dependence of Γ arises entirely through the running coupling $g(t)$. Introducing now the usual notation

$$\alpha_s = \frac{g^2}{4\pi} \qquad (6)$$

one can expand the beta-function in a power series in α_s

$$\beta(\alpha_s) = \mu\frac{\partial \alpha_s}{\partial \mu} = -\frac{\beta_0}{2\pi}\alpha_s^2 - \frac{\beta_1}{4\pi^2}\alpha_s^3 - \ldots \qquad (7)$$

where it results [6] that

$$\beta_0 = 11 - \frac{2}{3}N_f \qquad \beta_1 = 51 - \frac{19}{3}N_f \qquad (8)$$

Here N_f indicates the number of flavours that can be excited (with mass less than μ) at the scale μ.

It is clear that the solution of the differential equation for α_s introduces a constant, called Λ_{QCD}, which has to be fixed by using experimental data. The resulting α_s can be written as

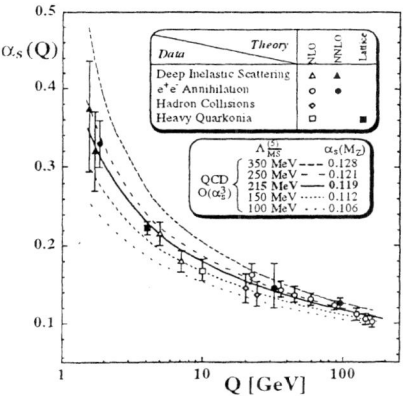

FIGURE 2. QCD running coupling.

$$\alpha_s(\mu, \Lambda_{QCD}) = \frac{4\pi}{\beta_0 \ln(\mu^2/\Lambda_{QCD}^2)} \left\{ 1 - \frac{2\beta_1}{\beta_0^2} \frac{\ln[\ln(\mu^2/\Lambda_{QCD}^2)]}{\ln(\mu^2/\Lambda_{QCD}^2)} \right\} \quad (9)$$

This expression for the running coupling shows clearly the property of asymptotic freedom of QCD, i.e., the coupling vanishes when the scale becomes asymptotic, namely $\mu \to \infty$. Consequently, in this momentum regime, perturbation theory is valid.

A very clear quantitative test of perturbative QCD is provided by the measurement of α_s in different processes at different scales Q^2. In Figure 2 there is a summary of the various determinations of α_s [7].

The present world average [8] for the coupling at the Z^0 mass is

$$\alpha_s(M_Z) = 0.119 \pm 0.002$$

which implies

$$\Lambda_{QCD}^{\overline{MS}} = 220 + 78 - 63 \, MeV$$

corresponding to five flavours excited and in the conventional $\overline{MS}$ prescription commonly used.

PROTON

Having reviewed the essentials of QCD, we now proceed with the analysis of one of the most rich benchmarks of perturbative QCD which is the short distance structure of the proton.

We start this section giving a set of definitions of the commonly used relativistic invariants related to Deep Inelastic Scattering, DIS, one of the fundamental experimental tools for hadron analysis, together with the corresponding formulae for the neutral and charged current cross-sections, where the structure functions are introduced. These functions will be latter expressed in terms of the quark-parton model including QCD corrections. For a more detailed treatment see [5]

Lorentz Invariants

The scattering of a lepton from a proton (or in general a nucleon, a hadron or a nucleus), at high enough Q^2, can be viewed as the (elastic) scattering of the lepton from a quark or antiquark inside the proton mediated by the exchange of the corresponding virtual vector boson γ^*, W^* or Z^*. Consequently, when the process is totally inclusive (one integrates over all final hadronic activity), can be fully described by two Lorentz invariants. With the momentum assignment of Figure 3 one can construct the following invariants:

$$s = (p+k)^2, \quad t = (p-p')^2, \quad W^2 = (p+q)^2, \quad Q^2 = -(k'-k)^2 = -q^2 \qquad (10)$$

defined as the centre of mass energy squared, the square of the four-momentum transfer between the proton and the final hadronic state, the invariant mass squared of the final hadronic state, and four momentum transfer squared between the lepton and the proton, respectively. It is also convenient to introduce dimensionless invariants (scaling variables)

$$y = \frac{p \cdot q}{p \cdot k}, \quad x = \frac{Q^2}{2 p \cdot q} \qquad (11)$$

i.e. the inelasticity of the scattered lepton, and fraction of the proton momentum carried by the struck quark, or Bjorken variable, respectively.

Notice finally that, ignoring masses,

$$Q^2 = s\,x\,y \qquad W^2 = Q^2 \frac{1-x}{x} \qquad (12)$$

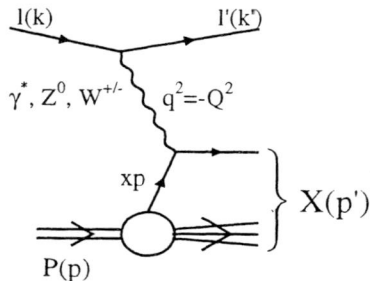

FIGURE 3. DIS kinematics.

so that $Q^2_{max} = s$, the center of mass energy squared, and small values of x imply increasing W. As we have already said, in principle one is allowed to use any pair of these invariants to describe totally inclusive DIS. Usually x and Q^2 are the preferred ones.

Another interesting point to remark concerns the resolving power of a DIS experiment. Clearly, the size d one can resolve inside the nucleon becomes smaller for large photon, or gauge boson in general, virtuality, namely

$$d \sim \frac{\hbar c}{Q} \simeq 0.2 \frac{GeV\,fm}{Q}$$

The 'magnifying power' is then $d \simeq 10^{-14}\,cm$ for $Q^2 = 4\,GeV^2$, and $d \simeq 10^{-16}\,cm$ and for $Q^2 = 40,000\,GeV^2$.

Experimental Reconstruction

Different DIS experiments cover different regions of the kinematical plane (x, Q^2) as shown in Figure 4 [9]. It is particularly interesting the very extended range covered by HERA: $Q^2 \simeq 0.2 - 10^4\,GeV^2$ and $x \simeq 10^{-5} - 10^{-1}$.

The kinematics of the DIS events, namely the two invariants required to specify DIS processes, can be determined (particularly at HERA) from measurements on the electron (the lepton) alone, on the final hadronic state corresponding to the struck quark alone or on a mixture of both. In general, the preferred method depends on the particular kinematical region of interest, strongly correlated with the detector performance.

The electron method: The input are the energy E'_e of the final electron and the electron scattering angle θ_e measured with respect to the proton beam direction

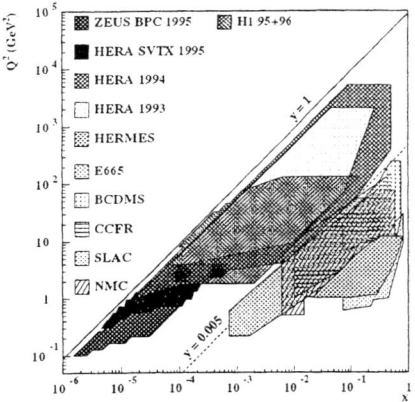

FIGURE 4. Coverage of the (x, Q^2) plane by the various DIS experiments

($\theta_e = 180°$ means zero electron scattering angle)

$$y_e = 1 - \frac{E'_e}{E_e} \sin^2 \frac{\theta_e}{2} \qquad Q_e^2 = 4 E_e E'_e \cos^2 \frac{\theta_e}{2}$$

The hadron method: The input is the hadronic energy E_h and the three-momentum components of the hadronic system. They are computed as the sum over all final state hadrons h

$$y_h = \frac{\sum_h (E_h - p_{z,h})}{2 E_e} \qquad Q_h^2 = \frac{p_{x,h}^2 + p_{y,h}^2}{1 - y_h}$$

This method is also known with the name Jacquet-Blondel, because these authors were able to show that the contribution from hadrons lost in the beam pipe is negligible.

The sigma method: This method is based on both electron and hadron measurements. The denominator of y_h is replaced by a sum over all final state particles including the scattered electron (certainly equal to $2 E_e$ due to energy-momentum conservation). The name of the method comes from the introduction of the variable $\Sigma = \sum_h (E_h - p_{z,h})$ that appears in

$$y_\Sigma = \frac{\Sigma}{\Sigma + E'_e (1 - \cos \theta_e)} \qquad Q_\Sigma^2 = \frac{E'^2_e \sin^2 \theta_e)}{1 - y_\Sigma}$$

Notice that the denominator in y_Σ is twice the true incident electron energy in the eventual case when the electron radiates photons that are not detected.

There are other more sophisticated methods, well adapted to particular situations, like the so-called double angle method or the PT method used by ZEUS [3].

Cross-sections

The fundamental measurement in DIS experiments concerns the totally inclusive cross-section for

$$\ell(k) + N(p) \rightarrow \ell'(k') + X(p')$$

as a function of the kinematical variables defined above. For charged lepton-nucleon scattering mediated by the neutral current, the spin averaged cross-section is given in terms of the structure functions F_2, F_L and F_3

$$\frac{d^2 \sigma_\pm}{dx\, dQ^2} = \frac{4 \pi \alpha^2}{x Q^4} \qquad (13)$$

$$\left[\left(1 - y + \frac{y^2}{2}\right) F_2(x, Q^2) - \frac{y^2}{2} F_L(x, Q^2) \mp \left(y - \frac{y^2}{2}\right) F_3(x, Q^2) \right]$$

here α is the QED coupling constant. For Q^2 values below the Z^0 scale, the parity violating effects related to F_3 are negligible and all the process is due to

γ^* exchange. Remember also that the longitudinal structure function F_L, a QCD correction important for large y, is defined in terms of the standard F_1 and F_2, related to the transverse and longitudinal $\gamma^* N$ cross sections respectively, as

$$F_L(x,Q^2) = F_2(x,Q^2)\left(1 + \frac{4\,M_N^2\,x^2}{Q^2}\right) - 2\,x\,F_1(x,Q^2) \qquad (14)$$

and that in the naive quark-parton model, really valid at extremely high Q^2, where quarks are considered free, massless, having spin 1/2 and without any p_T developed, F_L is zero because the Callan-Gross relation

$$2\,x\,F_1(x) = F_2(x)$$

is satisfied. Under this assumptions, the so called Bjorken scaling is fully satisfied, namely, structure functions are only function of the x variable.

The ratio R is defined by

$$R(x,Q^2) = \frac{F_L(x,Q^2)}{F_2(x,Q^2) - F_L(x,Q^2)} = \frac{\sigma_L}{\sigma_T} \qquad (15)$$

which is obviously zero in the Callan-Gross limit and can be interpreted as the ratio of cross-sections for the absorption of transversely and longitudinally polarized virtual photons on nucleon. The differential cross-section (13) can be rewritten in terms of $R(x,Q^2)$ as

$$\frac{d^2\sigma}{dx\,dQ^2} = \frac{4\,\pi\alpha^2}{x\,Q^4}\left[1 - y - \frac{M_N^2\,x^2\,y^2}{Q^2} + \frac{y^2}{2}\frac{1 + \frac{4\,M_N^2\,x^2}{Q^2}}{1 + R(x,Q^2)}\right]F_2(x,Q^2) \qquad (16)$$

where the parity violating contribution was discarded.

It is possible to specify the structure function F_2 in terms of the partons (or better quarks) within the nucleon as follows

$$F_2(x,Q^2) = x\sum_f e_{q_f}^2\left[q_f(x,Q^2) + \bar{q}_f(x,Q^2)\right] \qquad (17)$$

where the sum runs over all momentum distributions of quarks of the different flavours f excited ($x\,q_f(x,Q^2)$) and antiquarks ($x\,\bar{q}_f(x,Q^2)$) contained in the nucleon and e_{q_f} stands for the the different parton-photon couplings, namely their electric charges. Notice that the previous expression (17) is valid only in leading order of perturbation theory. In fact, there one has included the explicit Q^2 dependence which is the effect of implementing first order perturbative QCD. However, there is a particular renormalization and factorization scheme that can be used in higher order QCD, the so called DIS-scheme, where one can retain that expression at any order of perturbation theory. In other schemes, like the usual $\overline{MS}$ one, the expressions for the structure functions are more involved, including in general $g(x, Q^2)$, the gluon density in the nucleon.

Let us now refer for a moment to the inclusive reaction that goes via charged current

$$\nu_\mu + N \to \mu^- + X \tag{18}$$

where as usual we write N for an isoscalar nucleon. In the QCD improved parton model, the corresponding cross section reads

$$\frac{d^2\sigma}{dx\,dy} = \frac{2\,G_F^2\,M_N\,E_\nu}{\pi} \left(\frac{M_W^2}{Q^2 + M_W^2}\right)^2 \left[x\,q(x, Q^2) + x\,\bar{q}(x, Q^2)(1-y^2)\right] \tag{19}$$

M_W stands for the charged intermediate boson mass. G_F is the Fermi constant. The quark and antiquark distribution functions can be written in terms of the corresponding valence and sea flavour distributions in a proton as

$$2\,q(x, Q^2) = u_v(x, Q^2) + d_v(x, Q^2) + u_s(x, Q^2) + d_s(x, Q^2) \tag{20}$$
$$+ 2\,s_s(x, Q^2) + 2\,b_s(x, Q^2)$$
$$2\,\bar{q}(x, Q^2) = u_s(x, Q^2) + d_s(x, Q^2) + 2\,c_s(x, Q^2) + 2\,t_s(x, Q^2) \tag{21}$$

One can introduce immediately the corresponding structure function F_2 for neutrino scattering. It is of interest to compare this F_2^ν with the one for charged lepton DIS written above. Taking into account the sea corrections, the relation between both F_2 functions results

$$\frac{F_2^\ell}{F_2^\nu} = \frac{5}{18}\left(1 - \frac{3}{5}\frac{s + \bar{s} - c - \bar{c}}{q + \bar{q}}\right) \tag{22}$$

This relation compares well with experimental data [10].

QCD evolution equations

All the functions introduced above include an explicit Q^2 dependence. This fact reflects the presence of interactions among partons. Clearly, the parton model, that implies exact Bjorken scaling, has to be modified to include these interactions. Moreover, quarks and gluons are finally confined within hadrons by the non-Abelian gauge interaction in QCD. Even if QCD is with us for more than 25 years, it has not yet been possible to calculate in detail the structure of hadrons starting from quarks, gluons and their gauge interactions, making compatible the short distance almost free partons with the long distance confinement. This fact, related to the running character of the strong interaction strength that we have discussed above, is on the origin of our continuous use of structure (and fragmentation and fracture) functions in describing DIS. Fortunately enough, some *factorization theorems*, which are valid in QCD, allows one to split the problem into a perturbative part and a non-perturbative part. In other words, the cross-section can be expressed as the folding of the initial parton distribution function $q_{f/N}$, a non-perturbative input given the density of partons f in the nucleon, with a lepton-parton cross-section that can be computed by means of perturbation theory inside QCD. This fact can be schematically written as

$$\sigma_{\ell N} = \sum_f q_{f/N} \otimes \sigma_{\ell q}$$

and consequently, it can be sketched as in Figure 5

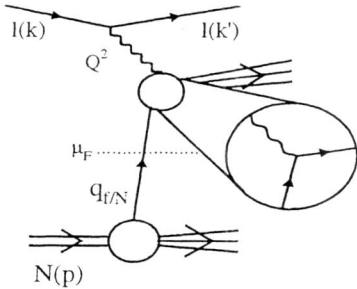

FIGURE 5. Factorization in QCD

It is clear that the non-perturbative part, namely, the part related to structure functions, has to be determined by fitting experimental data. Nevertheless, they are universal in the sense that once obtained from a given particular process, they can be used in connection with any other one.

Regarding QCD as an improvement of the quark-parton model, we can say that the nucleon is not simply composed of three point-like quarks, the so called *valence* quarks but as soon as Q^2 increases, the vector boson (γ^*, Z^* or W^*) increasingly resolves the composition of the nucleon. One of the targets that the vector boson could find is one of the quarks, called *sea* quarks, which originate from a gluon, namely through

$$g \to q\bar{q}$$

being this gluon itself radiated from one of the valence quarks. In other words, increasing Q^2 means that the resolution of our "view" of the nucleon increases, so that the effective number of partons sharing the nucleon momentum also increases. Consequently, the probability of finding partons with small x has increased while the corresponding one to large x has decreased. This process clearly entails a violation of Bjorken scaling through a Q^2 dependence, that was born from QCD interactions. Consequently, to compare experimental data with QCD predictions, one has to compute perturbative QCD corrections to the fundamental process

$$V^\star q \to q \qquad (23)$$

Nowadays, these corrections, even if the calculation is very involved, are well known up to order α_s^2 [11]. We give here only an introductory discussion of the lowest order corrections.

The relevant diagrams corresponding to the first terms of the perturbative expansion in α_s for the process (23) are presented in Figure 6 (We have omitted the vertex and self energy gluonic corrections, which are associated to the ultraviolet divergences. These are removed through coupling constant renormalization, and lead to the running coupling $\alpha_s(Q^2)$ given in Equation (9))

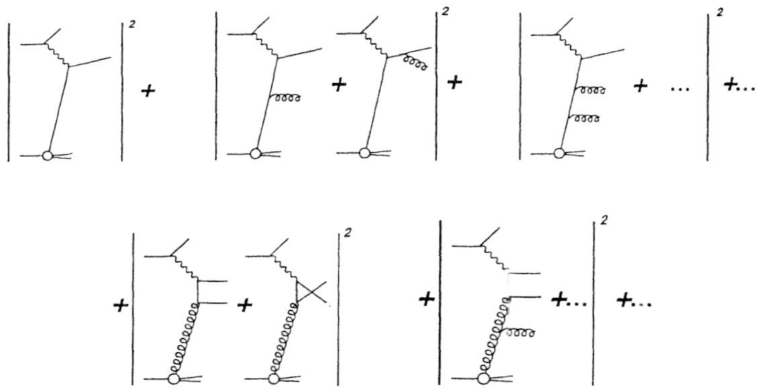

FIGURE 6. Lowest QCD corrections to $V^\star q \to q$

Because we are considering the inclusive case, the calculation implies phase space integrations which lead to divergent integrals when partons become collinear. They have to be regularized by means of dimensional regularization [12], a cut-off in the transverse momentum of the partons, or giving them masses. Any of these procedures introduce a regulator μ^2 which has the dimension of the energy scale Q^2.

For example, to order α_s, quarks contribute to the structure function F_2 through the following convolution integral

$$F_2^{quarks}(x, Q^2) = x \sum_{f,\bar{f}} e_f^2 \int_0^1 \frac{dy}{y} q_f(y) \left[\delta\left(1 - \frac{x}{y}\right) + \frac{\alpha_s}{2\pi} \left\{ P_{qq}\left(\frac{x}{y}\right) \ln \frac{Q^2}{\mu^2} + R\left(\frac{x}{y}\right) \right\} \right] \qquad (24)$$

where $q_f(y)$ is the 'bare' quark density. The delta function term comes from the lowest order diagram in Figure 6, and represents the corresponding contribution to F_2 as in Equation (25). The first term between the curly brackets includes the factor $\ln(Q^2/\mu^2)$ which diverges as the regulator μ^2 goes to zero and is associated to the collinear gluon emission diagrams, while the second term collects the finite contributions from those diagrams. $P_{qq}(x/y)$ and $R(x/y)$ are known calculable functions.

The collinear singularities of course threaten the validity of the parton model, however they can be consistently removed from the partonic subprocess absorbing them into the 'bare' parton densities $q_f(y)$ defining renormalized parton densities

$$q_f(x, M^2) = q_f(x) + \frac{\alpha_s}{2\pi} \int_0^1 \frac{dy}{y} q_f(y) \left\{ P_{qq}\left(\frac{x}{y}\right) \ln \frac{M^2}{\mu^2} + R'\left(\frac{x}{y}\right) \right\} \tag{25}$$

where M^2 is factorization scale chosen to separate the short distance ('partonic') effects from the long distance ('hadronic') ones. For the parton model to make sense, the renormalized parton densities must be process independent, i.e. must be the same for DIS, Drell-Yan, and any other process. Fortunately, this proves to be the case to all orders of perturbation theory, and one ends up with an expression for the (physical, and thus finite) structure function $F_2(x, Q^2)$ in terms of a finite renormalized parton density $q_f(x, M^2)$ and an also finite partonic cross section

$$F_2^{quarks}(x, Q^2) =$$
$$x \sum_{f,\bar{f}} e_f^2 \int_0^1 \frac{dy}{y} q_f(y, M^2) \left[\delta\left(1 - \frac{x}{y}\right) + \alpha_s C_2^q\left(\frac{x}{y}, \frac{Q^2}{M^2}\right) \right] \tag{26}$$

At this point, it is customary to chose the factorization scale M^2 equal to the energy scale Q^2, factorizing the scale dependence of the cross sections into the parton densities.

The crucial observation here is that although perturbation theory can not make an absolute prediction for $q_f(x, Q^2)$, from Equation (25) it follows

$$\frac{dq_i(x, M^2)}{d\log Q^2} = \frac{\alpha_s}{2\pi} \int_x^1 \frac{dy}{y} \left[q_i(y, Q^2) P_{qq}\left(\frac{x}{y}\right) \right] \tag{27}$$

which means that QCD actually gives the Q^2 dependence of the parton distributions.

By means of a similar procedure with the gluon densities, one can deal with the divergences in the diagrams initiated by gluons. However, due to the possibility of gluons emitting quark pairs which then interact with the photon probe, the evolution the evolution of the quark and gluon densities is no longer independent, but coupled. Taking into account all the $\mathcal{O}(\alpha_s)$ contributions, we end up with the so called DGLAP equations (or Altarelli-Parisi equations for short)

$$\frac{dq_i(x, Q^2)}{d \ln Q^2} = \frac{\alpha_s}{2\pi} \int_x^1 \frac{dy}{y} \left[q_i(y, Q^2) P_{qq}\left(\frac{x}{y}\right) + g(y, Q^2) P_{qg}\left(\frac{x}{y}\right) \right]$$
$$\frac{dg(x, Q^2)}{d \ln Q^2} = \frac{\alpha_s}{2\pi} \int_x^1 \frac{dy}{y} \left[\sum_i q_i(y, Q^2) P_{gq}\left(\frac{x}{y}\right) + g(y, Q^2) P_{gg}\left(\frac{x}{y}\right) \right] \tag{28}$$

Collecting all the $\mathcal{O}(\alpha_s)$ contributions to $F_2(x,Q^2)$, we finally arrive to the most usual expression for the structure function:

$$F_2(x,Q^2) = x \sum_f e_f^2 C_f(x) \otimes \left[q_f(x,Q^2) + \bar{q}_f(x,Q^2) \right] \quad (29)$$
$$+ C_g(x) \otimes g(x,Q^2)$$

where $\otimes$ indicates a convolution integral and the index g refers to gluons.

Deep Inelastic Scattering (DIS) experiments and hadron-hadron interactions provide information on the parton structure of the nucleon and constraints the dynamics of the quark-gluon interaction given by QCD. In Figure 4 the kinematic regions in x and Q^2 for cross section measurements in DIS ep scattering, ν scattering and jets in $p\bar{p}$ collisions were presented. Data from the last years [2] have a clear influence on parton distributions and have motivated the update of parton distribution function analysis. Thes most recent sets are CTEQ5 [13], GRV98 [14] and MRST

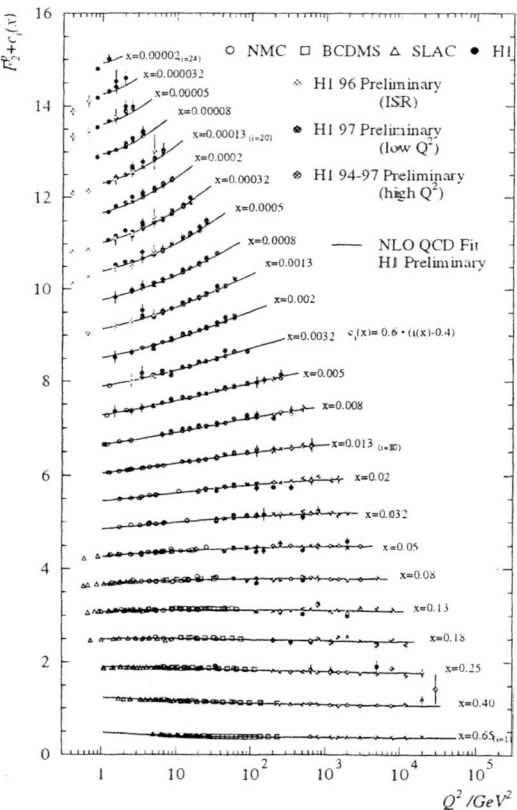

FIGURE 7. F_2 as a function of Q^2 and the corresponding QCD fit

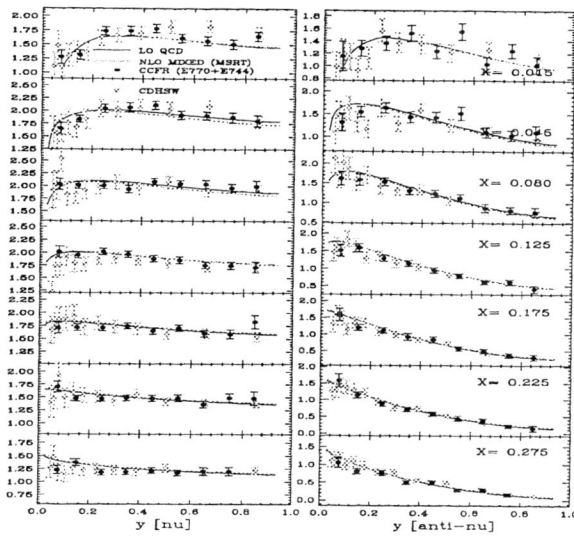

FIGURE 8. ν and $\bar{\nu}$ DIS as a function of y from CCFR experiment together with QCD calculations

[15]. The data playing a fundamental paper in these updates are the more precise ZEUS and H1 determinations of F_2^p including F_2^{charm}; the NMC and CCFR final muon-nucleon and neutrino data; the E866 pp and pd lepton pair production asymmetry; the W charge rapidity asymmetry; the D0 and CDF analysis of inclusive single jet production and the E706 direct photon production. It should be remarked that both approaches differ in the selection of data for the global analysis that are sensitive to gluons at large values of x. This means that a better understanding of the gluon behaviour at large x is in order.

The F_2 structure function measurements at HERA cover a wide range of five orders of magnitude in both x and Q^2 values as shown in Figure 4. One important point established by these data refers to the rise of the structure function with decreasing x. Moreover, QCD fits including NLO (next to leading order corrections) performed by the experimental collaborations are in very good agreement with the data even at low values of $Q^2 \simeq 1\,GeV$. In Figure 7, F_2 is presented as a function of Q^2 for fixed values of x from H1, ZEUS and the previous experiments, together with the QCD fit [2].

Figure 8 presents de neutrino and antineutrino cross section as a function of y and different values of x [2]. Data comes from the CCFR Fermilab experiment and is presented for $\langle E_\nu \rangle = 150\,GeV$. A good agreement with previous data is observed and the NLO QCD analysis is satisfactory.

We finish this paragraph with a brief comment about some QCD technical details.

The use of the renormalization group result for α_s in the AP approach, specifically the replacement of α_s by $\alpha_s(Q^2)$ in Equation (28), has an important consequence. For each order of perturbation in α_s where the factorization procedure is performed, a whole series of perturbative contributions is efectively re-summed, in addition to the contributions comming form the corresponding diagrams shown in Figure 6. This kind of contributions are depicted in Figure 9, the so called *ladder* diagrams.

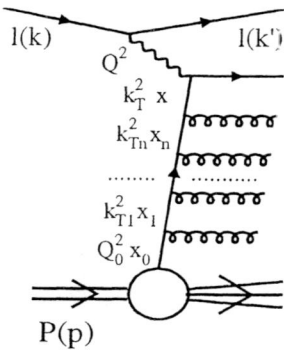

FIGURE 9. A typical ladder diagram

It can be shown that the improved AP approach takes into account contributions coming from the sum of all ladder diagrams in which the transverse momenta along the sides of the ladder are strongly ordered, namely: $Q_0^2 \ll k_{T1}^2 \ll \cdots \ll k_{Ti}^2 \ll \cdots \ll Q^2$. This condition, of course implies also a strong ordering for the momenta of the emitted partons.

Notice that the calculation of such a n-th order ladder diagram implies integrations over the internal momenta. This nested integration can be carried out because of the k_T ordering imposed, giving rise to results proportional to powers of $\alpha_s(Q^2)$ and $\ln(Q^2/Q_0^2)$.

Now, it is clear that large logarithms in Q^2 compensate the small values of $\alpha_s(Q^2)$, that as we have seen, decreases logarithmically. Consequently, all graph with rungs up to $n \to \infty$ would have to be summed up. The so called *leading log* approximation (LO), which is equivalent to follow the renormalization group improved AP approach up to order α_s, collects all the contributions proportional to

$$\alpha_s^n(Q^2) \left[\ln\left(\frac{Q^2}{Q_0^2}\right)\right]^n$$

while the *next to leading log* approximation (NLO), equivalent to the second order extension of the AP approach, also includes contributions proportional to

$$\alpha_s^n(Q^2) \left[\ln\left(\frac{Q^2}{Q_0^2}\right)\right]^{n-1}$$

The AP approximation is expected to be valid for Q^2 values sufficiently large but also for x not too small, just to ensure that small values of x do not give rise to other large logarithms. In other words,

$$\alpha_s(Q^2) \ln\left(\frac{1}{x}\right) \ll \alpha_s(Q^2) \ln\left(\frac{Q^2}{Q_0^2}\right) < 1$$

In is worth remarking that the AP equations can be solved analytically when a strong ordering in x is also required. In this way one ends with the *double leading log approximation*, (DLL), where the large logarithmic terms are of the form [16]

$$\alpha_s^n(Q^2) \left[\ln\left(\frac{Q^2}{Q_0^2}\right) \ln\left(\frac{1}{x}\right)\right]^n$$

This approximation is expected to be valid for large Q^2 and small x.

At small values of x, the parton content of the proton is gluon dominated. In this case the AP DLL can be obtained with the result

$$x\, g(x, Q^2) \approx x\, g_0(x, Q_0^2)\, exp\sqrt{\frac{144}{25} \ln\left[\frac{\ln(Q^2/\Lambda^2)}{\ln(Q_0^2/\Lambda^2)}\right] \ln(1/x)} \quad (30)$$

that shows that the gluon density increases faster than a power of $\ln(1/x)$.

In some cases of interest, x is small enough but Q^2 is not sufficiently large to be inside the DLL regime. Under these circumstances, the AP approximation is not more valid. The BFKL (Balitsky, Fadin, Kuraev, Lipatov) equation has been proposed to tackle the limit behaviour of large $1/x$ and Q^2 finite and fixed. In this scheme, the x_i variables in the ladder are stongly ordered, namely: $x_0 \ll x_1 \ll \cdots \ll x_i \ll \cdots \ll x$; while there is no order on k_T imposed. This approach ends with the so called leading log approximation in $\ln(1/x)$. The region of validity being

$$\alpha_s(Q^2) \ln\left(\frac{Q^2}{Q_0^2}\right) \ll \alpha_s(Q^2) \ln\left(\frac{1}{x}\right) < 1$$

Due to technical reasons, the BFKL equation is written in terms of the function $f(x, k_T^2)$, related to the usual gluon density $g(x, Q^2)$ that dominates at very small x by

$$x\, g(x, Q^2) = \int_0^{Q^2} dk_T^2 \frac{f(x, k_T^2)}{k_T^2} \quad (31)$$

The standard form of the equation is

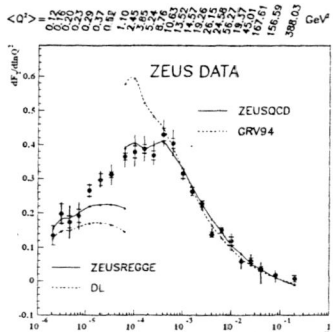

FIGURE 10. λ_{eff} and $dF_2/d\ln Q^2$.

$$\frac{\partial f(x, k_T^2)}{\partial \ln(1/x)} = \frac{3\,\alpha_s}{\pi} k_T^2 \int_0^\infty \frac{dk_T'^2}{k_T'^2} \left[\frac{f(x, k_T'^2) - f(x, k_T^2)}{|k_T'^2 - k_T^2|} + \frac{f(x, k_T^2)}{\sqrt{4\,k_T'^4 + k_T^4}} \right] \quad (32)$$

This approximate BFKL equation can be solved analytically for fixed α_s. The solution behaves like

$$f(x, k_T^2) \propto \left(\frac{x}{x_0}\right)^{-\lambda}$$

with

$$\lambda = \frac{N_c\,\alpha_s}{\pi} 4 \ln 2 \approx 0.5 \quad (33)$$

for $N_c = 3$ and $\alpha_s = 0.19$.

Consequently, the gluon density rises like a power of $1/x$ for decreasing x, namely

$$x\,g(x, Q^2) \propto x^{-\lambda}$$

clearly faster than the AP DLL prediction (30). It should be noticed that a running α_s and higher order corrections decrease the value of λ.

Assuming that gluons dominate at low x one can expect an approximate behaviour for the structure function given by

$$F_2(x, Q^2)\,|_{Q^2} = c\,x^{-\lambda_{eff}}\,|_{Q^2}$$

Figure 10 (a) shows the values obtained for λ_{eff} as a function of Q^2 from fits to ZEUS and E665 data. Figure 10 (b) shows $dF_2/d\ln Q^2$ as a function of x [17]. In both figures QCD and Regge inspired fits have been included. The schematic map

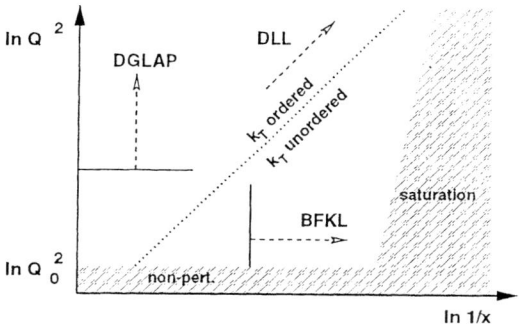

FIGURE 11. The schematic map of the plane $\ln(Q^2), \ln(1/x)$

of the plane $\ln(Q^2), \ln(1/x)$ shown in Figure 11 intents to divide the regions where each of the two approaches, DGLAP and BFKL, are in order [9].

In the figure, a *saturation* region is indicated, where one waits the gluon density to be so high as to prevent its continuous growing.

Semi-inclusive

In the analysis of hadron structure, more information can be obtained if one goes one step beyond totally inclusive DIS, namely to semi-inclusive DIS, where one of the final hadrons is also measured

$$\ell + h \rightarrow \ell' + h' + X$$

In describing semi-inclusive processes, in addition to quark distributions, the so called *fragmentation functions* are necesary. These functions [18] describe, or better parametrize, a given parton decay into a final hadron.

It is clear that, in principle, the best process to analyze fragmentation functions is the semi-inclusive e^+e^- annihilation, when a single hadron is fully detected in the final state

$$e^+e^- \rightarrow h(P) + X$$

The corresponding kinematics is depicted in Figure 12.

The relevant variables for this case are

$$z = \frac{2Pq}{Q^2} \quad \text{and} \quad \tilde{z} = \frac{2pq}{Q^2} \tag{34}$$

It is also convenient to introduce ζ that measures the fraction of the parton momentum carried by the final hadron. The corresponding cross section reads

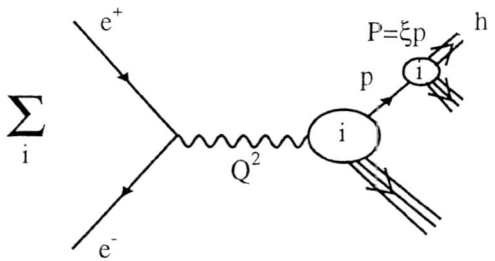

FIGURE 12. Schematic view of semi-inclusive e^+e^-

$$\frac{d\sigma_h(z,Q^2)}{dz\,dQ^2} = \frac{4\pi\alpha^2}{Q^2}\sum_i \int d\tilde{z} \int d\zeta\, \delta(z-\tilde{z}\zeta)\,\tilde{\sigma}_p^i(\tilde{z},Q^2)\,\zeta\, D_i^h(\zeta,Q^2) \quad (35)$$

$$= \sum_i \int_z^1 \frac{d\zeta}{\zeta}\, \tilde{\sigma}_p^i\left(\frac{z}{\zeta},Q^2\right)\,\zeta\, D_i^h(\zeta,Q^2)$$

Here $\tilde{\sigma}_p^i$ is the parton i production cross-section, that in the parton model results

$$\tilde{\sigma}_p^i(\tilde{z},Q^2) = e_i^2\,\delta(1-\tilde{z}) \quad (36)$$

if $i=q,\bar{q}$ and vanishes if $i=g$. The function $D_i^h(\zeta,Q^2)$ is the *fragmentation function* or probability for a parton i to decay into a hadron h carrying the fraction ζ of the parton momentum.

When QCD corrections are in order, the fragmentation functions acquire a Q^2 dependence. The evolution of these functions with the scale

$$t = \ln\left(\frac{Q^2}{\mu^2}\right) \quad (37)$$

is given by the AP-like integrodifferential equations

$$\frac{dD_q^h}{dt}(z,t) = \frac{\alpha_s(Q^2)}{2\pi}\int_z^1 \frac{dy}{y}\left[D_q^h dt(y,t)\, P_{qq}(z/y) + 2f\, D_g^h dt(y,t)\, P_{gq}(z/y)\right] \quad (38)$$

$$\frac{dD_g^h}{dt}(z,t) = \frac{\alpha_s(Q^2)}{2\pi}\int_z^1 \frac{dy}{y}\left[D_q^h dt(y,t)\, P_{qg}(z/y) + D_g^h dt(y,t)\, P_{gg}(z/y)\right] \quad (39)$$

The functions P_{ij} are exactly the same as those appearing in the AP equations for the structure functions (28). The corresponding graphical interpretation in terms of the basic QCD processes is included in Figure 13.

It is worth noticing that the properties of the splitting functions are enough to guarantee the momentum sum rule

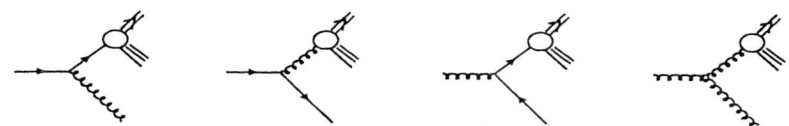

FIGURE 13. QCD processes providing the evolution of fragmentation functions

$$\sum_h \int_0^1 z\, D_q^h(z, Q^2) = 1 \qquad (40)$$

and the analogous one for the gluon fragmentation function.

To solve the AP equations it is necessary to know the fragmentation functions at some value Q_0^2. Exactly as in the case of structure functions, they have to be obtained from experimental data. In general one proposes a given form for the initial D-function, in agreement with data, and then AP equations provide the evolution with Q^2.

Fracture Functions

For those who are not acquainted with fracture functions, let us briefly summarize that the main idea behind fracture functions is the realization that the most familiar perturbative description for semi-inclusive processes, based on parton distributions and fragmentation functions is, at least, incomplete [19].

In the usual approach to semi-inclusive DIS, the corresponding cross section is expressed by a convolution between parton distributions and fragmentation functions accounting the process in which the struck parton hadronizes into a detected final state particle.

$$\frac{d^3\sigma_{curr.}}{dx\, dy\, dz} \simeq \frac{4\pi\alpha^2}{x\,(p+k)^2} \frac{1+(1-y)^2}{2y^2} \sum_i e_i^2\, q_i(x, Q^2) \times D_i^h(z, Q^2) \qquad (41)$$

Obviously, this approach only takes into account hadrons produced in the current fragmentation region, and what is more, within this approximation, and in leading order, hadrons can only be produced in the backward direction. Going to higher orders one finds a breakdown of hard factorization, as there are collinear singularities that can not be substracted in the parton distributions or in the fragmentation functions.

All this means that there are additional contributions missing, which are mainly target fragmentation processes and are included in the so called fracture functions M_i^h, defined by

$$\frac{d^3\sigma_{targ.}}{dx\,dy\,dz} \simeq \frac{4\pi\alpha^2}{x\,(p+k)^2}\frac{1+(1-y)^2}{2y^2}\sum_i e_i^2\,(1-x)\,M_i^h(x,z,Q^2) \qquad (42)$$

These functions, that can be thought as the probabilities to find a parton of a given flavour in an already fragmentated target, straightforwardly solve the factorization problem and also allow a LO description of hadrons produced in the forward direction. The only subtlety regarding them is that they obey slightly different evolution equations; their scale dependence not only depends on their shape at a given scale but also on that of ordinary structure functions and fragmentation functions, reflecting the fact that current and target fragmentation are not truly independent of each other. This is usually refered to as 'non homogeneus' evolution

$$\frac{\partial M_i^h(x,z,Q^2)}{\partial \ln Q^2} = \frac{\alpha_s(Q^2)}{2\pi}\int\frac{dy}{y}P_{ij}(y)M_j^h(\frac{x}{y},z,Q^2) \qquad (43)$$
$$+\frac{\alpha_s(Q^2)}{2\pi}\int\frac{dy}{x(1-y)}\hat{P}_{ijl}(y)q_j(\frac{x}{y},Q^2)D_l^h(\frac{zy}{x(1-y)},Q^2)$$

As for structure functions in totally inclusive deep inelastic scattering, QCD does not predict the shape of fracture functions unless it is known at a given initial scale. This nonperturbative information have to be obtained from the experiment, and, eventually, can be parametrized finding inspiration in nonperturbative models, as it is the case for ordinary structure and fragmentation functions.

Fracture functions have been succesfully applied to the description of leading baryon production [20], have been extended to spin dependent processes [21], and as we shall see, are relevant for the description of diffractive DIS.

PHOTON

It may seem strange that the next step in our study of parton structure relies in a particle, the photon, that it is not a hadron and of course has no structure by itself, as we very well know. However, a high energy photon can 'develop' structure when interacting with another object. In fact, as the electromagnetic field couples to all particles carrying the electromagnetic current, a photon can fluctuate into particle-antiparticle virtual states, particularly into quark-antiquark pairs or even more complex hadronic objects with the same quantum numbers. As long a the fluctuation time is longer than the interaction time we can talk about the structure of the the photon, and deal with it as a 'very special' hadron. Notice that at high energies, the fluctuation of a photon into a state of invariant mass M can persist for a time of the order of $\tau \sim 2\,E_\gamma/M$ until the virtual state materializes by a collision or annihilation with another system [22].

There are many reasons for this peculiar hadronic structure to be considered 'matter' of study. First of all, it gives the opportunity to study how QCD works in a completely different scenario. Theoretical studies of photonic parton distributions

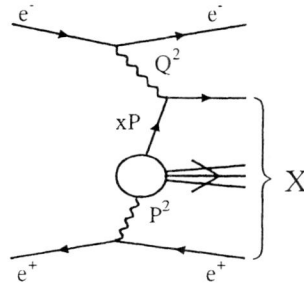

FIGURE 14. Photon Structure.

of real, i.e. on shell, photons have a long history initiated by Witten's work [23]. This description have been well established through measurements in $\gamma\gamma$ collisions at e^+e^- colliders (PETRA, PEP, LEP) however, extending the idea of photon structure to high virtualities as in ep processes like dijet production, a new insight has been gained as we link DIS, photoproduction, and $\gamma\gamma$ interactions [24].

The concept of photon structure functions for real and virtual photons can be defined and understood in close analogy to deep inelastic lepton nucleon scattering, via the subproccess $\gamma^*(Q^2)\gamma(P^2) \longrightarrow X$ in $e^+e^- \longrightarrow e^{\pm}X$ as in Figure 14, where we denote the probed target photon with virtuality $P^2 = -p_\gamma^2$ by $\gamma(P^2)$ and reserve $\gamma^*(Q^2)$ for the highly virtual one. The relevant differential cross section can be expressed, as in the hadronic case, in terms of the usual scaling variables x and y as

$$\frac{d^2\sigma(e\gamma(P^2) \longrightarrow eX)}{dxdy} = \frac{2\pi\alpha^2 s_{e\gamma}}{Q^4}\left[(1+(1-y)^2)F_2^{\gamma(P^2)}(x,Q^2) \right. \quad (44)$$
$$\left. -y^2 F_L^{\gamma(P^2)}(x,Q^2)\right]$$

with $F_{2,L}^{\gamma(P^2)}(x,Q^2)$ denoting the photonic structure functions. The measured e^+e^- cross section is obtained by convoluting Equation(44) with the photon flux for the target photon $\gamma(P^2)$ [25]. The range of photon virtualities explored in the above mentioned experiment is given by

$$m_e^2 y^2/(1-y) \leq P_{min}^2 \leq P^2 \leq P_{max}^2 \leq \frac{s_{e\gamma}}{2}(1-y)(1-\cos\theta_{max}) \quad (45)$$

where m_e is the mass of the electron, $s_{e\gamma}$ is the square of the c.m.s. energy, y is the energy fraction taken by the photon, and θ_{max} is the maximum scattering angle of the electron in the c.m.s. frame. $P_{min,max}^2$ are further determined by detector specifications and experimental settings.

It is worthwhile noticing that if one could neglect photon fluctuations into complex hadronic states, then the dependence of $F_{2,L}^{\gamma(P^2)}(x,Q^2)$ in both x and Q^2 would

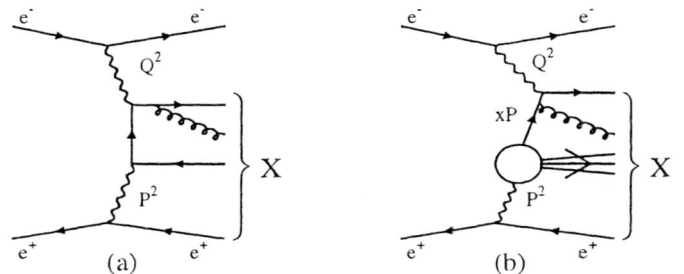

FIGURE 15. Direct (a) and Resolved (b) Contributions to $F_{2,L}^{\gamma(P^2)}(x,Q^2)$.

be fully predictable in perturbative QCD. The pointlike process $\gamma^*(Q^2)\gamma(P^2) \longrightarrow q\bar{q}$, corrected by gluon radiation effects, yields a definite prediction for $F_{2,L}^{\gamma(P^2)}(x,Q^2)$. However, as it was anticipated, this is only the so called 'direct' contribution to the total process, named in opposition to the 'resolved' one, where the virtual photon $\gamma^*(Q^2)$ strikes the nonperturbative 'dressing' of the photon fluctuation (Fig.15).

QCD corrections to the 'direct' component not only take into account the splitting of partons into partons as in the ordinary AP evolution equations, but also the possibility of a photon to split into a quark-antiquark pair. This implies an inhomogeneous term in the evolution equations and leads to a logarithmic enhancement in $F_{2,L}^{\gamma(P^2)}(x,Q^2)$ which means positive scaling violations for all values of x (Fig.16) [26].

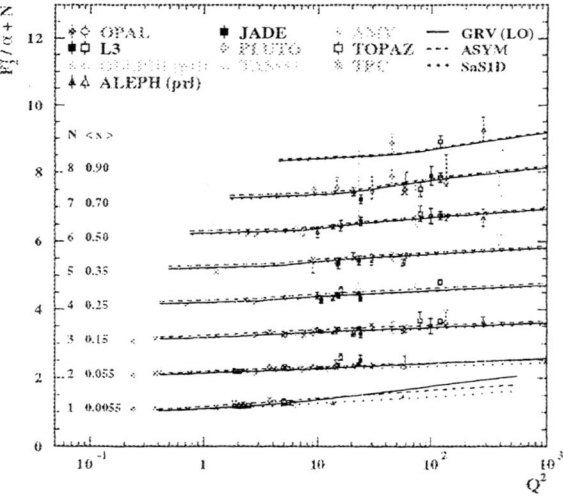

FIGURE 16. $F_2^\gamma(x,Q^2)$ as a function of Q^2.

The 'resolved' component obeys the same evolution equations typical of the lepton-hadron interactions. In a Vector Meson Dominance approach for the photon fluctuations, the resolved component is expected to vanish like $(1/P^2)^2$ in connection to the vector meson propagator, however it is not clear up to which values of P^2 the nonperturbative contributions are relevant. In recent years several sets of parton distributions for real and virtual photons have been proposed [27,28]. For virtual photons, different approaches have been followed in LO y NLO global fits to the available LEP data (Figure 17) [26].

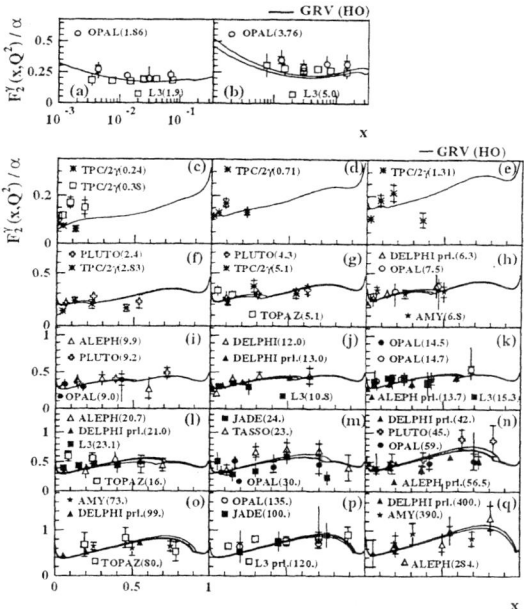

FIGURE 17. $F_2^\gamma(x, Q^2)$ as a function of x.

LEP data constraint reasonably well the quark distributions in the photon, however, in order to constrain the gluon densitiy it is much more suitable to analyze dijet production processes in e^+p collisions as measured at HERA. As we shall see, in this context the picture of 'direct' and 'resolved' is also more clearly illustrated. In this kind of process the parton content of the proton is used to probe the iqi partonic structure of the photon as in a hadron-hadron collision. The hard scale invoked to allow such inspection and guarantee the validity of a perturbative treatment, is the large transverse momentum of the final state jets. Figure 18 depicts LO 'direct' and 'resolved' contributions to dijet production e^+p collisions. As it can be seen, even at the lowest order, photonic gluons contribute to the cross section, at variance with what happens in e^+e^- collisions.

In leading order the differential cross section for two jet production in ep collisions

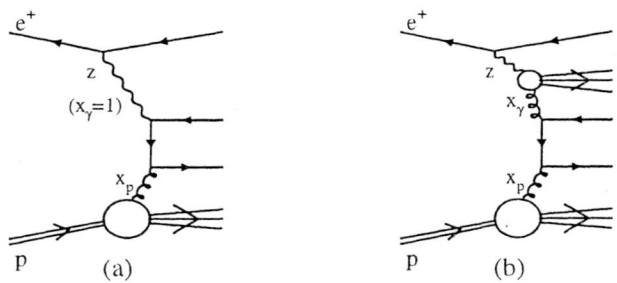

FIGURE 18. Direct (a) and Resolved (b) Contributions to dijet production.

takes a very simple form when written in terms of the fraction of the photon energy intervening in the hard process, x_γ, the fraction of the proton energy carried by the participating parton, x_p, and that of the electron carried by the photon, z [29]

$$\frac{d\sigma}{dx_\gamma \, dx_p \, dz \, dp_T \, dP^2} = \tilde{f}_{\gamma/e}(z, P^2) \, q^\gamma(x_\gamma, Q^2, P^2) \, q^p(x_p, Q^2) \frac{d\hat{\sigma}}{dp_T} \qquad (46)$$

Here, p_T is the transverse momentum of the jets, P^2 is the photon virtuality, and Q^2 is the relevant energy scale of the proccess, taken in this case equal to p_T^2. $d\hat{\sigma}/dp_T$ represents the hard parton-parton and parton-photon cross sections [30].

The functions $q^\gamma(x_\gamma, Q^2, P^2)$ and $q^p(x_p, Q^2)$ denote the parton distribution functions for the photon and the proton, respectively. The first one reduces to $\delta(1 - x_\gamma)$ -the probability for finding a photon in a photon- for direct contributions, i.e. those in which the photon participates as such in the hard process. $\tilde{f}_{\gamma/e}(z, P^2)$ is the unintegrated Weizsäcker-Williams distribution [31]

$$\tilde{f}_{\gamma/e}(z, P^2) = \frac{\alpha}{2\pi} \frac{1}{P^2} \frac{1 + (1 - z)^2}{z} \qquad (47)$$

which has been shown to be a very good approximation for the distribution of photons in the electron, provided the photon virtuality is much smaller than the relevant energy scale [27].

The differential cross section in Equation(46) can also be written in terms of the parton pseudorapidities η_1 and η_2, which are constrained by the experimental settings, and the electron and proton energies E_e and E_p [32]. Both pairs of variables are related to the energy fractions by

$$x_p = \frac{p_T}{2E_p} \left(e^{\eta_1} + e^{\eta_2} \right) \quad x_\gamma = \frac{p_T}{2zE_e} \left(e^{-\eta_1} + e^{-\eta_2} \right) \qquad (48)$$

Kinematical restrictions constrain x_γ to lay in the interval $[p_T^2/(x_p z E_e E_p), x_\gamma^{max}]$, x_p in $[p_T^2/(z E_e E_p x_\gamma^{max}), 1]$ and z in $[p_T^2/(E_e E_p), 1]$.

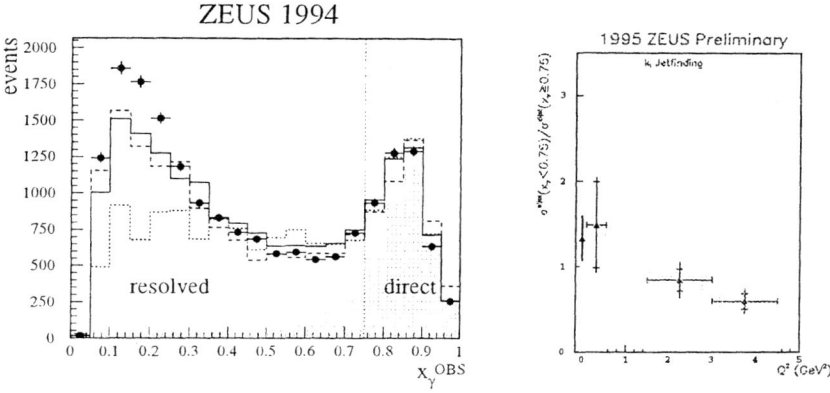

FIGURE 19. a) Event distribution measured in dijet photoproduction. b) Ratio of 'direct' and 'resolved' photon cross sections as a function of the photon virtuality.

In the lowest order of QCD the 'direct' contribution is characterized by $x_\gamma = 1$, which means that all the energy of the photon participates in the hard interaction. Allowing gluon corrections, 'direct' contributions may come from $x_\gamma \neq 1$, however one should expect them concentrated arround a peak at high x_γ. Conversely, 'resolved' contributions should be more copious al low x_γ as for ordinary hadrons. In fact, this trend is shown by the measurements, as can be seen in Figure 19a [33]. The variable x_γ^{obs} used there to approximate x_γ is the jet level equivalent of Equation (48)

$$x_\gamma^{obs} = \frac{p_T^{(1)} e^{-\eta_1} + p_T^{(2)} e^{-\eta_2}}{2zE_e} \qquad (49)$$

where $p_T^{(i)}$ and η_i refer to the transverse energies and pseudorapidities of the mea-

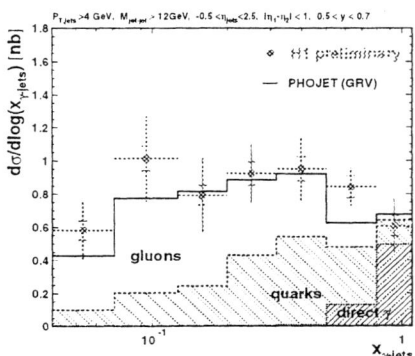

FIGURE 20. Dijet cross sections for $Q^2 < 0.01 GeV^2$ as a function of x_γ.

sured jets instead of those of the partons. Of course, the relative abundance of 'direct' and 'resolved' events depends strongly on the virtuality of the photon as can be seen in Figure 19b in accordance to the expectation about the suppression of the 'resolved' component as the virtuality increases [33].

Finally, it is quite instructive to anlyze the relative abundance of quark and gluon initiated contributions as a function of x_γ in a typical e^+p dijet cross section. In Figure 20 data obtained by H1 is compared to theoretical expectations based on a Montecarlo analysis using GRV parton distribution for the photon [27] and for the proton [34,35]. Notice that at large x_γ the 'resolved' component is clearly dominated by quarks, however as x_γ decreases, the photonic gluons become dominant.

COLOUR SINGLET (POMERON?)

In this section we draw our attention to a much more unfamiliar object, at least for the younger generation of physicists, although for some of us it may be an old acquaintance. We are talking about the *pomeron*, or in more modern language, about the colour singlet object which is exchanged when particles undergo strong interactions while preserving its nature. The possibility of studying the partonic structure of this old friend has been made real by a very recent generation of experiments and has added an extra quota of excitement in perturbative QCD.

Before we plunge ourselves into the partonic structure of the pomeron, it would be very helpful to first fetch some of the tools used in past to sail in those waters.

Peripheral model

Before the advent of QCD, the presence of structures in the differential cross-sections was mainly explained by means of dynamical exchange mechanisms like in the so called *peripheral model*. In this approach it is found that there is a clear enhancement near the forward direction whenever the crossed channel of the reaction has the quantum numbers of a known particle. Then, peripheralism proposes to express the scattering amplitude as a sum of contributions corresponding to the exchange of a particle. Namely

$$F = F_\pi + F_V + F_B + ...$$

with

$$F_i \propto \frac{1}{m_i^2 - t}$$

where it is clear that for small, physical (negative), values of t, there are peaks in the corresponding cross-section in s-channel, proportional to $|F|^2$. The indices $\pi, V, ...$ represent pions, vector bosons, etc., and their masses hierarchize each contribution. The explanation of peripheral events cannot be found in QCD, since the exchanged particle is not a gauge boson that can be treated perturbatively. In fact the particles involved are composite objects such mesons or baryons.

When the exchanged particle has a spin J, the previous expression for the amplitude has to be replaced by

$$F_i \propto \frac{P_J(\cos\theta_t)}{m_i^2 - t}$$

and, if one remembers that the scattering angle in the t-channel reads

$$\cos\theta_t = 1 + \frac{2s}{t - 4m^2}$$

it is found that the numerator of F_i behaves asymptotically as s^J for fixed t. This is not tenable because the Froissart bound is not satisfied for $J > 1$. Moreover, that expression is purely real and does not satisfy the principle of analyticity.

There is a way out of the above mentioned difficulties, while maintaining the spirit of an exchange mechanism, now of an entire family of related particles. This proposal is based on the Regge poles ideas. These ideas emerge naturally in potential scattering theory through the analyticity properties of the amplitude. The Regge pole model [36] is based then, upon the assumption of the following behaviour for the scattering amplitude

$$\lim_{s\to\infty, fixed\, t} F(s,t) \simeq \sum_i \beta_i(t) \frac{1 + \xi_i\, e^{-i\pi\alpha_i(t)}}{\sin[\pi\alpha_i(t)]} \left(\frac{s}{s_0}\right)^{\alpha_i(t)} \qquad (50)$$

here $\xi_i = \pm 1$ is called the signature of the corresponding Regge pole. The functions $\alpha_i(t)$ are called *Regge trajectories* and $\beta_i(t)$ residues of the poles that in general factorize between the initial and the final channels. Both functions $\alpha_i(t)$ and $\beta_i(t)$ are analytic functions of t.

The high energy model so introduced is understood as an exchange model where the Regge trajectories $\alpha_i(t)$ are such that they pass through the spin values of the particles, or better the resonances, each time positive t takes the values of their mass squared. In this way, the collective effect of the exchange of all members of the family is taken into account. Clearly, the Froissart bound (total cross-sections cannot increase faster than $\ln^2 s$) is satisfied as soon as

$$\alpha_i(t \leq 0) \leq 1$$

in the space-like region of negative t, relevant for the sacttering process.

Let us take as an example the case of pion-nucleon charge-exchange

$$\pi^- p \to \pi^0 n$$

that has always been the paradigmatic case of reggeology. The t-channel of this reaction is

$$\pi^- \pi^0 \to \bar{p} n$$

having the quantum numbers; $Q = 1$, $I = 1$, $Y = 0$. $G = +1$ that correspond to the well known meson resonances $\rho\,(770\,MeV)$ of spin 1 and $g\,(1680\,MeV)$ of spin 3. This sequence of resonances defines precisely the meson Regge trajectory of isospin 1, G-parity +1 and positive signature that is noted α_ρ. The continuation of this trajectory to negative t values, takes us into the physical region corresponding to the s-channel charge-exchange reaction. The parameters of a standard linear trajectory are the intercept at $t = 0$ and the slope. In the present case, as it is shown in Figure 21, they are

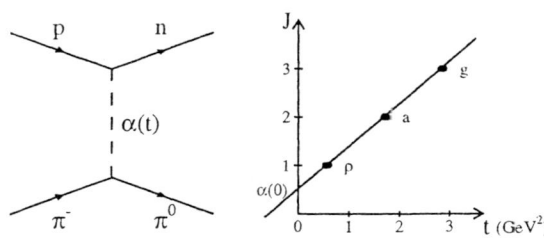

FIGURE 21. ρ-Regge trajectory and $\pi^- p \to \pi^0 n$ data

$$\alpha_\rho(0) \simeq 0.55 \quad \alpha' \simeq 0.9\,GeV^{-2}$$

respectively. Consequently, the differential cross section has the behaviour

$$\left.\frac{d\sigma}{dt}\right|_{t=0} \simeq s^{2(\alpha_\rho(0)-1)} \sim s^{-1}$$

But what happens if the crossed channel of the reaction, and correspondingly the eventual exchanged object, has the quantum numbers of the vacuum? The answer arrives in the following paragraphs.

Diffraction

Landau and his collaborators, in the fifties, introduced the term diffraction in high energy physics used in complete analogy with the well known phenomenon in Optics that occurs when light interacts with obstacles or holes whose dimensions are of the order of the electromagnetic radiation wavelength. The interaction of a hadron could be think as the absorption of its wave function caused by the different channels open at high energy and consequently the name diffraction seems to be intuitive. For a recent review of this topic see [37].

In the Fraunhofer limit, namely when the product of the wave number times the area Σ of the obstacle is of the order of the observation distance, the energy distribution at the observation point, given by the Kirchoff formula, can be written as

$$T(x,y,z) \approx \frac{k}{2\pi i} \frac{e^{ikr_0}}{r_0} \int_\Sigma d^2b\, S(\vec{b})\, e^{i\vec{q}\cdot\vec{b}}$$

where $\vec{b}$ stands for the impact parameter, r_0 is the position of Σ, $|\vec{q}| = k\sin\theta$ is the 2-D momentum transfer and the scattering matrix is expressed as

$$S(\vec{b}) = 1 - \Gamma(\vec{b})$$

in terms of the so called profile function $\Gamma(\vec{b})$. From the expression for T, one can immediately obtain the scattering amplitude

$$f(\vec{q}) = \frac{ik}{2\pi} \int d^2b\, \Gamma(\vec{b})\, e^{i\vec{q}\cdot\vec{b}} \tag{51}$$

i.e.: given as the 2-D Fourier transform of the profile function.

When the function Γ is spherically symmetric, the integral becomes a Bessel transform, namely

$$f(\vec{q}) = ik \int_0^\infty b\, db\, \Gamma(\vec{b})\, J_0(q\,b)$$

meaning that if the profile function is merely a disk of radius R, the amplitude reduces to the black-disk form

$$f(\vec{q}) = ik R^2 \frac{J_1(qR)}{qR}$$

The prediction coming from this very simple model is a series of diffractive maxima and minima, entirely similar to the case in Optics, and have been clearly observed in several experiments

In the specific field of particle physics, diffraction is said to be the dominant process of scattering at high energy if no quantum numbers are exchanged between the colliding particles. In other words, diffraction dominates asymptotically as soon as the particles in the final state have the same quantum numbers of the incident ones. This sort of definition of diffraction clearly includes, for the two body scattering, three cases: elastic scattering, single diffraction and double diffraction. In the first process the outgoing particles are exactly the same as the incident ones. In the second case, one incident particles goes out unmodified while the second one gives rise to a resonance, or to a bunch of final particles, with total quantum numbers coincident with its own ones. Finally, when double diffraction occurs, each incident particle gives rise to a resonance, or to a bunch of final particles, with the same quantum numbers of the initial ones.

Pomeron

It is possible to discuss diffraction from the viewpoint of an exchange model, in particular within the framework of reggeology. Remember that we have found precisely that an enhancement in the differential cross section can be expected whenever the quantum numbers of the crossed channel correspond to an existing particle. However, in the elastic case at high energies, all the hadronic systems show the diffraction peak while the exchange of a given particle gives rise to different contributions in different systems. Moreover, via the optical theorem, the forward elastic amplitude is connected to the total cross section and this fact cannot be understood in terms of only one exchange process. Nevertheless, the diffraction peak can be interpreted in terms of a special trajectory that summarizes all the diffractive contributions and has the vacuum quantum numbers: the *pomeron*. Its contribution to the amplitude is

$$F_{dif}(s,t) = -\beta_{I\!P}(t)\frac{1+e^{-i\pi\alpha_{I\!P}(t)}}{\sin[\pi\alpha_{I\!P}(t)]}\left(\frac{s}{s_0}\right)^{\alpha_{I\!P}(t)} \tag{52}$$

Consequently, the asymptotic behaviour of the total cross section (via optical theorem) is given by

$$\sigma_{tot} \simeq \frac{\beta_{I\!P}(0)}{s_0}$$

if the intercept of the pomeron trajectory is the maximum value allowed by the Froissart bound, namely $\alpha_{I\!P}(0) = 1$. This fact implies that the total cross section in this approximation behaves asymptotically as a constant. It is clear that logarithmic corrections are always allowed. Moreover, they are necessary in order to cope with experimental data.

The pomeron concept answer the question at the end of the previous section using the reggeon jargon: a diffraction process is dominated by the exchange of a *pomeron*, that in fact means exchange of no quantum numbers.

It is interesting to mention that Donnachie and Landshoff [38] were able to describe all available total cross section data on $\bar{p}p$, pp, πp and Kp by a very simple Regge-Pomeron inspired parametization of the form

$$\sigma_{tot} = X\,s^{0.0808} + Y\,s^{-0.4525}$$

where X and Y are reaction dependent parameters. Clearly the first exponent of the center of mass energy squared s correspond to a pomeron intercept of $\alpha_{I\!P}(0) = 1.0808$, while the second exponent comes from a typical Regge intercept $\alpha_{I\!R}(0) = 0.5475$. The successful fit is shown in Figure 22.

Trying to connect the pomeron with the language and understanding of QCD, one can imagine it as a colour singlet combination of partons such as the simplest picture of a pair of gluons proposed by Low and Nussinov [39]. Two-gluon exchange is compatible with all soft phenomenology except that this simple model gives a constant, not a rising, total cross-section. Clearly in order to analyze the parton content of pomeron, a hard or high-momentum transfer interaction analogous to the one used to find the quark-gluon content of the proton is necessary. This kind of processes is called hard diffraction making reference to the hard scale that ultimately allows the pertubative description.

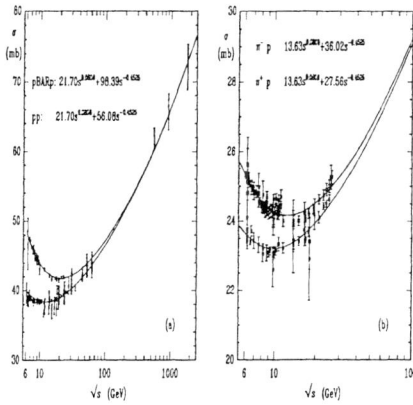

FIGURE 22. Donnachie and Landshoff parametrization of total cross sections

Nowadays, people is trying to go beyond in the understanding of pomeron properties following different theoretical approaches [40], although there remain several open questions. The main point to be solved is related to the precise relation between hard and soft diffraction. As we said, to study the pomeron parton content, small distances and high-momentum transfer are necessary, while the natural environment of the soft pomeron is at large distances and low-momentum transfer. In other words, the clear notion of a pomeron is still missing.

Most of the recent renewal of interest in diffraction was triggered mainly by a theoretical proposal of Bjorken [41], who pointed out a new signature of diffraction related to the presence of large rapidity gaps. These rapidity gaps were found both at HERA and at the Tevatron.

The presence of diffractive processes at HERA should be expected since the photon behaves like a hadron in several circumstances. In fact, about 40 % of the photoproduction events are of diffractive character. In this kind of processes, the pomeron is exchanged and the incident proton remains as a proton or is diffractively dissociated in a state with the same quantum numbers. Consequently, as it will be discussed below, there appears a large rapidity gap between the proton, or the diffractive system, and the hadrons coming from the system into which the photon was diffracted. The quite unexpected fact observed at HERA was the presence of large rapidity gap events also in the DIS domain [42]. The observation that also a highly virtual photon can participate in a diffractive process, opened the road to the analysis of the pomeron structure. In fact, if Q^2, the virtuality of the photon, is larger than say $4\,GeV$, the scattering γ^*-pomeron can be treated perturbatively. The impact of this DIS diffractive data is also evident because one is facing in a DIS process characterized by a large Q^2 scale, diffractive properties which were been expected at a soft scale.

Large rapidity gaps

As it was predicted by Bjorken [41], the most evident signal for diffractive physics at high energy, is the presence of large rapidity gaps. Just to get some insight on this jargon, remember that a single inclusive diffractive reaction, noted as

$$A(k_A) + B(k_B) \to A'(k'_A) + X(k_X)$$

implies no exchange of quantum numbers different from those of vacuum. In this inclusive case, at variance with the elastic reaction, besides the two variables (k and θ or s and t) a third variable is needed to describe the process. Generally it is used

$$M_X^2 = (k_A + k_B - k'_A)^2 = E_X^2 - \vec{k}_X^2 \tag{53}$$

or, alternatively, the Feynman x_F defined as

$$x_F \equiv \frac{|k'_{A\,long}|}{k'_{A\,long}} \approx 1 - \frac{M_X^2}{s} \tag{54}$$

Another useful variable is rapidity, defined by

$$y = \ln\left[\frac{E'_A + k'_{A\,long}}{E'_A - k'_{A\,long}}\right] \tag{55}$$

or equivalently, the pseudorapidity which is its limit for large energy

$$\eta = -\ln\left(\tan\frac{\theta_{A'}}{2}\right) \tag{56}$$

In a typical DIS event, the struck parton of the proton, emerges forming an angle $\theta_{A'}$ with the proton remnant direction, as it is shown in Figure 23.

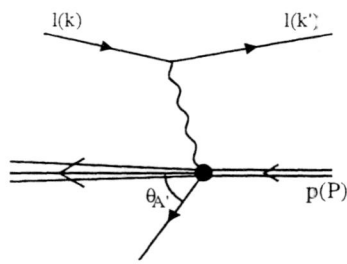

FIGURE 23. DIS event diagramm

This angle can be expressed in terms of the difference in total pseudorapidity η between these directions, namely

$$\triangle \eta = \eta_{remnant} - \eta_{parton}$$

If the center of mass energy squared of the system is s, the pseudarapidity interval covered is

$$\triangle \eta \sim \ln\left(\frac{s}{m_p^2}\right)$$

Consequently,

$$\triangle \eta \sim \ln\left(\frac{W}{m_p}\right) - \ln\left(\frac{xW}{m_p}\right) \sim \ln\left(\frac{1}{x}\right)$$

where the first term correspond to the pseudorapidity covered by the $\gamma^* - p$ system and the second to that covered by the $\gamma^* - quark$ system respectively. Clearly x is the Bjorken variable that measures the amount of momentum of the proton carried by the quark.

The confinement property of QCD can be rephrased by saying that the struck quark and the proton remnant are connected via a colour string in order to end with colourless hadrons. For this reason, the mentioned pseudorapidity gap $\triangle \eta$ is filled with particles during the hadronization process. Moreover, when the value of x decreases, the average hadron multiplicity increases, making less likely the visibility of any rapidity gap in the DIS event. In principle, the rapidity gap between the remnant proton, and the struck quark jet is then exponentially suppressed.

After this kinematical preface, we briefly discuss the so called *leading particle effect*, characteristic of high energy hadronic interactions. Around 10% of the inclusive hadron scattering events present a Lab system configuration where the incident particle flies apart essentially unscattered in almost the forward direction leaving behind a stream of slow moving produced particles. The experimentally determined cross section looks entirely similar to the elastic case. Namely, between the initial particle and the final fast one, there is no change of quantum numbers and the reaction is clearly of diffractive type. There is only a minor loss of momentum to produce the slow particles. This effect requires the fast particle to be exactly the same as the incident one.

When the leading particle, the one we called above A', is produced diffractively, and because the process has this characteristic, one has to expect the remaining cluster X to be produced at the opposite end of the rapidity spectrum. That is usually expressed by saying that the reaction presents a large rapidity gap from A'. Diffraction definition makes evident this concept. In fact, as A' has exactly the same quantum numbers as A, no quantum numbers at all can be exchanged until X is produced. If a particle were produced in between, it would mean that some quantum number has been exchanged and the process is no longer diffractive. Consequently, in this large rapidity gaps events, as no particle is produced between A' and X, they are clearly separated in rapidity.

Large rapidity gap events can be generated in soft processes as elastic and single diffractive hadron-hadron scattering, where the momentum transfer t between the scattered particles is small, and the center of mass energy s large, i.e., peripheral scattering.

It should be stressed that nowadays, diffraction is considered synonym of large rapidity gap processes. In other words, the evidence of diffractive hadronic events comes mainly from large rapidity gaps as have been observed in D0 and CDF Collaborations at the Tevatron [43].

On the other hand, at HERA, the electron-proton collider, diffraction has also been extensively observed using both the rapidity gap criteria for diffraction and also the detection of final state protons. In semi-inclusive DIS, whenever the detected final hadron coincides with the incident one, one is dealing with diffraction. In order to follow the outgoing proton a Leading Proton Spectrometer (LPS) was designed.

Hard Diffraction at HERA

Before discussing the interesting HERA DIS diffractive events, let us consider the particular semi-inclusive processes where the detected final hadron exactly coincides with the initial one, namely

$$\ell(k_\ell) + N(k_N) \to \ell(k'_\ell) + N(k'_N) + X(k_X)$$

In this case the scattering diagram looks like the one in Figure 24

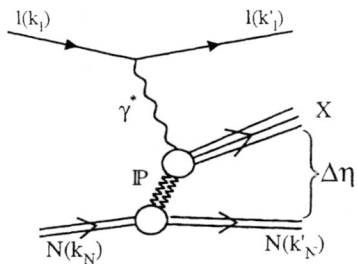

FIGURE 24. Diffractive DIS event diagramm

The process is clearly diffractive since no quantum numbers are exchanged between the γ^* and the nucleon N. As a consequence, one has for the remnant X the quantum numbers $J_X^{PC} = 1^{--}$ exactly equal to those of the incoming photon. Notice by the way that the production of any vector boson, the explicit replacement of X by any V in the reaction above, is a diffractive process because the V meson quantum numbers are precisely those of γ^*.

The kinematics related to Figure 24 needs further variables to be introduced. They are usually defined as

$$t = (k_N - k'_N)^2 \quad , \quad M_X^2 = (k_N - k'_N + q)^2 \quad , \quad M_Y^2 = k'^2_N \tag{57}$$

$$x_{I\!P} = \frac{(k_N - k'_N) \cdot q}{k_N \cdot k'_N} = \frac{M_X^2 + Q^2 - t}{W^2 + Q^2 - m_p^2} \quad , \quad \beta = \frac{Q^2}{M_X^2 + Q^2 - t} = \frac{x}{x_{I\!P}} \tag{58}$$

Notice that frequently ξ is used instead of $x_{I\!P}$ and that $x_F = 1 - x_{I\!P}$. Clearly, the range of values of $x, x_{I\!P}, x_F$ and β is $(0,1)$. Moreover, β could be understood as the momentum fraction carried by the parton directly coupled to γ^*.

In the context of diffractive DIS, it is customary to define *diffractive structure functions*, in analogy with the ordinary ones, through the expression of the corresponding cross section

$$\frac{d^4\sigma(\ell N \to \ell X Y)}{dx\, dQ^2\, dx_{I\!P}\, dt} = \frac{4\pi\alpha_{em}^2}{x\, Q^4}\left[\left(1 - y + \frac{y^2}{2}\right) F_2^{D(4)} - \frac{y^2}{2} F_L^{D(4)}\right] \tag{59}$$

where

$$F_2^{D(4)} = F_T^{D(4)} + F_L^{D(4)}$$

It is also of interest the t-integrated diffractive structure function defined en each case by

$$F_i^{D(3)}(x_{I\!P}, \beta, Q^2) = \int_{|t|_{min}}^{|t|_{max}} d|t|\, F_i^{D(4)}(x_{I\!P}, \beta, Q^2, t) \tag{60}$$

with $i = 2, T, L$. The superindices 3 and 4 refer to the number of variables in the structure functions. In the integral $|t|_{min}$ is the lower kinematic limit of $|t|$ and $|t|_{max}$ has to be specified in each case.

In the most naive Regge inspired approach, the diffractive structure function is assumed to be given by the product of the probability $f_{I\!P/p}(x_{I\!P}, t)$ to find a pomeron in the incoming proton, which only depends on the variables $x_{I\!P}$ and t, and a *pomeron structure function* $F_2^{I\!P}(\beta, Q^2)$, given by parton densities which are assumed to behave according to Altarelli-Parisi evolution equations and factorize as ordinary parton distributions [44].

In recent years different theoretical approaches have been proposed for the description of hard diffraction and specifically for the diffractive structure functions. See for example [40] and references therein.

Some of these approaches modelize the diffractive interaction in the proton rest frame as an exchange of two gluons between the proton and the hadronic system into which the photon has evolved as depicted by the upper line diagramas in Figure 25. Others, also in the proton rest frame, evaluate within a semiclassical analysis the effect of the colour field of the proton into the photon hadronic system as in the lower line diagrams of Figure 25. The main difference between the above

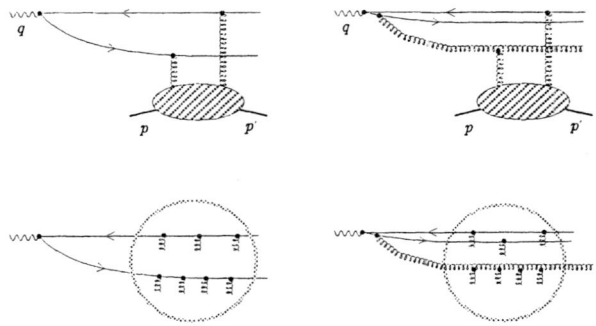

FIGURE 25. Different approaches for diffractive DIS

mentioned models is that in the former only two gluons are exchanged while in the latter a multiple exchange of soft gluons is assumed [40].

A third strategy consists in dealing with the diffractive interaction as a special kinematical limit of a more general semi-inclusive process, i.e. regarding the diffractive structure function just as the low $x_{I\!P}$ limit of the fracture function of protons into protons [19]. Doing this, the whole perturbative techniques can be rigurously applied without need to make additional assumptions in a program similar to what has been done for structure and fragmentation functions [45]. Another advantage of this last strategy is that in this framework the large $x_{I\!P}$ behaviour of the diffractive structure function can be explored with leading proton production experiments. The fracture function approach can be in some sense related to the two gluon exchange model by a frame transformation as it is sketched in Figure 26.

A remarkable thing to notice regarding these kinds of models for diffraction and also the factorization approach, is that although they seem to differ in rather strong assumptions, they all give a reasonably good account of the data suggesting a large gluon content in the diffractive structure function with gluons concentrated at high

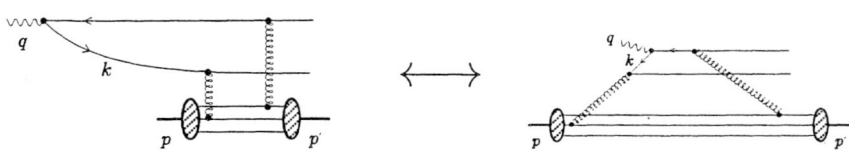

FIGURE 26. Correspondence between the two gluon exchange models and the factorization approach.

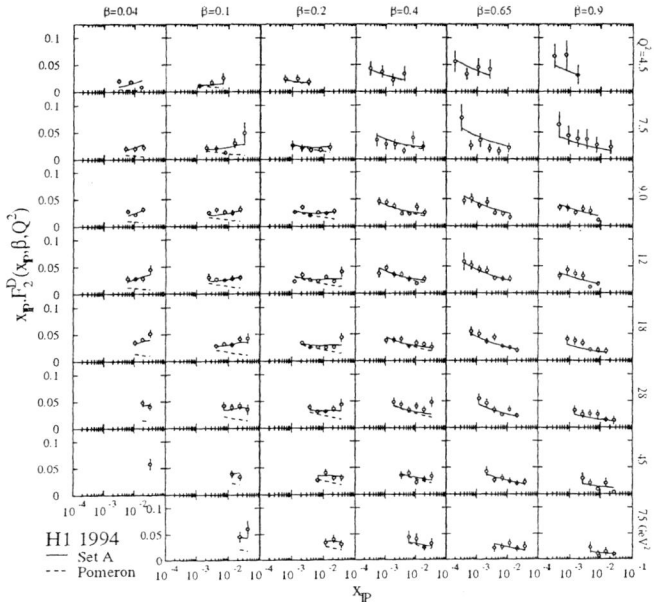

FIGURE 27. Diffractive DIS data.

β.

In Figure 27 measurements of the diffractive structure function $F_2(x_\mathbb{P}, \beta, Q^2)$ as a function of $x_\mathbb{P}$ obtained by H1 at HERA are compared with the fit coming

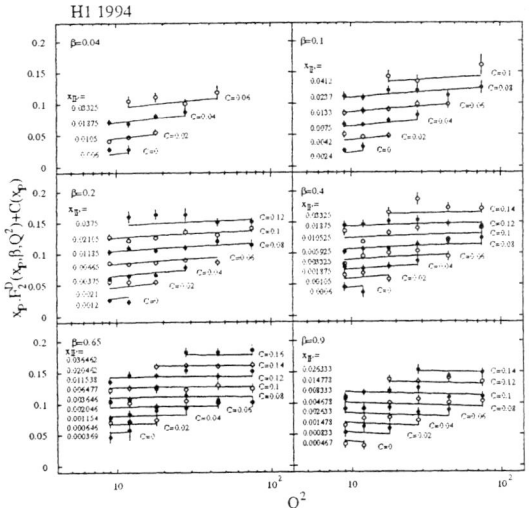

FIGURE 28. Scale dependence of $F_2(x_\mathbb{P}, \beta, Q^2)$.

from a factorization approach [45]. As for ordinary structure functions, QCD predicts the scale dependence of the diffractive structure function. Figure 28 shows the agreement between the data and the behaviour expected in the factorization approach.

Having established a rigorous and precise description of diffractive DIS we can go back to the most naive Regge inspired approach and see which of their hypothesis or assumptions may survive. The first thing to notice is that although in principle diffractive parton distributions or fracture functions obey non homogeneus AP evolution equations, in the low $x_{I\!P}$ limit the non homogeneus contributions are numerically negligible so the standard assumption is a very good approximation. Instead, what fails, at least in the unrestricted kinematical range accesed by HERA, is the flux factorization hypothesis, i.e. the posibility to factorize the $x_{I\!P}$-dependence of the structure function as a simple power of $x_{I\!P}$. Global analysis of the data shows not only deviations form this behaviour, suggesting the admixture of Regge exchanges, but also a β-dependent level of admixture [45].

HERA also measures other diffractive procceses like diffractive production of vector mesons and diffractive photoproduction of jets [46]. This last kind of processes, depicted in Figure 29, is particularly exciting as it represents a combined test of our knowledge of the parton structure of the photon and the colour singlet. ¿From the theoretical point of view it is also very important because the resolved photon contribution to the process is in fact a hadron-hadron diffractive process for which hard QCD factorization may be broken.

Indeed, although hard factorization can be proven to be valid for diffractive DIS, the proof fails for hadron-hadron interactions. The possibility of soft interactions taking place before the hard scattering occurs may spoil factorization in hadron-hadron collisions [48], although there are no model-independent estimates of how large the factorization breaking effects may be or under which circumstances can be found.

Hard factorization has been a very active theoretical topic in the three preceding years, and it is particularly relevant for the discussion of Tevatron diffractive data.

Figure 30 shows a comparison between H1 dijet photoproduction data and the

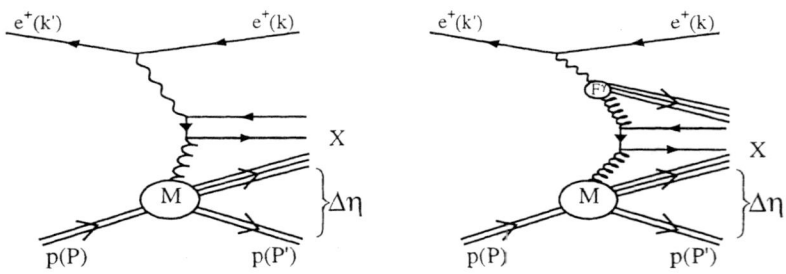

FIGURE 29. Direct and Resolved Diffractive Dijet Photoproduction

corresponding prediction computed with photonic and diffractive parton distributions [47]. The comparison suggests that for diffractive photoproduction the factorization breaking mechanisms are either neglible or beyond the accuracy of the present data.

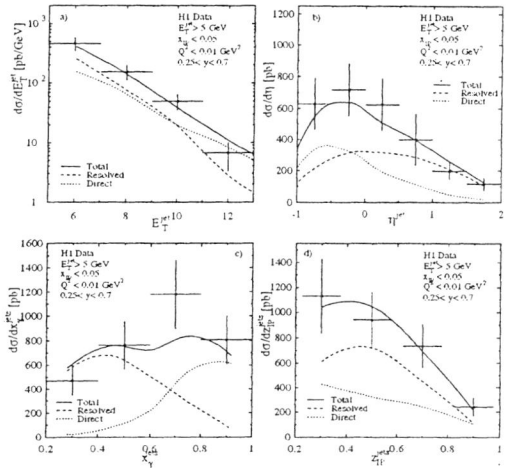

FIGURE 30. H1 dijet photoproduction data and the prediction computed with photonic and diffractive parton distributions

Hard Diffraction at Tevatron

As we mentioned earlier, large rapidity gaps in proton-antiproton collisions have been observed by both D0 and CDF collaborations at Tevatron. The typical rapidity gap events observed at Tevatron are classified acording to three categories: hard single diffraction, hard double pomeron exchange, and hard colour singlet.

In the first case, Figure 31a, a large rapidity gap in the forward direction is found between the outgoing antiproton and the debris produced by the proton-pomeron interaction. In the double pomeron exchange, Figure 31b, both the proton and the antiproton survive the interaction and emerge leaving rapidity gaps between each of them and the debris of the pomeron-pomeron interaction. In the third case, both the proton and the antiproton disociate but their corresponding remnants leave a large rapidity gap between them, Figure 31c.

For a straightforward comparison with HERA data the most simple topology to analyze is hard single diffraction, as the only change to be made in the conceptual framework developed in the previous section is the replacement of the lepton probe (the positron) used at HERA by an hadronic one (the proton) that takes its place at the Tevatron. The diffractive structure function for the antiproton is simply

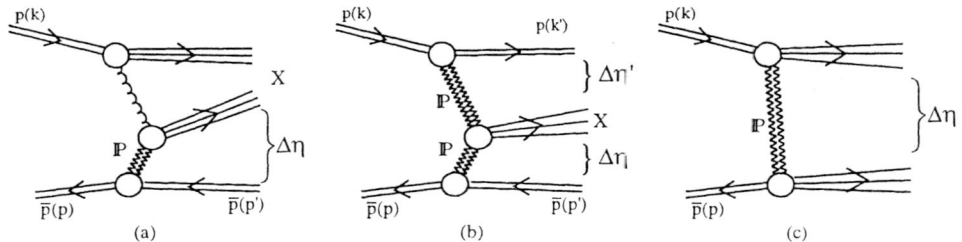

FIGURE 31. Rapidity gap topologies at Tevatron.

related by charge conjugation to that for the proton, and which is measured at HERA. Whithin this topology, several observables can be measured such as dijet, $W^{\pm}$, and Z^0 production.

Performing this kind of analysis what has been found is that even though the Tevatron data allow a description in terms of diffractive parton densities, the predictions computed with those coming from HERA largely overshot Tevatron data signaling a breakdown of factorization [49], which remains to be fully understood.

As we have already pointed out in the previous section, the factorization breaking seen Tevatron is not at all unexpected, since QCD factorization may be broken in hadron-hadron diffraction due to soft exchanges. In addition to these, there are also some other effects which conspire against factorization and need some consideration.

For the analysis of these effects it is crucial to take into account that the standard criterium used to identify diffractive events both at HERA and at Tevatron, and thus the correspoinding contributions to the diffractive parton distributions and the Tevatron observables, respectively, is the presence of rapidity gaps of a given size. Obviously, changing the size of these rapidity gaps, the number of events selected also changes, effect that in fact has been observed and studied by the different experiments. In Figure 32 we show as an example dijet photoproduction events generated with a Monte Carlo in the kinematical regimes of H1 and ZEUS experiments as a function of the pseudo-rapidity η_{had}^{max} of the most forward particle belonging to the system X [47]. Events to the left of the thick vertical line (rapidity gap limit) are those taken into account while those to the right are discarded.

It is worthwhile noticing that even for the same kind of process, the different kinematical regimes covered by both experiments given for example by cuts in the transverse energy E_T and pseudo-rapidity η of the two most energetic jets, etc., yield rather different distributions in η_{max}^{had} and thus, are affected in a different way even for the same pseudo-rapidity gap definition. For ZEUS, approximately 25 % of the total number of events are concentrated at $\eta_{max}^{had} < 1.8$, while for H1 kinematics approximately 10 % of them survives the same constraint.

Clearly, the situation is much more involved if we try to relate pseudo-rapidity gap data from different processes as we are doing in the HERA-Tevatron comparison.,In other words, is not evident what is the relation between DIS events with

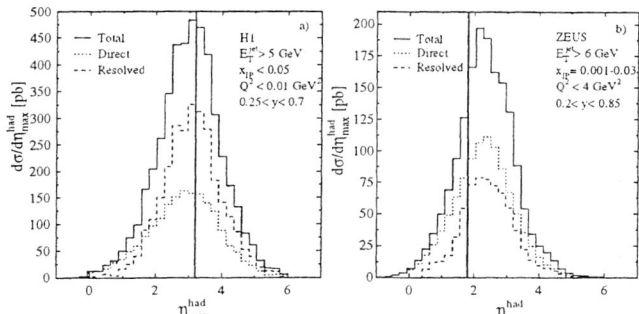

FIGURE 32. Dijet photoproduction events generated in the kinematical regimes of the a) H1 and b) ZEUS experiments

a pseudo-rapidity gap of given size, and proton-antiproton events with pseudo-rapidity gaps of a different size, and obtained in a completly different kinematical regime.

It is somewhat paradoxical that the same rapidity gaps that helped so much the development of hard diffraction, seem at this point to be a burden. In any case, ongoing experimental programs based on the Roman Pot technique, which allows a positive identification of the final state hadron as signature of the colour singlet exchange, will surely allow a much more deep and complete picture of these exciting phenomena.

CONNECTIONS

Throughout the present lectures we have review the most significant features of partons, the fundamental protagonists of QCD, as seen in the three rather different enviroments, the proton, the photon, and the pomeron (colour singlet exchange). As we have seen, and in fact was anticipated in the introduction, these objects, or more precisely the high energy processes involving them, are three of the main benchmarks of perturbative QCD.

In the process of developing the predictive power of QCD we have introduced structure, fragmentation, and fracture functions, objects that link intimately both our knowledge and our ignorance about the partonic structure.

The fundamental prediction of QCD in these enviroments, the energy scale dependence of the corresponding cross sections or structure functions, is beautifully illustrated in Figure 33 [50]. The observed behaviour in each case can be traced back to a basic parton interaction mechanism which alternatively dominates over the others: gluon radiation, quark pair creation from gluons, and from photons, respectively, and to the respective parton compositions.

At intermediate x, the case in the figure, the proton structure function is dominated by valence quarks, however increasing the resolving power of our probe, gluon

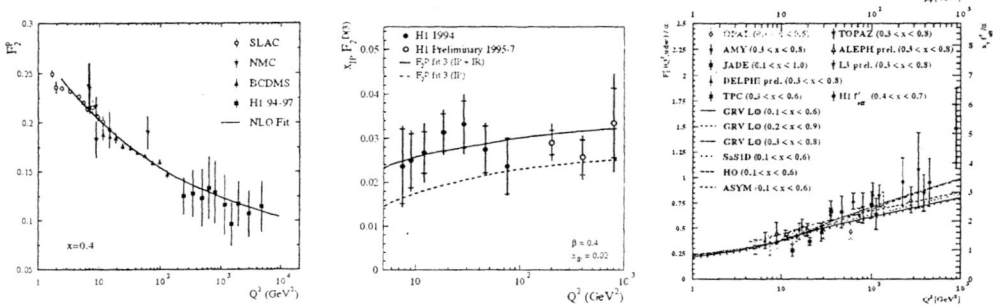

FIGURE 33. Q^2 Dependence of the proton, the color singlet, and the photon.

radiation depletes dramatically these densities through the corresponding term in the AP evolution equation. At variance with this scenario, the diffractive structure function is dominated by gluons at intermediate β, and consequently our lepton probes mainly quarks produced by pair creation from gluons, and the corresponding evolution. The logarithmic increase in the photon structure function is caused by quark pair creation from photons, which in the lowest order approximation appears in the AP equations as Q^2 independent term.

Finally, the partonic structure as seen in our benchmarks in some way relate the physics made in the three main high energy physics laboratories. LEP provides us with the most precise determinations of photon structure functions and also fragmentation functions, using electron-positron collisions. Tevatron in turn, contribute to our understanding of the proton structure and also to diffraction by means proton-antiproton interactions. Positron-proton collisions at Hera, not only test the structure of the proton and diffraction by themselves, but also brings toghether the proton and the photon structure functions in inclusive photoproduction, relates the latter with fracture functions in the diffractive photoproduction, and explores the relation between fragmentation and fracture functions in leading baryon production.

REFERENCES

1. For a recent comprehensive review see: B. Lampe and E. Reya, Spin Physics and Polarized Structure Functions, hep-ph/9810270. See also E. Leader et al., Phys, Rep. 261, 1 (1995).
2. U. Bassler et al. hep-ph/9906027.
3. H. Abramowicz, A. Caldwell, DESY 98-192 (1998).
4. J. Huston, 29th Intern. Conf. on High-Energy Physics, Vancouver, Canada 1998, hep-ph/9901352.
5. E.Leader, E.Predazzi, *An Introduction to Gauge Theories and the New Physics*,

Cambridge University Press (1996), F. Halzen, A. D. Martin, *Quarks and Leptons: An Introductory Course in Modern Particle Physiscs*, John Wiley & Sons, (1984).
6. F. J. Yndurain, *The Theory of Quark and Gluon Interactions* Springer Verlag (1992).
7. S. Bethke, QCD Euroconference 97, Montpelier, 1997 hep-ph/9710030.
8. Review of Particle Properties, The European Physical Journal, C3, 1 (1998).
9. Michael Kuhlen, hep-ph/9712505
10. V. Chekelian in 18th International Symposium on Lepton-Photon Interactions, Hamburg, July 1997.
11. W. L. van Neerven in *Physics at Hera*, F. Barreiro, L. Hervás, L. Labarga, Editors. World Scientific (1994).
12. C.G.Bollini, J.J. Giambiagi, Nuovo Cimento 12B, 20 (1972),
    G. 't Hooft, M. Veltman, Nucl. Phys. B44, 189 (1972). B409, 271 (1997).
13. H.L. Lai et al. hep-ph/9903282.
14. M. Glück, E. Reya, A. Vogt, hep-ph/9806404.
15. A.D. Martin et al. hep-ph/9803445.
16. A.D. Martin in *Physics at Hera*, F. Barreiro, L. Hervás, L. Labarga, Editors. World Scientific (1994).
17. A. T. Doyle hep-ex/9812029
18. R. Feynman in 'Photon-Hadron Interactions', Benjamin, New York (1972).
19. L. Trentadue and G. Veneziano, Phys. Lett. **B323**, 201 (1993).
20. D. de Florian and R. Sassot, Phys. Rev. **D56**, 426 (1997).
21. D. de Florian, C. A. García Canal and R. Sassot, Nuc. Phys. B470, 195 (1996).
22. S.Brodsky, P. Zerwas, Nucl. Inst. and Methods, A355, 19 (1995).
23. E. Witten, Nuc. Phys. B120, 189 (1997).
24. F. Barreiro, International Europhysics Conference on High Energy Physics, Tampere 1999.
25. C. F. von Weizäcker, Z. Phys. 88, 612 (1934), E. J. Williams, Phys. Rev. 45, 729 (1934), S. Frixione *et al.*, Phys. Lett. B319, 339 (1993).
26. R. Nisius, Photon 99, Freiburg 1999 hep-ex/9811024.
27. M.Gluck, E.Reya, M.Stratmann, Phys. Rev. D51, 3220 (1995).
28. T.Uematsu, T.F. Walsh, Phys.Lett. B101, (1981) 263, Nucl.Phys.B199, (1982) 93; M.Drees, R.Godbole, Phys. Rev. D50, 3124 (1994); G.Schuler, T.Sjöstrand, Z. Phys. C68, 607 (1995), Phys. Lett. B376, 193 (1996).
29. M.Drees, R.Godbole, Phys.Rev.D39, (1984) 169.
30. B.L.Combridge et al., Phys. Lett. B70, (1977) 234; D.W.Duke, J.Owens, Phys.Rev.D26, (1982) 1600.
31. S.Frixione, et al., Phys. Lett. B319, (1993) 339.
32. M.Klasen, G.Kramer, DESY 95-226 (1995).
33. M. Derrick et al. Eur. Phys. J. C1 109 (1998).
34. M. Glück, E. Reya, A. Vogt, Z. Phys. C67 (1995) 433.
35. H1 Collab., contribution # 549 to the 29th Intern. Conf. on High-Energy Physics, Vancouver, Canada (1998).
36. P. D. B. Collins, *An Introduction to Regge Theory & High Energy Interactions*, Cambridge University Press, Cambridge 1977
37. E. Predazzi, hep-ph/9809454.

38. A. Donnachie, P.V. Landshoff, Phys. Lett. B296, 227 (1992)
39. F.E. Low, Phys. Rev. D12, 163 (1975); S. Nussinov, Phys. Rev. Lett 34, 1286 (1975).
40. M. Diehl, hep-ph/9906518; M. Wüsthoff, A. D. Martin, hep-ph/9909362; A. Hebeker, hep-ph/9905226
41. J.D. Bjorken, Phys. Rev. D47, 101 (1993).
42. ZEUS Collab., Phys. Lett. B269, 481 (1993); H1 Collab., Nucl. Phys. B429, 477 (1994).
43. D0 Collab., Phys. Rev. Lett. 72, 2332 (1994); CDF Collab., Phys. Rev. Lett 74, 855 (1995).
44. T. Gehrman and W. J. Stirling, Z. Phys. **C70**, 89 (1996); Z. Kunszt, W. J. Stirling hep-ph/9609245 (1996).
45. D. de Florian and R. Sassot, Phys. Rev. **D58** 054003 (1998).
46. D. M. Jansen, M. Albrow, R. Brugnera, hep-ph/9905537.
47. R. Sassot, Phys. Rev. **D** (to be published), hep-ph/9908348
48. J. Collins, L. Frankfurt, M. Strickman, Phys. Lett. B307, 161 (1993); A. Berrera, D. E. Sopper, Phys. Rev. D50 4328 (1994); A. Berrera. D. E. Sopper, Phys. Rev. D53 6162 (1996).
49. L. Alvero et al., Phys. Rev. **D59** 074022 (1999).
50. H1 Collab., contribution # 533 and 571 to the 29th Intern. Conf. on High-Energy Physics, Vancouver, Canada (1998).

SESSION IN HONOR OF LEON M. LEDERMAN

Welcome Address

Rene Asomoza
Academic Secretary
CINVESTAV

Ladies and Gentlemen:

I am here in this special occasion on behalf of Prof. Adolfo Martínez Palomo, Director General of the Center for Research and Advanced Studies of the National Polytechnic Institute - Cinvestav. He welcomes you.

I am very pleased to participate in this meeting, specially because Prof. Lederman is at Cinvestav.

Most of you are familiar with his career and achievements, nevertheless, I will give you a brief biographical sketch in order to emphasize the importance of the interaction that Mexican groups have had with him.

Leon M. Lederman, internationally known specialist in high energy physics, was born in New York City, in 1922. He received a B.S. from City College in 1943. Columbia awarded him a Master degree in 1948 and a Ph. D. in 1951. He was appointed full professor there in 1958.

In 1956, working with a Columbia team at the Brookhaven cosmotron accelerator, Dr. Lederman discovered the long-lived neutral K-meson, which had been predicted from fundamental symmetry ideas.

In 1957, in conjunction with Dr. Richard L. Garwin and R. Marcel Weinrich, Dr. Lederman carried out the crucial experiment on muon decay at the Nevis synchrocyclotron. This work verified the startling prediction of the non-conservation of parity by Professors Lee and Yang, who won the 1957 Nobel Prize in Physics for this theoretical work.

Dr. Lederman's fundamental experiments on the interactions of neutrinos and high energy muons were carried out primarily as part of the Nevis Laboratory program at the Brookhaven National Laboratory. There in 1961, the group discovered the second neutrino, that, associated with a muon. This discovery was recognized by the award of the 1988 Nobel Prize in Physics to Lederman, Schwartz and Steinberger. It was crucial in establishing the doublet structure of lepton currents and the classification of fundamental particles into what became known as "The Standard Model". It also initiated the program of high energy neutrino physics which has dominated programs at the major accelerators since that time.

In 1963, he proposed the idea that eventually became the Fermi National Accelerator Laboratory in Batavia, Illinois.

Following a series of experiments at the Brookhaven 30 GeV AGS accelerator which included a search for free quarks, the first observation of antideuterons and a study of nuclear Fermi motion, Dr. Lederman's team in 1968 initiated a measurement of pairs of muons emerging from primary proton collisions. A suspicious "shoulder" in the effective mass distribution of these dimuons was clarified six years later as the J/Ψ resonance. The dimuon data stimulated the "Drell-Yan" model which has been a rich source of information on the quark structure of hadrons.

Dr. Lederman worked with a team of physicists from Columbia and Rockefeller Universities on a large experiment at the Intersecting Storage Rings at CERN, the European Center for Nuclear Research in Geneva, Switzerland. In 1972, this group established that pions were copiously produced at large transverse momentum, which is evidence for hard collisions of constituents of the protons. This initiated a large program of research in many laboratories and has become one of the most incisive tools in the probing of the substructures of nuclear particles.

In 1973, he worked with a separate team from Columbia, Fermilab and Stony Brook at Fermilab on the production of electrons and muons at high transverse momentum to study the fundamental structure of matter. The group carried out a series of experiments culminating in 1977, with the discovery of the Upsilon particles, a series of three narrow and closely spaced levels generally interpreted as evidence for a new quark called "bottom" or "beauty". This interpretation was confirmed in subsequent experiments carried out at electron-positron colliders. Dr. Lederman led a team, which carried out one of these experiments at the Cornell Electron Storage Ring facility and is the co-discoverer of the fourth Upsilon state. In 1995, the "top" quark was found at Fermilab's proton-antiproton collider, culminating a 17-year search for the quark, companion to "bottom". It is interesting to note that in the late 1990's two of the most active fields of particle physics research have to do with the properties of neutrinos and the properties of the beauty quark.

Dr. Lederman, retired in July 1989 after ten years, as the Director of Fermi National Accelerator Laboratory. That year he became the Frank L. Sulzberger Professor of Physics at the University of Chicago and is currently Pritzker Professor of Physics at the Illinois Institute of Technology. He was previously the Eugene

Higgins Professor at Columbia University. He was associated with Columbia as a student and faculty member for more than 30 years, and was Director of Nevis Laboratories in Irvington, which is the Columbia Physics Department's center for experimental research in high energy physics, from 1961 to 1979. With colleagues and students from Nevis, he has led an intensive and wide-ranging series of experiments which have provided major advances in the understanding of weak interactions. In 1998, he became Resident Scholar at the Illinois Mathematics and Science Academy, a three-year residence public school for gifted Illinois high school students, which he helped to create in 1986. He has continued to write and teach and has dedicated a great effort to improve science education in the United States.

He has published over 200 papers and has written books that are plenty of humor, metaphor and storytelling to delve into the mysteries of matter, discussing particle accelerators and high energy physics. He has received numerous awards including the Nobel Prize which was awarded to him and two collaborators for "transforming the ghostly neutrino into an active tool of research".

His interest in developing science education is not limited to the United States, some Mexican groups have been beneficiary of his interest to participate in consolidating research groups in our Country. The "Academia Mexicana de Ciencias" has recognized the impact of his activities on the Mexican community of physicists and from last Tuesday he has become a corresponding Member of the "Academia Mexicana de Ciencias".

In particular, Cinvestav has taken advantage of his willingness and help to allow our researchers and students to participate in collaboration programs with groups at Fermilab.

Prof. Lederman welcome to Mexico. We are very happy to have you here again. I hope you will enjoy your stay in this splendid city and I take advantage of this occasion to invite you to visit Cinvestav.

Leon M. Lederman and the High Energy Physics in Mexico

G. Herrera

Physics Department
CINVESTAV
Apdo. Postal 14 740, Mexico 07300, D.F.

It was a great idea to have a Session in honor to Leon Lederman during our Workhsop. Not only for his well known achievements in modern physics but for the role he has played in the start of experimental high energy physics in latinamerica.

Two days ago (on Nov. the 9th), Prof. Lederman became *Miembro Correspondiente* of the Mexican Academy of Sciences in Mexico City.

In 1981 in collaboration with Prof. Jorge Flores and Clicerio Avilés (UNAM), Leon organized the First Latinamerican Symposium on High Energy Physics and Technology. It was celebrated in Cocoyoc, Morelos not far from Mexico City. This meeting was the starting point for the creation of the first experimental groups in high energy physics in Argentina, Brazil, Colombia and Mexico. All of them take part in experimental collaborations in the most important laboratories of the world.

As director of Fermilab, Leon supported in a decisive way the first Mexican group in this area conducted by Clicerio Aviles who was at that time at the National University (UNAM) and moved later on to the Physics Institute of the University of Guanajuato.

Leon supported a different initiative started by Augusto García, Miguel A. Pérez and Arnulfo Zepeda from the Centro de Investigaci'on y de Estudios Avanzados (CINVESTAV). He accepted and gave financial support to PHD students from CINVESTAV working in their thesis at Fermilab. As a result we have now high energy physics groups in CINVESTAV-Zacatenco, CINVESTAV-Merida, Physics Institute of the University of San Luis Potosí, University of Puebla, University of Guanjuato and University of Michoacan.

Another very important initiative was the creation of an NSF-Fermilab Foundation to support Latinamerican physicists during the critical years 1984-1993 and that enable people to travel to institutions in the USA and to buy equipment and accesories. This support was very important to sustain scientific activity of many collegues (including myself) in hard times.

The proposal presented by Miguel A. Pérez to the Academy was supported by Prof. Jorge Flores (UNAM), Mauricio Fortes (IF-UNAM), Gerardo Herrera (Division of Particles and Fields of the Mexican Physical Society) and Arnulfo Zepeda (CINVESTAV) and it reflects the wish of the whole comunity of physicist in Mexico which is honored to have Leon as a member of the Mexican Academy of Sciences.

I had the opportunity to meet Prof. Lederman in Cartagena, Colombia in 1992 during a version of the Panamerican Symposium that he started years before in 1981. There he explained why he considered worthwhile investing time and effort in the latinamerican region. He said he believed that somewhere between Rio Grande and the Patagonia a children may have the cure to cancer, or a clue to solve the problems of our society. This thought shows Leon's vision, a vision that moves us, physicists in latinamerica, to keep science alive.

Gerardo Herrera
President of Division of Particles and Fields
Mexican Physical Society

Origin of experimental high-energy physics at UNAM

Jorge Flores

Centro de Ciencias Físicas, Universidad Nacional Autónoma de México, Avenida Universidad s.n., Chamilpa 62210, Cuernavaca Morelos, México.

Abstract. At the early eighties mexican physicists received a great help from Leon Lederman, at that time director of Fermi Lab, so I dedicate this note to him.

In this note, I will try to put the origin of the effort to establish experimental high-energy-physics in the National University of Mexico (UNAM) in a historical perspective.

Physics in Mexico is a very young enterprise. The Instituto de Fsica of UNAM was the first research center in which physics could be done in a professional way. And it was founded as late as 1938! This means that physics research in Mexico is only sixty years old. At the beginning, the Institute occupied only a large room at the Palacio de Mineria, the old School of Mines, which was the site of the School of Engineering. Only a few physicists, mostly engineers converted into physics, were the members of the Institute. In the next 15 years people like Sandoval Vallarta, Baos, Graef and Moshinsky, all educated abroad, came to the Institute. It was not until 1952, when the University moved to the University City, far from downtown Mexico City, that real experimental physics began, since up to that time mostly theoretical work, in gravitation and nuclear physics, had been done. A 2 MeV Van de Graaff accelerator was acquired and some good work in nuclear spectroscopy was started. This was done in collaboration with the MIT group. The Latin American School of Physics was founded by Moshinsky, Leite Lopes and Giambiaggi in 1959, and this started the collaboration within latinamerican groups, then only existent in Brazil, Argentina and Mexico.

Up to the sixties there was only the Instituto de Fisica. The Instituto Politecnico Nacional then founded its School of Physics and Mathematics, a center for advanced studies (CINVESTAV), and initial seeds of physics were laid in San Luis Potosi and Puebla. Also at UNAM, the Nuclear Science Center and the Materials Research Center started working. Solid state and elementary particles physics began to be studied and in the early seventies groups in statistical mechanics and atomic physics were established. The Nuclear Energy Commission and the Mexican Oil Institute played crucial roles at those days, supporting research in an important way.

By that time, people could already obtain a reasonable Ph. D. in physics in Mexico. The first one to obtain it was Fernando Alba in 1959! When I became director of the Institute in 1974, we had 60 researchers of which 30 had a Ph. D. By the end of the seventies, it had 90 Ph. D members and the other large center, CINVESTAV, had of the order of 25 Ph. D's in physics. In 1971 CONACYT was founded and in 1974 the Universidad Autnoma Metropolitana was established. A good physics department in Iztapalapa was created, and in 1980 it had of the order of 40 people working there. At the same time, Puebla and San Luis had small but serious groups, and another one in Sonora was already going on. Experimental physics in low-energy nuclear physics, solid state and atomic collisions had matured.

Around 1975 we realized that nuclear physics was moving into heavy-ion accelerators and that these would be difficult for us to acquire. The group in UNAM decided to do experiments in the US, mainly at Oak Ridge and Berkeley. We therefore learned how to become users of large facilities.

It is at this point that Prof. Lederman enters into the game. Around 1980 there was a small group -Cocho, Avilez, Moreno, Mondragon and Jacobs -doing work in elementary particles from the theoretical standpoint. Leon was visiting several latinamerican countries, promoting high-energy physics. In his first visit to the Instituto de Fisica we decided to organize the first Panamerican Workshop on High-Energy Physics, and he promised to have several Nobel Prize winners attending it.

The meeting took place in Cocoyoc and indeed several Nobel Prizes came as well as many physicists from other latin countries. I think this meeting, which afterwards has taken place several times more, played and important role in developing HEP in our region.

But Prof. Lederman went further. He convinced Clicerio Avilez to became an experimentalist! Clic, as we all called him, went with several young students to Brook- haven and with groups from Amherst and Columbia started working in a real HEP experiment related to strange particles. The detector they were using in Brookhaven would be eventually taken to Fermi Lab, where experiment EO76 was to be conducted. Some of the instrumentation was built at the Instituto de Fsica and several students were attracted to the plan. Even people from Argentina and Venezuela joined the enterprise through UNAM, and then CINVESTAV got also interested in this project. This will be referred to by Augusto Garca in the next talk.

Clicerio worked at the Instituto de Fsica for a few years, with frequent trips to Batavia. Then he decided to form his own group at the University of Guanajuato, where with great enthusiasm he established an Institute of Physics, devoted to experimental high-energy physics. He got great help from Leon and bought a nice computer to analize the data they were getting at Fermi Lab. Unfortunately Clic died, at his early forties, but his work remained in Guanajuato where the Institute is doing nice work, not only in the field that was originally the main subject there, but also in some other fields.

There are now several latinamerican universities doing this type of work. Curiously enough experimental HEP did not flourished at UNAM, where 300' Ph D's in physics work, mostly in experimental and applied physics, but none of them in this field. Maybe, again with the help of Leon Lederman, we should make another big effort to establish such a field in our National University.

SELEX

Antonio Morelos

Instituto de Física, Universidad Autónoma de San Luis Potosí,
Álvaro Obregón 64, Zona Centro, 78000 San Luis Potosí, México
morelos@ifisica.uaslp.mx

Abstract. A summary of the path which lead to a high energy physics group at Instituto de Física de la Universidad Autónoma de San Luis Potosí is presented. This group is the result of the initial push made by Leon Lederman at the beginning of the 80's.

By mid 80's I started as a graduate student at experiment E761-Fermilab. The first experiment where San Luis Potosí participated is SELEX, experiment dedicated to charm baryon and hyperon studies. In parallel with SELEX another professor joined San Luis Potosí, together with him the group enters into a new challenge in experiment CKM-Fermilab including Ring Imaging Cherenkov technology.

THE HISTORY: E761

Experiment E761, "An Electroweak Enigma: Hyperon Radiative Decays" [1] was my first encounter with experimental high energy physics.

I got experience on silicon micro strip and wire chamber detectors, magnet spectrometers for momentum precision measurements, data acquisition and trigger systems based on NIM and CAMAC, reconstruction and analysis packages, and performed high precision physics measurements of the Σ^+ magnetic moment and of the Σ^+ and $\bar{\Sigma}^-$ production polarization [2-4].

This experiment now represents the starting push for the particle physics group at San Luis Potosí.

THE PRESENT: SELEX

There are several aspects to talk about SELEX, recent hyperon and charm physics results, I only highlight a few of them.

Also prior to talk about physics I should mention that E781 or SELEX is the first experiment where San Luis Potosí is a formal collaborating institution. Two students have gotten their master degree thesis related to SELEX, Ricardo López Fernández working in the RICH group and analyzing the beam composition using the RICH, and Galileo Domínguez Zacarías working with the K_s^0 sample looking into π asymmetry [5,6]. Personally I worked in the smart crate controller in the CAMAC setup and trigger installation. An impact at San Luis Potosí also happen in relation to this collaboration: Jürgen Engelfried, member of SELEX and previously WA89 at CERN, accepted to work as professor at San Luis Potosí, his expertise brings more life and conforms the local group and opens more physics opportunities as I'll discuss in the section "The near future".

SELEX is a new fixed target experiment designed to enhance charm - strange baryon over meson data. The data taking lasted from summer 1996 to fall 1997. It includes a tagged hadron beam on π, Σ and proton using a TRD detector; a micro-vertex detector, and particle id using TRD, lead glass, and RICH detectors; all these detectors distributed among three magnet spectrometers [7].

SELEX was designed to be a high x_F charm baryon spectrometer, and in fact this can be appreciated by the high acceptance at $x_F > 0.5$. In the three modes, $\Lambda_c^+ \to p K^- \pi^+$, $D^0 \to K^- \pi^+$, and $D^+ \to K^- \pi^- \pi^+$, the acceptance is grater than 6 %, and identical for particle and antiparticle decays. The control on the acceptance gives the opportunity to study charm baryon production as a function of x_F with very good precision for challenging theoretical models and other experiments [8–10].

Charm baryon and meson lifetime measurements are also under control in the experiment. SELEX was designed with a trigger to enhance all charm baryon production and decay modes, as a result it has the ability to look for unseen decay modes, right now SELEX is reporting the first observation of Cabibbo suppressed $\Xi_c^+ \to p K^- \pi^+$ decay [11].

The fact of having a Σ^- beam to study charm production also provides by its own nature the tool to study properties of the hyperon itself. At present there are preliminary reports on Σ^- radius and total cross section of Σ^- beam on different target material [12–15].

THE NEAR FUTURE: CKM & INSTRUMENTATION

On April 26 1996, Fermilab invited physicists for a workshop towards the use of a 120 GeV/c proton beam on collider and fixed target mode. All mexican groups had the opportunity for joining this activity. The event really represented a great opportunity for young mexican groups since the enterprises are small in size and with a lot of opportunities to start at the zero point of the design and construction for leading a project or subproject. Of all the mexican existing groups only San Luis Potosí, so far, has taken the challenge to participate actively in one of these experiments.

CKM "Charged Kaons at the Main Injector", a proposal for a Precision Measurement of the Decay $K^+ \to \pi^+ \nu \bar{\nu}$ and Other Rare K^+ Processes at Fermilab Using the Main Injector, is one of the experiments which were born after the April 96 workshop [16].

The experiment will measure the branching ratio of the ultra-rare charged kaon decay $K^+ \to \pi^+ \nu \bar{\nu}$ by observing a large sample of those decays with small background. The physics goal is to measure the magnitude of the Cabibbo, Kobayashi, Maskawa matrix element V_{td} with a statistical precision of about 5% based upon a ~ 100 event sample with total backgrounds of less than 10 events. This decay mode is known to be theoretically clean. The only significant theoretical uncertainty in the calculation of this branching ratio is due to the charm contribution. A 10% measurement of the branching ratio will yield a 10% total uncertainty on the magnitude of V_{td}.

In this experiment IF-UASLP is in charge for testing parts and the whole design of two Ring Imaging Cherenkov Counters, RICH's [17]. Experience on this technology came from the participation of Jürgen Engelfried on the design, construction, operation, and analysis of two previous RICH's, one in WA89 and another in SELEX. Also, in SELEX, Ricardo López Fernández, graduate student from IF-UASLP, worked on and used the RICH as part of his M.Sc. thesis.

CKM marks the near future experimental enterprise we are working on at IF-UASLP. We are initiating high energy physics instrumentation at IF-UASLP, aimed right now at the RICH technology applied to the CKM experiment.

HEP: A GROUP AT IF-UASLP

Experimental high energy physics evolves with projects and this aspect is also reflected on the IF-UASLP group which has seen the pass of experiments E761, WA89, E781 and now CKM. Presently, the group is creating a high energy instrumentation laboratory towards detector research and development. The basic idea of the laboratory is to target user defined detectors at world wide experiments, right now we have the RICH design and testing for CKM. In 1999, IF-UASLP just hired Ruben Flores Mendieta to strength particle physics theory and phenomenology research. Looking backwards from the beginning of the 80's to the end of the 90's a spawn of close to 20 years has happened for initiating an experimental group at San Luis Potosí, that is a positive result from an initial kick.

ACKNOWLEDGEMENT

This work was partly financed by IF-UASLP and CONACyT.

REFERENCES

1. E. Jastrzembski et al., Fermilab Proposal 761, An Electroweak Enigma: Hyperon Radiative Decays, April 3, 1985.
2. A. Morelos et al., $p_\perp$ and x_F Dependence of the Polarization of Σ^+ Hyperons Produced by 800 GeV/c Protons, Phys. Rev. D, 52, 3777, (1995).
3. A. Morelos et al., Measurement of the Magnetic Moments of Σ^+ and $\bar{\Sigma}^-$ Hyperons, Phys. Rev. Lett., 71, 3417, (1993).
4. A. Morelos et al., Polarization of Σ^+ and $\bar{\Sigma}^-$ Hyperons Produced By 800 GeV/c Protons, Phys. Rev. Lett., 71, 2172, (1993).
5. Ricardo López Fernández, Identificación de Partículas producidas en Interacción p-N mediante el E781 RICH, M. of Sc. Thesis, IF-UASLP, January, 1997, unpublished.
6. Galileo Domínguez Zacarías, Distribución Angular del $K_s^0 \to \pi^- \pi^+$ en E781, M. of Sc. Thesis, IF-UASLP, September 21., 1998, unpublished.
7. R. Edelstein et al., A Proposal to Construct - SELEX - Segmented Large-x Baryon Spectrometer, November 8, 1987.
8. Fernanda G. Garcia, Talk: Wine & Cheese, Fermilab, Batavia, Il, August 27, 1999.
9. Fernanda Garcia, The SELEX Collaboration, First Charm Baryon Physics from SELEX, Proceedings of the 1999 Division of Particle and Fields, UCLA, Los Angeles, California USA, January 5-10, 1999, FERMILAB-Conf-99/070-E.
10. J. Russ et al., First Charm Hadroproduction Results from SELEX, Proceedings of the International Conference High energy physics, Vancouver 1998, vol. 2 1259-1262, hep-ex/9812031.
11. S.Y. Jun et al., Observation of the Cabibbo Suppressed Decay $\Xi_c^+ \to p K^- \pi^+$, FERMILAB-PUB-99-217-E, hep-ex/9907062, accepted for publication en Phys. Rev. Lett. 2000.
12. U. Dersch et al., Total Cross-Section Measurements with π^-, Σ^- and protons on Nuclei and Nucleons Around 600 GeV/C, FERMILAB-PUB-99-325-E, hep-ex/9910052, submitted to Nuclear Physics B, Oct 20, 1999.
13. M.A. Moinester et al., Inelastic Electron Pion Scattering at FNAL (SELEX), Proceedings of the 8th International Conference on the Structure of Baryons (BARYONS '98) Bonn, Sept 22-26, 1998, TAUP-2568-99, hep-ex/9903039.
14. V. Kubarovsky et al., Radiative Width or the a(2) Meson, Proceedings of the International Conference High energy physics, Vancouver 1998, vol. 2 1296-1299, hep-ex/9901014.
15. I. Eschrich et al., Hyperon Physics Results from SELEX, Proceedings of the Heavy Quarks at Fixed Target, Batavia 1998, 303-313, hep-ex/9812019.
16. R. Coleman et al., Charged Kaons at the Main Injector, A Proposal for a Precision Measurement of the Decay $K^+ \to \pi^+ \nu \bar{\nu}$ and Other Rare K^+ Processes at Fermilab Using the Main Injector, submitted to Fermi National Accelerator Laboratory, Batavia, IL, USA, April, 1998.
17. Erik Ramberg et al., CKM Research and Development Project Plan, The CKM Collaboration, submitted to Fermi National Accelerator Laboratory, Batavia, IL, USA, September, 1998.

The University of Guanajuato Institute of Physics (IFUG). Leon Lederman, The Big Boss

J. Félix, G. Moreno

Instituto de Física de la Universidad de Guanajuato, León GTO. 37150 México

Abstract. This paper is to honor Leon Lederman trajectory in experimental high energy physics and, in particular and most, his invaluable help to start experimental high energy physics in Mexico and in Latin America in general. That began around 1982. Now, in 1999, there are many active groups in Latin America that devote their best efforts to experimental high energy physics. In Mexico there are six groups.

INTRODUCTION

I have asked many times who, and when, began to think about the possibility of cultivating experimental high energy physics in Mexico, and in general, in Latin America. I have obtained many sort of answers. No one satisfactory. All different. Honest.

One version is this:

That first story tells that people from UNAM and CINVESTAV considered that idea seriously back to 1981. Those groups dedicated to theoretical high energy physics and were unsure if either to convert to experimentalists or help young people to get Ph.D in experimental high energy physics. The results was that each group followed different road. They only agreed in one thing: Start experimental high energy physics in Mexico. The group from CINVESTAV sent abroad young people to study experimental high energy physics and the group from UNAM transited to experimentalist. Boths approaches worked well.

Another version is as follows:

I asked the same question to Leon Lederman. He was visiting the State of Guanajuato invited by the Council of Science & Technology of the State of Guanajuato. That was November 26, 1998. And in the way from the airport of León Guanajuato to Celaya, I got this answer in a joke tone to Kam Biu Luk, who was visiting the IFUG, and to me:

"When I was the big boss at Fermilab, I had nothing important to do. So, I decided something: To look at Latin America to see what was happening there...".

And in the same joke tone, I continue:
Those were the moments of the big bang. When the Universe of the experimental high energy physics in Mexico, and in Latin America, was created.

THE BIG BANG

According to Lederman himself -around 1981 at Fermilab USA-, during the creation he was not thinking solely in high energy physics, but in physics in general. He was considering applied physics, solid state physics, high energy physics, experimental high energy physics, etc. However, experimental high energy physics began to be cultivated. The reason is obvious. Experimental high energy physics was not studied in those days in Mexico. The others were.

THE FIRST THREE MINUTES

Also the initial idea was to involve the Americas as a whole. This means, from Alaska to Tierra del Fuego. Inside this spirit, the first Pan American Symposium on Particle Physics and Technology was held in Cocoyoc Morelos, Mexico from January 5-7, 1982. It was co-organized and co-sponsored by Instituto de Física de la Universidad Nacional Autónoma de México and Fermi National Accelerator Laboratory, USA. At that time Leon M. Lederman was the general director of Fermilab; and, of course, he was standing behind this new movement all the time.

Many personalities in physics, from USA and whole America, participated: J.D. Bjorken, G. Cocho, J.C. Cronin, R.P. Feynman, C. García-Canal, S. Glashow, L. Lederman, B. Richten, R. Salmeron, and R.R. Wilson.

The symposium poster exhibits a composition of the Aztec Calendar. That 24 tons and three meter diameter monolit shows not the face of the warrior in its center, as originally does, but the Fermilab logo. Also its shows some symbols from experimental high energy physics like $\nu, W, P, e^-, \psi, K, X, Z, q, \alpha, N$. The designer was trying to tell us something. I interpreted this composition as to record in hard stone, and forever, the scientific relationship between Fermilab and Latin America.

THE FIRST COLLABORATION IS BORN

Inside this purpose of joining the Americans, the first Mexican collaboration in experimental high energy physics was born, the Brookhaven National Laboratory 766 experiment, under the leadership of Clicerio Avilez. This experiment has the goal of studying the production of charm in exclusive pp reactions. However it proved to be very fruitful in other physical topics like hyperon polarization, $\pi\pi$ correlations, precise mass particle determination, etc.

THE YOUNG UNIVERSE. THE FIRST INSTITUTE. THE IFUG

As a consequence of those meetings, the IFUG was born, originally at the Universidad Nacional Autónoma de México. Around 1986 was moved to the Universidad de Guanajuato at León Guanajuato. It was the first institution devoted exclusively to experimental high energy physics. Two years later, the FNAL-690 experiment, a continuation of the BNL-766 experiment was planned and run at Fermilab. The guest book of the IFUG registers the first BNL-766-FNAL-690 experiment meeting at León Guanajuato. M.M. Kreisler from the University of Massachusetts at Amherst and Bruce C. Knapp from Columbia University participated, as well as the entire crew; that was December 17, 1988.

Leon M. Lederman visited the IFUG, for the first time, in October of 1989. The year before he got the Nobel prize in physics. In the guest book of the IFUG, he wrote: "We wish this Institute great success and prompt realization of the vision and energy of its founder Clicerio Avilez. (sings Leon Lederman)".

LEON LEDERMAN AT GUANAJUATO

The second visit of Leon M. Lederman to the IFUG occurred in November 28-29, 1998. In that occasion he wrote in the guest book of the IFUG, "My second great visit to León (signs Leon Lederman)".

During that second visit he dictated the conference *"Physics and education"* at the general auditorium of the IFUG. He analyzed the education on physics in the United States of America and showed the new plans to structure again the scientific education in United States of America.

THE ADULT UNIVERSE. THE NEW IFUG

The IFUG has evolved into a new dynamic Institute. Now it includes, Mathematical Physics and gravitation, theoretical high energy physics, statistical physics, applied physics -photoacustics-, astronomy, and experimental high energy physics.

The original experiments in high energy physics -BNL-766 and FNAL-690- are still producing physics. The last two publications are: J. Félix et al, Phys. Rev. Lett. 82, 5213 (1999) and M. Sosa et al, Phys. Rev. Lett. 83, 913 (1999). The IFUG experimental group in high energy physics has moved to the hyper CP (FNAL-871 experiment) collaboration. This collaboration includes Leon M. Lederman as a distinguished member.

If those days were days of planting, these days are of harvesting and more planting. And more work. The first scout that was send out, to learn experimental high energy physics, retorned to Mexico in 1989 as a young experimentalist. Up to 1999, the IFUG has graduated three young Ph. D. experimentalists. Now there are 13 Ph. D's in experimental high energy physics working in Mexico.

ACKNOWLEDGEMENTS

We are thankful to many persons and institutions. From north: M.N. Kreisler, from the University of Massachusetts and Bruce P. Knapp from Colombia University. Special thanks to D.C. Christian and E. Gottschalk and G. Gutiérrez from Fermilab and E.P. Hartouni from the Laurence Livermore National Laboratory and to the BNL-766 and FNAL-690 crew: J. Félix, M.C. Berisso, C. Avilez, D.C. Christian, A. Gara, E.E. Gottschalk, G. Gutiérrez, E.P. Hartouni, B.C. Knapp, M.N. Kreisler, S. Lee, K. Markianos, G. Moreno, M.A. Reyes, M.H.L.S. Wang, H. Whemann, D. Wesson. And from the south thanks to: Augusto García, Matias Moreno, Octavio Novaro, Jorge Flores, Miguel A. Pérez A., Arnulfo Zepeda y Octavio Obregón.

Let me back to the north, to thank Leon M. Lederman. Thanks Leon, thanks for the big bang. Thanks for the experimental high energy physics big bang in Mexico and Latin America.

The Latin American Collaboration in DØ

A. Sánchez-Hernández

Depto. de Física, CINVESTAV
Apdo. postal 14-740, 07000 México, D.F.
asanchez@fis.cinvestav.mx

Abstract. A short description of the impact of the Latin American collaborators in DØ is given in the paper. The DØ collaboration has become an example on how Latin American countries can participate in high energy physics.

INTRODUCTION

DØ had its origins in a call for proposal for a "small, clever" experiment at the DØ interaction region at the Tevatron in about 1981. There were about 8 proposals that were considered and reviewed by the Physics Advisory Committee (PAC) over the next year and half. Several of these focused on high p_T physics topics such as W boson mass, search for new physics, high p_T jets, etc.

In June of 1983, the PAC met and felt that what was needed was something more ambitious and rejected all proposals. Then, P. Grannis formed a new collaboration to attack the high p_T physics issues. The new collaboration grew over the next few months to include about 40 physicists. In November 1983, a new proposal was submitted to the PAC, incorporating the main ideas of the detector which was built: uranium–liquid argon calorimetry, no solenoid, thick muon detection, and inner tracking/TRD. The institutions involved at that time were: University of Arizona, Brookhaven Laboratory, Brown University, Columbia University, Fermilab, Florida State University, Indiana University, Michigan State University, Northwestern University, Saclay, State University of New York, Stony Brook, and Virginia Institute of Technology.

Approval was given in two stages: Stage I was the approval to continue to develop the proposal to a technical design report and cost estimate. Stage II approval involved a review by the Department of Energy (DOE) and permitted funds to be applied. Stage I approval came in December 1983. The DOE review with costs, management, technical design, etc. was in December 1984. Stage II approval by the laboratory came not too long after that. Subsequently, DOE reviewed the project at roughly 9 month intervals. Funding in 1984 was very small due to the pressure of other projects, so DØ did not begin spending until 1985 and the real years of major funding were 1988 – 1991.

The first couple of years were spent mostly on Radiation and Design : DØ had test beam runs to study the muon system design, and several major runs to prototype and test liquid argon calorimetry. A premodule of the central tracker was put into the DØ collision region during the 1989 run for test and saw first collisions in DØ.

By the time of roll–in (in February 1992) the collaboration had grown to about 300 physicist, LAFEX/CBPF from Brazil, UNIANDES from Colombia and CINVESTAV from Mexico were already in. First collisions were seen on May 12 of 1992 and the commissioning run lasted until August of that year. After a short shutdown for the accelerator, DØ started data taking in September of that year.

Nowadays DØ has collaborators from several Latin American institutions: University of San Francisco de Quito from Ecuador, State University of Rio de Janeiro, and University of the State of São Paulo from Brazil, apart from CBPF, UNIANDES and CINVESTAV. The number of collaborators from these institutions is approximately 7.2% of the total amount of collaborators. In the following sections I will describe briefly the contributions of the different Latin American institutions to the DØ collaboration.

THE BRAZILIAN COLLABORATION

The Brazilian collaboration in DØ has its origins around 1986 when A. Santoro (from LAFEX/CBPF) leading a group of about half dozen of physics from Brazilian institutions joined DØ. However, in February of 1987 due to the economic situation in Brazil they had to stop temporarily. By the end of 1989 and beginning of 1990 LAFEX returned to DØ with a larger group which included Ph. D. students from the State University of Rio de Janeiro (UERJ). Since then four Ph. D. students have defended their thesis using DØ data. Since 1996 CBPF/LAFEX has been forming a regional collaboration which in the beginning included physicists from the Federal University of Rio de Janeiro (UFRJ) and from the Federal University of Bahia (UFBa). By 1997, another physicist from the University of Campinas (UNICAMP) joined to the CBPF group. And in 1999, three physicist from the University of the State of São Paulo (UNESP) joined DØ too .

At present the Brazilian collaboration comprised the larger Latin American group working in a single high energy physics experiment. They are around 18 workers, including engineers, software experts and physicists. This consortium of physicists and engineers come from mainly six Brazilian institutions: LAFEX/CBPF, UERJ, UFRJ, UNESP, UNICAMP and LNLS (National Synchroton Laboratory).

The contributions of the Brazilian groups include several projects in Software (including Monte Carlo Production) Hardware and Electronics for the so called Run I. For the Run II, they are contributing in the FPD (forward proton detector), the Level 3 trigger development and Monte Carlo Production. Almost all of them have done B-Physics using the Run I data, and for the Run II they are also planning to do Diffractive physics. We have to note that the contributions of the Brazilian groups have been very important for the experiment, and at the moment their major project is the design, and construction of the FPD (Roman Pots).

THE COLOMBIAN COLLABORATION

The Colombian collaboration (UNIANDES) was admitted in DØ in May of 1991. It started with Juan Pablo Negret and Bernardo Gomez among others. Later on, Bruce Hoeneisen of the University of San Francisco de Quito (USFQ), Ecuador was invited to participated as adjunct professor of UNIANDES. In 1998, USFQ was given its own place in the Collaboration.

Since 1991, UNIANDES has participated with about a dozen of members: professors, postdocs, students, and engineers. Nowadays there are ten active members in the collaboration, from which four sign papers. We have to mention that The UNIANDES group is not just a Colombian group. They had have participants from all the Andinian region: Colombia, Ecuador, Peru, Chile, and Argentina.

Among the UNIANDES contributions, we have to mention that in at least 50 (out of 4000) internal physics and technical reports, called DØNotes members of the group have participated. Since 1991, engineers from UNIANDES had collaborated in the design, construction and tests of the SVX chips, used in the silicon trackers and optical fibers. Another projects where the group has participated are: hardware triggers for Run I and Run II, data acquisition read-out for silicon trackers, studies of the radiation damaged in the silicon detectors, and calibration of the muon system detectors. Their physics contributions have mostly been in B-physics.

THE MEXICAN COLLABORATION

CINVESTAV joined DØ in February of 1992 with professors Castilla and Herrera as members. In 1994 Herrera leave the collaboration, but many students (Masters and Ph. D.) started working with the group. By 1996, the CINVESTAV group had five active collaborators in the experiment: Castilla, Hernandez, González, Magaña and Sánchez. All of them, except Castilla, as PhD students. After the students finished their thesis only Castilla, and Gonzalez remained as active members. At the moment CINVESTAV has three collaborator in the experiment.

The group with Castilla as leader has awarded: 4 master's thesis and 4 Ph D's thesis. All of them in physics analysis. The group is principal author of one PRL paper: PRL 78, 1441, (1997), and has had direct participation or their results have being included in at least a dozen of DØ papers, those related with properties of the W boson.

The group has participated in electroweak physics topics like: Search for anomalous production of Trilinear Gauge Boson Couplings, Mass of the W Boson, Width of the W Boson, p_T distribution of the Z boson, Measurement of the α_s, and measurements of Jet cross sections. In hardware, the group has been involved in the Silicon tracker group responsible of the Upgrade of the DØ detector. The group has also participated in the maintaining and shielding of the muon system. Another software projects includes the development and implementation of the Silicon detector raw data algorithms for Run II.

THE ARGENTINIAN COLLABORATION

In about 1992, two students (Daniel Elvira and Cecilia Gerber) went to work on DØ as Ph.D. students, officially from University of Buenos Aires, but under the supervision of people in Fermilab. They completed their degrees in about 1995, having returned to Argentina in 1994. In September of 1995 University of Buenos Aires joined the collaboration officially. At present, University of Buenos Aires has five active members: one professor (R. Piegaia) and four graduated students (S. Grinstein, M. Mostafa, V. Sorin, and A. Schwartzman).

Since they joined to the collaboration they have worked mainly in QCD physics topics, like: Measurement of the Inclusive jet cross sections, Measurement of jet resolutions with the cone algorithm, Measurement of the p_T spectrum of the W boson, and Measurement of event topological variables. In hardware, they are participating in: setting of the burning and chip testing stand for the silicon detector, the alignment and calibration of the superconducting magnet, and calibration and measurement of the attenuation in the optical fibers. They are also involved in other software projects like: the development and implementation of the Kalman filter technique in object oriented language for secondary vertices reconstruction, and development and implementation of the topological vertex finding algorithms for Run II.

SUMMARY

In summary about 7.2% of the DØ collaborators are from Latin American institution, and about 8.2% of the DØ Ph. D. awarded have been from any of these institutions. The contributions of these institutions have been very significant and valuable for the collaboration.

ACKNOWLEDGMENTS

I am grateful to P. Grannis (former spokesperson of the DØ Collaboration) for the detailed information provided about the origins of the DØ collaboration, as well as to A. Santoro, J.P. Negret, H. Castilla, and R. Piegaia (leaders of the their respective Institutions) for all the information provided about their groups.

LEON LEDERMAN AND 15 YEARS OF FERMILAB - CBPF COLLABORATION

J. C. Anjos[1]

Centro Brasileiro de Pesquisas Físicas, CBPF
Rua Dr. Xavier Sigaud 150, 22290-180 Rio de Janeiro Brazil

Abstract. The contribution of Leon Lederman to the creation of an experimental High Energy Physics community in Brazil is stressed and the present situation is summarized.

I am very honored to talk in this session in honor of Leon Lederman and I would like to stress his contribution to the creation of an Experimental High Energy Physics community in Brazil and more specifically at the Centro Brasileiro de Pesquisas Físicas, CBPF. The other part of my talk, with recent results on charm physics, is included elsewhere in these proceedings.

Back in 1982 Lederman, along with Clicerio Avilez, organized the 1st Pan American Symposium in High Energy Physics, held at Cocoyoc, Mexico and invited Latin American physicists to participate in experiments at Fermilab. Jayme Tiomno, one of the founding fathers of brazilian physics, was at the meeting and back in Rio de Janeiro spread the good news, inviting three of his former students, Santoro, Moacyr Souza and myself, to form a group to work in experimental high energy physics at Fermilab. The CBPF group was very enthusiastic about the possibility of working in the frontiers of experimental physics and a project was quickly submitted to CNPq, the Brazilian Research Council, asking for postdoctoral fellowships to support the stay of the group at Fermilab. We never got any answer.

In the meantime Lederman proposed holding the 2nd Pan American Symposium in Rio de Janeiro, and it subsequently took place there in 1993. During the Symposium Tiomno told him the difficulties we had to get financial support from CNPq and Lederman decided to go ahead and invited the four physicists, as Carlos Escobar from Sao Paulo had joined the group, to come as guest scientists to Fermilab for the period of two years for the purpose of establishing an experimental collaboration.

---

[1] janjos@lafex.cbpf.br

Thus at the beginning of 1984 we were at Fermilab, and after looking at the experimental fixed target program and talking with several collaboration leaders we decided to stay together forming a group and to join E691, a charm photoproduction experiment led by Mike Witherell. It was a good choice. The experiment made a breakthrough in charm physics, with the successful use of silicon microstrip detectors, and was able to collect, for the first time, a high statistic sample with more than 10,000 charm particles, making the deep exploration of the charm sector possible.

In 1986, just before coming back to Brazil, during a lunch at the Fermilab cafeteria I asked Steve Braker, responsible for the data acquisition system, what kind of computer we should buy in order to be able to do some physics analysis at Rio. He said we should buy ACP's, the parallel computing system that was being developed by the Advanced Computer Program group at Fermilab, lead by Tom Nash, one of the members of the E691 collaboration. "Main frames will soon be old technology for high energy physics", he anticipated.

Back at Rio de Janeiro in l986, the main challenge with my colleagues was to set up a laboratory of experimental high energy physics at CBPF. When we arrived we had just an empty room, no money and a lot of enthusiasm. But we knew that good computing capabilities were essential to be able to analyze the data we had collected at Fermilab. In order to buy a Microvax II, needed to run the ACP's, Leite Lopes, the CBPF Director at that time, gave us the money he had reserved for painting the CBPF building, and with the help of Jayme Tiomno, Department Head, we got money from CNPq to buy the ACP farm. Santoro, then the Head of the Laboratory, sent several engineers to participate in the ACP Project at Fermilab and also PhD students to work in the experiment E769, aimed at studying charm hadroproduction and successor of E691 at the Tagged Photon Lab. We spent two years creating the lab infrastructure. Finally in 1988 the laboratory of high energy Physics, LAFEX, was inaugurated, with the presence of the Minister of Science and Technology and with the first parallel computing farm in Brazil, the ACP-I, running with 13 microprocessor nodes.

This was the beginning of the CBPF-FERMILAB Collaboration. I could continue talking on the subject but there are time and space limitations to this presentation. It is better to see what we have accomplished in the last fifteen years and where we are now.

Since returning to São Paulo in 1986 Carlos Escobar has also created a strong group in experimental high energy physics at the University of São Paulo and more recently at the University of Campinas, working in the experiments E781, SELEX, KTEV and in the Auger observatory.

CBPF groups have participated in various experiments: E691, E769, E791, FOCUS, SELEX and DZERO at Fermilab. One important part of the collaboration is the formation of students. CBPF graduate students typically spend two years at Fermilab on scholarships provided by CNPq working on preparing the hardware for data taking runs, manning shifts during data taking, and assisting in the software effort needed for analysis of these data. They begin work on a thesis topic at this time and complete the analysis at CBPF, with supervision from their advisors. Many CBPF students had already obtained their degrees analyzing the experimental results of these experiments: six had PhDs based on E769 results, four on E791 and four on DZERO. Two had MSc degrees based on E691 results, one on E769, and three on E791. All students have led or contributed to physics analysis that gave rise to publications.

Another important part of the collaborative work is the data analysis and the publication of the experimental results. In this regard the contribution of the CBPF group has also increased over the years. The experiment E791, for example, will publish a total of 28 papers in several topics on charm physics. Although representing only 10% of the collaboration members, the CBPF group is leading the analysis of 7 research topics, that will represent about 20% of the number of publications. This was possible thanks to the very stable participation of the CBPF group in E791 for almost ten years!

In 1995, ten years after the start of the HEP program in Brazil, the groups started to build hardware locally as part of their commitments to the experiments. The group from the University of São Paulo, lead by Carlos Escobar, provided the Vee Drift Chambers for SELEX, (three stations with a total of 9 planes) a project that cost a few hundred thousand dollars. At CBPF we provided three very high rate multiwire proportional chambers for FOCUS and now the DZERO group, lead by Alberto Santoro, is building a Roman pot and a very small angle detector to study diffractive physics at the proton antiproton collider. At the AUGER Observatory the CBPF physicists and engineers are responsible for the installation of the first 40 tanks of the surface detector in Argentina.

In addition to the recognition it has received for its contribution to experiments in particle physics, the CBPF group has established its reputation on the international scene by organizing and successfully managing a number of conferences and schools, like the ICFA Instrumentation School held at CBPF in 1990 and the LISHEP International School, held in 1993, 1995 and 1998. In the year 2000 CBPF will host the workshop on "Heavy Quarks at Fixed Target" HQ2K.

Another activity of the CBPF group in the international arena is the cooperation with other Latin-American groups. During a stay at Fermilab in 1994, I commented with Gerardo Herrera from CINVESTAV, Mexico, that we should try to establish a Latin-American collaboration in order to increase the impact of our groups in the experiment E791. He accepted an invitation to come to CBPF for a one-year stay and in 1995 we started a collaboration in experimental high energy physics and also in phenomenology with the participation of Javier Magnin, postdoc from Argentina who was also at CBPF. Since 1995 with the help of CLAF, Centro Latino Americano de Física, the collaboration has increased with the stay of a number of Mexican students and postdocs and more recently with the participation of a group from the University of Montevideo, Uruguay. Several papers in phenomenology were published in topics related to charm physics and to the measurements we have made at Fermilab. Several MSc and PhD theses were also generated at CBPF and at CINVESTAV.

Lederman's initiative has opened new possibilities of collaboration in experimental high energy physics. In 1987, Carlo Rubia, then the Director of CERN, following the steps of Leon Lederman, invited Brazilian physicists to join experiments at CERN. Consequently a group of physicists and engineers from CBPF and from the Federal University of Rio de Janeiro joined the DELPHI collaboration at LEP.

Presently there are brazilian groups working in the following international collaborations:
FOCUS - CBPF
SELEX - Univ. of São Paulo, Univ. of Campinas, CBPF
KTEV - Univ. of São Paulo, Univ. of Campinas
AUGER - Univ. of São Paulo, Univ. of Campinas, CBPF
DZERO - CBPF, State Univ. of Rio de Janeiro, Federal Univ. of Rio de Janeiro
DELPHI - CBPF, State Univ. of Rio de Janeiro, Federal Univ. of Rio de Janeiro
LHC-b - Federal Univ. of Rio de Janeiro

Additionally, there are contacts to form groups to work on CMS and ATLAS at the Large Hadron Collider at CERN.

There are about 40 physicists working in experimental high energy physics with permanent positions at these universities and a lot of PhD students: A factor 10 increase in the last 15 years, and all of this thanks to Leon Lederman!

A Challenge to Join the CDF Experiment

Jacobo Konigsberg

University of Florida, Department of Physics, Gainesville, FL 32611, USA
E-mail: konigsberg@phys.ufl.edu

Abstract. The CDF experiment was established at around 1977 and has now become a major international enterprise with about five hundred physicists from about fifty institutions from eleven different countries. Unfortunately no Latin-American institutions have so far become official members of the CDF experiment and even the contribution from individual Latin- American physicists has been limited. Here I describe briefly the status of the CDF experiment, the challenges it faces and outline the possibilities for new groups to help meet these challenges and partake in the very exciting future awaiting just around the corner.

I INTRODUCTION

The CDF experiment has established itself as one of the major endeavors in which particle physics research of the highest level is performed. Two major running periods, in 1988-89 and in 1992-96 ("Run I"), have produced an enormous harvest of physics results crowned by the discovery of the Top quark in 1995. Unfortunately only a handful of physicists that come from Latin-American countries have been part of this very successful enterprise. In an informal survey I found that, throughout more than twenty years of CDF history, less than ten such physicists have worked in CDF. These physicists came from Chile, Colombia, Argentina, Mexico and Brasil and worked with groups from various American institutions, typically as post-doctoral researchers. Only one Latin-American Ph.D. student has done his dissertation on CDF and no Latin-American institution has been an official member of the CDF experiment. Fortunately this is not the case in other Fermilab experiments such as D0 and the fixed target program. Other presentations in these proceedings will attest to the accomplishments of Latin-American groups in these experiments. In here, I will proceed to describe briefly the status of the CDF experiment and to challenge the Latin-American experimental physics community to participate in what is presently one of the most exciting and important high energy physics experiments in the world.

The year 2000 is a particularly relevant time for CDF, major upgrades of the detector are being finished and the detector is slowly being put back together after more than four years since it last took data. Practically every subsystem of the detector has been upgraded with the goal of being able to take data at a much faster rate and with greater robustness, precision, and coverage. The Tevatron accelerator complex has also been upgraded in order to deliver much higher instantaneous luminosities than in previous runs. The expectation is to reach in the upcoming "Run II" a luminosity of $2 \times 10^{32} cm^2 s^{-1}$ with 36×36 proton-antiproton bunches crossing every 132 ns and producing about six proton-antiproton interactions, on average, per crossing. Being able to take reliable physics data under these conditions will allow the CDF (and D0) experiment to gather datasets between twenty and one hundred times larger than those accumulated in Run I, depending on how long the run actually is. These datasets, corresponding to a total integrated luminosity of 2 to 20 fb^{-1}, will be an enormous source of physics research projects in almost every area of particle physics, and, until the advent of the LHC experiments, CDF will be not only be at the frontier of particle physics but will define that frontier. This is a unique time for Latin-American physicists to join the experiment and be part of this fantastic adventure.

II CDF'S PROGRAM

The ongoing physics program at CDF touches most of the major questions of HEP today:

- What is the origin of EWK symmetry breaking and mass generation ?

- What is beyond the Standard Model,

- Is QCD a correct theory at the smallest distance scales?

- Are quarks composite objects, what is the origin of CP violation ?.

- Is the top quark a Standard Model object?

Run I, which has yielded over two hundred publications, has been already a great success for CDF highlighted with new discoveries, precise measurements of Standard Model (SM) parameters and by searches for new phenomena. All these have moved us a few steps closer to being able to answer some of these questions. The list below exemplifies this incredibly rich harvest of results:

- Top quark discovery, mass measurement to $\sim 3\%$, measurement of the production cross- section, of production kinematics, of the CKM matrix element V_{tb} and of the helicity of the W- boson in top decays.

- B_c discovery, B_d mixing, indication of CP violation in $B_d \to J/\psi K_s$, lifetime measurements of b hadrons, cross-section measurements for b-quark and onium production.

- W-boson properties, mass measured to $\sim 0.1\%$, width, decay asymmetry, couplings and rare decays.
- High E_t jets, dijet probes to new limits ($\sim 10^{-17}$ cm), PDF's at high x etc.
- Tests of quark, lepton compositeness tests at scales up to a few TeV.
- Searches and limits of exotic processes in SUSY, leptoquark and technicolor theories.

Run I has also thought us an enormous amount regarding detection and measurements techniques, all of which we've tried to apply in our improvements to the CDF detector and infrastructure for Run II. The strengths of the CDF detector come about in the simultaneous, reliable and highly efficient detection, identification and measurement of all the objects that make up an event. Leptons, jets, missing transverse energy, photons, charged tracks, secondary vertices, etc. all are measured with well understood calibrations, resolutions, and with low fake rates. Additionally CDF has benefited from the work of excellent physicists from very good institutions.

For Run II, we have made significant improvements in almost every area:

- The front-end electronics, trigger and data acquisition systems have all been replaced by systems with technologies that provide much higher speed, handling and storage capabilities.
- The end-cap gas calorimeters have been replaced by faster scintillating tile detectors.
- The muon systems have been enhanced in coverage, shielding and triggering capabilities.
- The tracking detectors have all been replaced by larger, more segmented and more robust systems.
- The luminosity detectors will consist of gaseous Cherenkov tubes for more precise measurements.
- All online and offline software has been modernized.

III   CDF'S FUTURE AND POSSIBILITIES FOR PARTICIPATION

Run II is currently scheduled to start in earnest in March 2001, Fermilab, under its new director, has made a commitment to support CDF and D0 running until the LHC era is well under way. This is a commitment of the order of ten or so years. The CDF has experiment has established a strong team fully engaged in Run II preparations and is encouraging the full collaboration to take responsibilities in the commissioning, operation and future planning; all needed to make the next few years of data taking and data analysis as successful as possible. In all these areas there are projects which offer very good opportunities for newcomers' contributions, for example:

- In the Online Project: Data taking, data quality and monitoring tasks.

- In the Offline Project: Data storage, data access and data analysis.

- Subsystem Operations: Long term maintenance of CDF's detectors and infrastructure.

- Physics analyses and tools: Preparation of Monte Carlo and simulation datasets, analysis tools and algorithms etc.

- Study potential detector improvements: silicon replacement, trigger, calorimetry, muons, controls, monitoring etc.

Groups outside CDF in many cases are under the impression that it is not so easy to join the experiment. However, over the last three years CDF has accepted proposals from ten new institutions, from six different countries −not including the U.S.− that have now joined the experiment officially. The current list of countries that have groups working on CDF is: Canada, U.S., England, Scotland, Spain, Russia, Switzerland, Germany, Japan, Korea and Taiwan. The process that these groups have followed in order to become members of CDF is relatively straightforward. The first step is, of course, to gather a group of interested physicists that can write a proposal to join CDF. In this proposal, areas in which the group can help CDF should be identified. This task is facilitated by a-priori experience in the experiment from individuals in this group and/or by a series of discussions with the various CDF project leaders and management. This group can propose to work on CDF as visitors for a period in which they get to know the experiment and the areas in which they can contribute or directly propose to be accepted as members right away. Acceptance is granted by a majority vote in the CDF Executive Board after the group presents their case, based of their proposal. Usually, if the group has undergone the appropriate preparations, their proposal is very well matched to a specific CDF need and their acceptance is relatively straightforward. In some cases, the Executive Board will ask to clarify some issues and/or to strengthen the proposal with tasks that are more relevant to the experiment and/or with more personal.

An interesting model that Latin-American groups could follow is through individuals first visiting CDF for one or two years and then recruiting other physicists from their home institution and forming a group with a proposal. This group can be from a single institution or a collaboration of various Latin-American institutions. These models have worked well for the Spanish, British and some U.S. groups, for example.

IV CONCLUSIONS

Fermilab's Tevatron will continue to be the energy frontier for close to another decade and will effectively help us explore deeper into the structure of matter with one or two orders of magnitude more data than ever gathered before. CDF is well poised to make whatever new discoveries these explorations may yield. It is a very exciting place to be, there is plenty of work to do, and CDF is open to enthusiastic and committed groups. It will be great if Latin American physicists and institutions could join and reap the rewards that such an experiment will certainly yield. For this to happen, a coherent effort between physicists, institutions and funding agencies needs to be achieved. It is my hope that in the next few years this can become a reality.

Mexican participation in the H1 experiment,
A bit of history, a bit of physics

J. G. Contreras*

Departamento de Física Aplicada[1], CINVESTAV–IPN, Unidad Mérida, A. P. 73 Cordemex, 97310 Mérida, Yucatán, México

Abstract. This talk has been presented during a session to honor Leon Lederman for his long time commitment to support and impulse latinamerican groups in experimental high energy physics. It portraits the experience of the mexican participation within the H1 collaboration.

INTRODUCTION

The H1 collaboration is an international group formed by approximately 400 scientist form some 40 institutions and around 15 different countries. This group designed, built and maintain the H1 detector, which is located in the hall north of the HERA ring in the german laboratory DESY in Hamburg.

HERA is the first, and up to now only, collider of electrons on protons. It was desing to study the structure of the proton through the deep inelastic scattering (DIS) of electrons (or positrons) traveling with 27.5 GeV off protons with 820 (in 1999, 920) GeV. This process is described by only two kinematic variables. HERA extended the experimentally available phase space in each of these two variables in several orders of magnitude. It became thus a favorite playground to test the fundamental ideas behind our understanding of DIS.

The H1 detector is an assembly of several subdetectors arranged concentrically around the point where the electron collides with the proton. Seen from this interaction point the produced particules meet first a silicon tracker, then several drift chambers distributed in the central and forward region of the detector[2]. Surrounding the trackers there is a liquid argon calorimeter and a lead–scintillator spaghetti type of calorimeter. All this is inside a magnetic field parallel to the beam pipe of 1.15 T. Outside of it there is an instrumented return yoke to measure muons and leaked energy. The detector is completed by a luminosity detector, several tens of

---

[1] Partial support by Conacyt–Mexico
[2] the term forward in H1 refers to the flight direction of the proton

meters backwards from the main detector, consisting on an electron and a photon coalorimetrs, and forward detectors of protons and neutrons some hundred meters after the interaction point. A detailed description of the whole detector can be found in reference [1].

A BIT OF HISTORY

The mexican effort started in individual form and has been growing little by little. Among the institutions conforming the H1 collaboration, the Dortmund group lead by Professor Dietrich Wegener has played a key role in the birth and grow of the mexican participation in H1. Thanks to the infrastructure and the support of Prof. Wegener and his group the mexican individuals have had the chance to contribute in a more profitable way to the H1 collaboration, as if they had been left alone. This is important, because it provides a model through which latinamerican or other small institutions can become partners in this kind of experiments.

In 1987 Gerardo Herrera Corral, now Profesor at Cinvestav México, went to Dortmund to work for his Ph. D. under the direction of Prof. Wegener. Gerardo was assinged to work in the Argus experiment and finished his doctorate in 1991. Upon his return, he invited Prof. Wegener to give a short course in México [2]. There, I met him and he invited me to do my Ph. D. in H1. I did it from the end of 1993 to mid 1997. After that I stayed still one year as a postdoctoral fellow. Then I returned to México, but Gerado Herrera went for a sabatical leave to stay one year, from 1998 to 1999, working for H1. In 1998 Conacyt, the mexican funding agency, granted a proyect so that Gerado and myself, along with several students could continue collaborating in H1. On the german side, H1 accepted Cinvestav as a collaborating institute through the Dortmund group. These two developments allowed the adquisition of computer equipment in Mexico devoted to H1 physics, and the possibility to travel to Germany to perform shift and other duties for the collaboration.

In 1998 Miguel Mondragon started his Ph. D. in the Dortmund group working in H1. He is currently there and it is expected to obtain his degrees in one or two years. This means that currently the mexican group has two professors, one Ph. D. student in DESY and one starting his Ph. D. in México. There is also the subjet of mexican summer students at DESY and particularly at H1. This will be treated in the next section.

As already mentioned, Prof. Wegener came to México to hold a series of lectures [2]. He is not the only member of H1 who has came to teach here. In 1996 Tancredi Carli talked about Final States in DIA at HERA [3] and in this proceedings Eckhardt Elsen [4], current spokeperson of the collaboration, lectures on the experimental challenges at HERA and Albert de Roeck does the same on the subject of structure functions [5]. Finally it is expected that a german H1 postdoc will spend one year at Cinvestav Mérdia starting in the middle of the year 2000.

MEXICAN SUMMER STUDENTS IN H1

Starting in 1993, there has been eight summer students, up to 1999, working for H1. Next year (2000) it is expected that another two will join the club. The summer students and their contributions to H1 are listed here:

1993 Fabiola Vasquez. Reweighting in the hadronic LAr calorimeter.

1994 Javier Espinoza. Noise files as a tool to understand jet rates at high rapidity.

1995 Oscar Ramirez. Reweighting in the hadronic LAr calorimeter.

1996 Miguel Mondragon. Charge collection efficiency of the LAr calorimeter.

1997 Gabriela Murguía. WWW monitoring of the LAr calorimeter.

1998 Andrea Vargas. Upgrades to the Time of Flight system of H1.

1999 Manuel Rendón. Monitoring of the charge collection efficiency of the LAr calorimeter.

1999 J. L. Gamboa. Data quality monitoring of the SpaCal calorimeter.

Note that from the 8 summer students three are women. Also note that in the last year the number of mexican summer students in H1 increased to two. It is to hope that this tendency will persist (or improve!).

Also important is to know what are these students doing today. It is quite comforting to realize that all of them are still in science. Moreover only one has left HEP and he is working in synchrotron radiation as applied to material sciences which is also a topic of interest at DESY. Their current occupations are:

- F. Vasquez, Ph. D. in Cinvestav. Working in FOCUS.

- Javier Espinoza, post Doc in Seattle. Synchrotron Radiation.

- Oscar Ramirez, Ph. D. in Heidelberg. Theory of DIS diffraction.

- Miguel Mondragon, Ph. D. in Dortmund. Working in H1.

- Gabriela Murguía, M. Sc. in UNAM. Dirac particles and magnetic fields.

- Andrea Vargas, M. Sc. in Cinvestav. Working in AUGER.

- Manuel Rendón, finishing his B. Sc.. Interested in H1.

- J. L. Gamboa, M. Sc. in Cinvestav. Interested in H1.

A BIT OF PHYSICS

The mexican contribution to H1 has been as follows. In the technical side all summer students have done their share. Also the Ph. D. students, Miguel and before myself, had to perform some technical chore for H1. In physics we have been involved in three different groups. Gerardo and Miguel have worked in charm physics.

Myself, I have worked for the jet group investigating the possibility to describe the dynamics of the proton structure using the so called BFKL equations instead of the traditionally used DGLAP equations. I also was a member of the radiative corrections group and been invoved in the analysis performed by younger members of the Dortmund Group. Today we are interested in Charm physics, QCD dynamics and the newly revived skewed parton distributions, through the measurement of DVCS process.

OUTLOOK

The near future looks promising. The effort of sending sumer students is starting to pay off. Several young students are interested in eventually joining H1. Some of them already with the summer experience, and some interested because their friends and collegues were already at DESY and came back with a lot of enthusiasm. DESY itself, through the director Prof. Wagner, has expressed their interest in the mexican participation. Prof. Wagner was in México in November 1999 to discuss possible ways to improve the mexican collaboration at DESY. H1 and Dortmund continue their support of our effort. So everything looks OK. It is up to the new students to answer thess gestures of help and support, but at the same time of challenge.

REFERENCES

1. H1 Collaboration, I. Abt et al., Nucl. Instrum. Meth. **A386** (1997) 310 and 348.
2. D. Wegener *New results from experiments at the HERA storage ring and from ARGUS*. In *Guanajuato 1992, Proceedings, Particles and fields* 228-347.
3. T. Carli, *Hadronic final state in deep inelastic scattering at HERA*. DESY-97-010, Jun 1996. 24pp. Invited talk at 16th International Conference on Physics in Collision (PIC 96), Mexico City, Mexico, 19-21 Jun 1996. In *Mexico City 1996, Physics in collision* 415-438.
4. E. Elsen, *this proceedings*.
5. A. De Roeck, *this proceedings*.

The Pierre Auger Observatory

Arnulfo Zepeda*, for the Pierre Auger Collaboration[1]

*Centro de Investigación y de Estudios Avanzados del IPN (Cinvestav)
P.O. Box 14-740
07000 México D.F., México

Abstract. Construction of the southern site of the Pierre Auger Observatory is starting now in Argentina. It is expected that in two years the construction of the northern site will begin. The two sites together will comprise an area of 6000 km^2 covered with 3200 water Cherenkov detectors and 8 fluorescence detectors with an aperture and a duty cycle large enough to collect in few years several hundreds of cosmic rays with energies around and beyond the GZK cut-off, that is with energies $\geq 5 \times 10^{19}$ eV. This sample will allow the Pierre Auger Collaboration to establish with confidence the spectrum and the direction of the incoming cosmic rays as well as to identify statistically the type of particles that compose the flux of cosmic rays in this part of the spectrum.

INTRODUCTION

There are by now several written documents where one may find concise [1] or detailed [2] information on the scope and the design of the Pierre Auger Observatory. We will therefore concentrate, after a brief description of the objectives and the structure of the Observatory, on the latest developments in its construction. In the first part we give a short account on the incorporation of Mexico into the Pierre Auger Collaboration.

---

[1] **Argentina**: CRICyT Mendoza, Fisica & Ingenieria Universidad Nacional de La Plata, IAFE Buenos Aires, IAR Villa Elisa, Instituto Balseiro-CNEA Bariloche, Tandar-CNEA Buenos Aires, UTN Mendoza and San Rafael; **Armenia**: Yerevan Physics Institute; **Australia**: University of Adelaide; **Bolivia**: University of La Paz; **Brasil**: University of Campinas, University of Sao Paulo, CBPF-Lafex Rio de Janeiro, Federal University Ric de Janeiro; **China**: IHEP Beijing; **France**: ENST Paris, LTF Observatoire de Besancon, College de France, LPNHE Universite Paris 6; **Germany**: FZK-HPE and FZK-IK Karlshruhe, Universitat Karlsruhe; **Greece**: NTU Athens; **Italy**: INFN Catania, University of Milano, University of Roma 2, University of Torino; **Japan**: ICRR Tokyo; **Mexico**: BUAP Puebla, CINVESTAV, UNAM, UMSNH Morelia; **Poland**: INP Jagiellonian University Krakow, University of Lodz; bf Russia: MEPhI Moscow; **Slovenia**: Nova Gorica Polytechnic; **UK**: University of Leeds; **USA**: University of Colorado Boulder, EFI University of Chicago, FNAL, Louisiana State University. Michigan Technological University, University of New Mexico, Northeastern University, Pennsylvania State University, University of Utah; **Vietnam**: DNRI-VAEC Dalat, HINCST-VAEC,Hanoi.

MEXICO INTO THE PIERRE AUGER COLLABORATION

Since this part of the workshop is dedicated to honor the role that Leon Lederman and Fermilab have played in the development of experimental physics in Mexico, I would like to mention that Fermilab played a central role in the origin of the involvement of Mexico in the Pierre Auger Project. Indeed, at the occasion of a visit of one of my Ph.D. students, Umberto Cotti, to Fermilab, within the program of summer stays founded by Leon and supported by Fermilab and Cinvestav (now by Fermilab and the Mexican Academy of Sciences), he became aware of an important development taking place at Fermilab, namely the continuous workshop for the design of the Pierre Auger Observatory and he enthusiastically informed me about it. Motivated by this first approach and reinforced by recommendations from UNESCO, Professor Jim Cronin invited me to the final workshop for the design of the Pierre Auger Observatory that was held at Fermilab in May of 1995. These encounters lead to the formation in May of 1996 of a group of Mexican scientists, engineers and students from 5 institutions (BUAP, Cinvestav, INAOE, UNAM, and the University of Michoacan) which joined the then recently formed Pierre Auger Collaboration. The size of the Mexican group has been fluctuating during these years and at present it is of about 20 scientists and engineers and 10 students. More people would love to join this exciting project if sufficient funds will be provided by the Mexican agencies. In the meantime the Mexican group has accepted the suggestion of Professor Marcos Moshinsky and adopted for itself the name of "Manuel Sandoval Vallarta" in honor of the great late Mexican physicist who played a key role in the development of physics in Mexico and who participated in several important discoveries in cosmic rays in the 1930s. Another important development worth mentioning has been the incorporation of Rotoplas, a Mexican industry, to the team in charge of finalizing the design of the container of the water Cherenkov detector, a key component of the Pierre Auger Observatory.

THE DESIGN AND OBJECTIVES

The Pierre Auger Observatory is designed to detect with unprecedented statistics cosmic rays of the highest energy achievable in our Universe. The interest in these phenomena stems not only from their extreme nature but mostly from their puzzling existence. Indeed, it is believed that the change of slope in the spectrum of cosmic rays around 10^{14-15} eV, the "knee", shows that the higher energy cosmic rays are of extragalactic origin. Moreover no sources of ultra high energy cosmic rays (with E> 10^{18} eV) have been identified in the neighborhood of 100 Mpc around our galaxy. Several observatories (Volcano Ranch [3], Haverah Park [4], Yakutsk [5], Fly's Eye [6], and AGASA [7]) have however detected a handful of cosmic rays with energy bigger than 10^{20} eV, well beyond the so-called GZK cut-off [8], of about

5×10^{19} eV, for the energy of cosmic rays originating at distances greater than 50 Mpc.

Motivated by this puzzle Jim Cronin and Alan Watson steered an international collaboration, to which now belong scientists and engineers from 19 countries around the world. The aim of the collaboration is to design and construct the Pierre Auger Observatory.

To improve the resolution of the observatory it has been decided that its structure be of the hybrid type, that is composed of fluorescence and surface detectors; and to have a significant statistics of registered events in the part of the spectrum to which the Pierre Auger Observatory is aimed its size has been set up to be 60 times bigger than the largest of its predecessors, which consisted of either only surface detectors (Volcano Ranch, Haverah Park, Yakutsk, AGASA) or only fluorescence detectors (Fly's Eye). The Pierre Auger Observatory will be the first large hybrid detector (recently this technology has been tested with the HiRes/MIA experiment [9]). It will be divided in two sites, one in the northern hemisphere (Utah, USA) and one in the southern one (Malargüe, Argentina) covering each an area of 3,000 km^2 with 1,600 water Cherenkov detectors (as in the Haverah Park experiment) and four fluorescence detectors which have a range of view large enough to cover together the whole site.

The array of surface detectors will operate 100% of the time while the fluorescence detectors will need, as was the case of Fly's Eye and is the case of HiRes, dark cloudless nights and will therefore have only a 10% duty cycle. With this arrangement it will be possible to collect 100 events per year in the surface detector with a primary energy $> 10^{20}$ eV and 500 per year with an energy $> 4 \times 10^{19}$ eV. The designed energy resolution with the surface array alone is 15% at E = 10^{20} eV and 30% at E = 10^{19} eV. With the hybrid mode this numbers improve to 10% and 20%, respectively. A similar improvement is obtained for the angular resolution: from 1 to 0.20 deg. at E = 10^{20} eV, and from 2 to 0.35 deg. at E = 10^{19} eV. The hybrid mode is therefore a definite advantage. This follows from the complementary role that the surface array and the fluorescence system play. Ground arrays sample the lateral density profile of muons, electrons, and photons in the shower front while atmospheric fluorescence detectors observe the evolution of the air showers, their growth and attenuation, as they move longitudinally downward.

CONSTRUCTION STRATEGY

The ground breaking ceremony of the southern site was held in Malargüe on March 17, 1999. The southern part of the Pierre Auger Observatory will be constructed in two stages. In the first stage an "Engineering Array" will be installed consisting of 40 surface detectors and one fluorescence detector together with one central control unit, communication devices, and solar power systems for each water Cherenkov detector. When the Engineering Array is completed, in 2001, a good deal of experience with each component will have been accumulated and fine tuning of the design will be possible in order to increase the efficiency and maybe lower the cost of the second stage in which the southern observatory should be completed in two more years. Immediately after finishing the southern site, the construction of the northern part should start and it is expected that in three more years the Pierre Auger Observatory will be completed.

CONCLUSIONS

The Pierre Auger Observatory is on its way to construction and soon will shed light on the puzzle of the origin of the extremely high energy cosmic rays. In particular it is possible that the apparent uniformity of the few extreme high energy events will be resolved into point sources (in a way similar to how the Chandra X-ray telecope resolved the x-ray background into mostly emission from discrete sources). While no ordinary astrophysical objects may by the source of these events, several hypothesis have been advanced about an exotic origin, notably the possibility that they may be the signal of topological defects or of the decay of relic heavy particles [10] and that they may have a common origin with the Gamma Ray Burts [11].

REFERENCES

1. http://www.auger.org/ http://www.unicamp.br/~turtelli/www4.htm\#Auger
   http://www-lpnhep.in2p3.fr/auger/presentation.html
   http://lpnhp2.in2p3.fr/auger/press_book.html
   http://lpnhp2.in2p3.fr/auger/phys_aims.html
   http://cosmicray.dtc.millard.k12.ut.us/backgrnd/pabakgrn.htm;
   Pryke C., *Proc. Workshop on Observing Giant Cosmic Ray Air Showers from $10^{20}eV$ Particles from Space, College Park, Maryland, USA, November, 1997, AIP Conf.Proc.* **433**, 312 (1998); M.Boratav, *Proc. 25th Int. Cosmic-Ray Conference ICRC, Durban, South Africa, July 28 - August 8, 1997, World Scientific, Singapore* **5**, 205 (1997).
2. Pierre Auger Collaboration, Design Report, 1997;
   http://brenta.ijs.si/auger/private/DesignReport/; Zavrtanik D., in *Proc. 2nd Latin American Symposium on High Energy Physics, San Juan, Puerto Rico, April 1998, AIP Conf. Proc.* **444**, 95-104 (1998); Mantsch P.M., *Proc. of the Latin American Work. on Particles and Fields and Phenomenology of Fundamental Interactions, Puebla, Mexico, October 29 - November 3, 1995, Puebla* 370-382 (1996); Matthews J., in *Towards the Millennium in Astrophysics Problems and Prospects*, Erice, Sicily, 1996.
3. Linsley J., *Phys. Rev. Lett.* **10**, 146 (1963).
4. Brooke, G. et al., in *Proc. 19th ICRC* **2**, 150 (1985); Lawrence M.A., Reid R.J.O., Watson A.A., *J. Phys. G.* **17**, 773 (1991).

5. Efimov, N.N. et al., in *Proc. Int. Workshop on the astrophysical aspects of the most energetic cosmic rays*, ed. Nagano M. and Takahara F., World Scientific, p.20 (1991); Glushkov A.V. et al., *Astropart. Phys.* **4**, 15 (1995).
6. Bird, D.J. et al., *Astrophys. J.* **441**, 144 (1995).
7. Yoshida, S. et al., *Astropart. Phys.* **3**, 105 (1995).
8. Greisen, K., *Phys.Rev.Letters* **16**, 748 (1966); Zatsepin G.T. and Kuz'min V.A., *JETP Lett.* **4**, 78 (1966).
9. Abu-Zayyad T. et al. in 26th *International Cosmic Ray Conference, Contributed Papers*, ed. by Kieda D., Salamon M. & Dingus B., Vol. 3, p. 260-263, University of Utah, 199; ibid, Vol. 5, p. 365-368.
10. Berezinsky V., *Ultra High Energy Cosmic Rays from Cosmological Relics* hep-ph/0001163; L. Masperi and G. Silva, *Astropart. Phys* **8**. 173 (1998).
11. Berezinsky V., B. Hnatyk, and A. Vilenky *Superconducting cosmic strings as Gamma ray Busrts engines*, astro-ph/0001213.

A BRIEF INTERVIEW WITH LEON LEDERMAN

During the VII Mexican Workshop on Particles and Fields held from November 10 to November 17, 1999 in the city of Mérida in México, Leon Lederman conceded a short interview to the local television. The interview took place on November 11, 1999. We would like to thank Sistema Tele Yucatán Canal 13, who kindly made available to the editors of these proceedings a video of the interview, from where the following transcript was taken:

TV Why and when did you decide to study science?

LL Ah! May be I was three years old, may be older. But I read a book when I was very young, written by Einstein, for children. There Einstein said: "Doing science is like solving a detective story". Some clues and somewhere you find the solution to the problem. And that was very inspiring. Then I had very good teachers in school. Excellent teachers. So more and more I was attracted to the problem of trying to the understand the world in which we live.

TV Can you tell us your experiences about research and development in science, in physics?

LL Well I dont know exactly what you want to know. I know that we try to understand the world in which we live. Thats what physics does. And while we are doing this we have to invent all kinds of instruments and we find that those instruments turn out to be extremely useful in society, in medicine, in manufacturing; and these things pay for our research. In other words they make companies which make profits, and the profits pay taxes and we only use a small part of those taxes for our research.

TV What meaning does it have for you to have received the Nobel Prize?

LL Well I was surprised because I had already received many prizes. When I got the Nobel Prize I said: "Oh good, one more prize". But, the reputation of the Nobel Prize is enormous, so that encouraged me to use the reputation of the Nobel Prize to do things in education of children.

TV Why all this effort to support education for children?

LL Ah! thats very important. I think that we live in a changing world. Today we have Internet, we have huge computers, we have television. We have so many technological changes in our lives, that it is important for every human being on this planet to understand enough science, so he can make decisions, wise decisions for his own life and wise decisions for the future of the nation. I think, today there is no choice: either you will become a primitive country or you will join the high technology world and make use of the technology for the benefit of all citizens.

TV As a scientist how do you see in Mexico the technological development?

LL It takes two things. It take human resources, people, and it takes financial resources to become a scientific and technologically advanced country. In Mexico you have fantastic human resources. What you need is a decision to give those people financial resources. It is just like if you have oil underground and you say we cant afford to dig, so then you don't use the oil. In the same sense you must find the resources to exploit the human people. Very simple.

TV Based on the science development, what is the destiny of the human beings?

LL Well, again we have a choice. We can make use of these technological opportunities, which would enable us to lift the standard of living of all people of the world. We can do this, because we have the technology. Or we can waste the resources by military expending, or by making the rich people even richer and the poor poorer. We have a choice.

TV Thank you very much. It was an honor to talk a little bit with you.

SEMINARS

Heuristic Derivation of Weinberg's Angle from Space-Time and Gauge Symmetries Unification

J. Besprosvany

Instituto de Física, Universidad Nacional Autónoma de México, Apartado Postal 20-364, México 01000, D. F., México.

Abstract. We analyze modern theories' use of the concepts of the wave function, the field, and the underlying space-time in which they live, at the local and global levels and point out at incompatibilities involved. These are resolved by proposing a description in which these three items are considered inseparable. We implement this idea by assigning group generators to Lorentz scalars in an extended Clifford algebra. In the simplest application of the theory, we associate the vector carriers of the $SU(2)_L \times U(1)$ isospin and the hypercharge symmetries with linear combinations of elements in the Clifford algebra which reproduce the required quantum number of leptons in the interactions. We show that at symmetry breaking to an electromagnetic $U(1)$, the demand of the presence of a non-axial vector, to be linked to the electromagntic field, leads to a Weinberg's angle θ_W with $sin^2(\theta_W) = .25$.

PROBLEMS IN THE CONCEPTIONS OF SPACE

The multiplication of free parameters in the standard model motivates a search for ideas for a clue on their origin. A possible source is the conceptions of space on which all modern physical theories rely for the description of their objects. Although these theories use views which are well established, the views can differ and clash, even to the point that some of these theories remain incompatible. A fundamental difference of views appears in the relation of space to the matter that moves in it. In the "monistic" view space is inseparable and indistinguishable from the matter that exists in it, and the distinction between space and matter is simply a matter of convention. In the "reductionistic" view space is of a wholly different nature from matter, if at all, and is mainly the receptacle where bodies exist. While the first view is aesthetically and philosophically more appealing as it conforms to a unified view of nature, the second is unnatural but more intuitive and has been more successful and useful, by providing a simple framework to treat phenomena, as in Newtonian mechanics in constrast to Cartesian.

Today's accepted theories of nature suscribe to both views. This information is summarized on Table 1. In general relativity (GR) both points of view are present. It is inherent to this theory that locally space-time behaves as a Minkowskian framework, in which objects fall freely and physical phenomena are the same as in flat space, independently of the particular spot in which they occur (second view). On the other hand, globally space-time is a manifold representing the gravitational field and is coupled to matter (first view). Coordinates are labels to account for particular points in the manifold but any particular choice of coordinates lacks physical significance. (This is expressed by the coordinate invariance of GR). The inseparability of space, the gravitation field and matter gives space back its status as a physical object. The understanding of space as a field strongly suggests a link to electromagnetic phenomena. This possibility was explored by Kaluza and Klein who by extending space to more dimensions implied a picture which encompasses both four-dimensional space-time and another dimension for the electromagnetic field.

Classical (CM) and quantum mechanics (QM) suscribe to the second view as the wave function, which contains all the information of the matter it describes, is defined as a field. The same occurs in quantum field theory (QFT), which describes varied number of particles by allowing for an infinite number of degrees of freedom represented by its principal element, the field. This has space-time as a parameter and satifies causality constrains from special relativity. However, doubts emerge on the physicality of space. For example, Leibnitz argued in this direction by stating that space is only a system of relations between bodies. In QM the use of configuration space as a basis is by no means compulsory; indeed, momentum and configuration space are just bases (coordinates), which means none is more relevant than the other (this leads to Heisenberg's uncertainty relations). Moreover, the theoretical and now experimental implications of quantum mechanics as emerging from the Einstein-Podolsky-Rosen paradox [1] (EPR) and Bell's inequality [2] point at an inherent non-local behaviour.

Space enters modern theories as both a meaningful physical object and as a simple descriptive device without physical meaning. The conceptions of space as a "field" as in global GR or as a "stage" or the background of events in local GR, CM, QM, and QFT belong to the first class. However, QM and QFT, by denying any physical meaning to a well defined place where objects are, point at the relevance of the pure "coordinate" conception of space, which belongs to the second class.

In searching for a theory that should overcome the incommesurable views of space as "field", "stage", and "coordinate", we prefer the view of space as a field and the idea that space is not void but is a manifestation of a "space-matter" substance. We propose to identify the wave function with space itself in the "field" meaning and keep its "coordinate" use while dropping the "stage" meaning. Hence, by suscribing to the view that space is in fact filled up with the wave function we dispose of space as "stage" at the local level too.

¿From this proposal only the wave function remains a physical entity which encompasses space and matter. In this way, we are emulating the treatment of GR by interpreting the wave function (space) as the relevant field, whose coordinates do not have a physical meaning. Depriving space of its physical meaning as "stage" implies also a possible solution to the EPR paradox and Bell's implied non-locality. This idea therefore brings nearer QM and GR and allows for a removal of an inconsistency in an entirely new framework which gives space a new meaning locally. Through it the wave function is seen as the ripples of space and matter; from here, space, and its fields become aspects of the same entity. The possibility of having space, whose representing field, the graviton, is a boson, and fermions under the same footing implies a closer connection between fermions and bosons, and in general, interaction fields on the same footing as matter fields. In addition, we may expect that in a classical limit space regains its usual meaning of "stage".

TABLE 1. Two incompatible conceptions of space in modern physics, the theories based on each conception, and their pros and cons.

Space as receptacle where bodies move	Successful and useful view ✓ Reductionistic ×
Theories: Classical mechanics Quantum mechanics and quantum field theory General relativity (locally)	EPR paradox and Bell's inequality ×
Space as integral part of physical medium	Unified view ✓ Hard to quantize ×
Theories: General relativity (globally) Kaluza-Klein	

ELECTROWEAK AND SPACETIME UNIFIED TREATMENT

Without pretending to propose a theory which would encompass these ideas it is possible to use them as an inspiration, and conceive the possibility of a closer connection between the generators of the space-time symmetries and those of gauge symmetries, but otherwise using a conventional quantum mechanical framework.

A straightforward implementation of this assumption can be obtained with the electroweak symmetries $SU(2) \times U(1)$ and their breaking to the electromagnetic $U(1)$. On one hand we shall use a formalism that permits a description of vector fields through Dirac's equation

$$\gamma_0(i\partial_\mu \gamma^\mu - M)\Psi = 0, \tag{1}$$

which can be achieved by extending its space of solutions and assuming that Ψ is a matrix instead of a spinor. On the other hand, the simplest extension of the usual 4-d Hamiltonian is obtained by generalizing it to a 6×6 matrix. This is sufficient to permit that spacetime Lorentz scalars be assigned to electroweak generators at symmetry breaking. The following terms

$$H_R = P_{+-}\frac{1}{2}(1+\gamma_5)i\gamma_0\nabla\cdot\boldsymbol{\gamma}. \tag{2a}$$

$$H_{L1} = P_{-+}\frac{1}{2}(1-\gamma_5)i\gamma_0\nabla\cdot\boldsymbol{\gamma}, \tag{2b}$$

$$H_{L1} = P_{--}\frac{1}{2}(1-\gamma_5)i\gamma_0\nabla\cdot\boldsymbol{\gamma}, \tag{2c}$$

are assumed to be 2×2 matrices that can be arranged so that the projection operators P_{+-}, P_{-+} P_{--} signify the positions in a 6×6 matrix

$$\begin{pmatrix} H_R & \\ & \begin{pmatrix} H_{L1} & \\ & H_{L2} \end{pmatrix} \end{pmatrix}, \tag{3}$$

where all other components are null. At symmetry breaking, the only relevant generators to reckon with are one component of the isospin I_3 and the hypercharge Y. Consideration is also made of the vertex vector-lepton (a more elaborate argument could include quarks). The only possible assignement giving the correct quantum numbers to fermions are of the form

$$Y \propto P_{+-} + \frac{1}{2}(P_{-+} + P_{--}), \tag{4a}$$

$$I_3 \propto P_{-+} - P_{--}, \tag{4b}$$

where I_3 lives in the part acting on left-handed leptons and the different factors in Y give the correct hypercharge of left and right-handed parts. (The remaining combination $P_{++} + P_{-+} + P_{--}$ can be interpreted as the particle number).

It is enough to use the a schematic form of solutions, defined respectively as $H_R \to \hat{u}_{R\mu}$, $H_{L1} \to \hat{u}_{L1\mu}$, $H_{L1} \to \hat{u}_{L2\mu}$, where these are assumed normalized. We then employ these to construct the normalized carriers of the third component of the isospin and the hypercharge according to their assignement to combinations of projection operators in Eq. 4. We get

$$B_\mu = \sqrt{\frac{2}{3}}[\hat{u}_{R\mu} + \frac{1}{2}(\hat{u}_{L1\mu} + \hat{u}_{L2\mu})] \tag{5a}$$

$$W_\mu^0 = \sqrt{\frac{1}{2}}(\hat{u}_{L1\mu} - \hat{u}_{L2\mu}) \tag{5b}$$

If we are to compare with fields in the standard model [3], at symmetry breaking these should form

$$A_\mu = \frac{1}{\sqrt{g^2 + g'^2}}(gB_\mu + g'W^0_\mu) \tag{6}$$

The demand that this combination transform as a vector under parity leads to the requirement that each of the right and left-handed components have the same weight. This implies setting, say, the coefficient of $\hat{u}_{L1\mu}$ to zero. This leads immediately to the condition $\frac{g'}{g} = \frac{1}{\sqrt{3}}$. As in the SM $tan(\theta_W) = \frac{g'}{g}$, we find $sin^2(\theta_W) = .25$. Other consequences [4] of this theory as a flavor quantum number, vector-fermion vertices, and coupling constants can be also derived, making the connection between gauge and space-time structure symmetries a fruitful assumption.

REFERENCES

1. Einstein A., Podolsky B., and Rosen N., *Phys. Rev.* **47**, 777 (1935).
2. Bell J. S., *Speakable and Unspeakable in Quantum Mechanics*, Cambridge: Cambridge University Press, 1987.
3. Glashow S., *Nucl. Phys.* **22**, 579 (1961); Weinberg S., *Phys. Rev. Lett.* **19**, 1264 (1967); A. Salam in *Elementary Particle Theory*, Stockholm: Almquist and Wiskell, ed. W. Svartholm, 1968.
4. Besprosvany J., Submitted to *Nucl. Phys. B* (1999).

Zero Momentum Gluons and Perturbative QCD

Marcos Rigol* and Alejandro Cabo[†]

Instituto Superior de Ciencia y Tecnologia Nuclear, Habana, Cuba
[†] *Instituto de Cibernetica, Matematica y Fisica, Habana, Cuba*

Abstract. A modified initial state for the construction of the perturbative expansion of QCD is investigated. It is formed as a coherent superposition of zero momentum gluon pairs and shows Lorentz as well as global SU(3) symmetries. It follows that the gluon and ghost propagators determined by it, coincides with the ones used in an alternative of the usual perturbation theory proposed in a previous work. Therefore, the ability of such a procedure of producing a finite gluon condensation parameter already in the first orders of perturbation theory is naturally explained. It also follows that this state satisfies the physicality condition of the BRST procedure in its Kugo and Ojima formulation. Therefore, after assuming that the adiabatic connection of the interaction does not takes out the state from the interacting physical space, the predictions of the perturbation expansion, for the physical quantities should have meaning.

INTRODUCTION

Quantum Chromodynamics (QCD) was discovered in the seventies and up to this time it is considered as the fundamental theory for the strong interactions as a consequence it has been deeply investigated [1].

In a previous work of one of the authors (A.C.) [2], a modified perturbation theory for QCD was proposed. This expansion retains the main invariance of the theory (the Lorentz and $SU(3)$ ones), and is also able to reproduce main physical predictions of the chromomagnetic field models. It seems possible to us that this procedure could produce a reasonable if not good description of the low energy physics. If it is the case, then, the low and high energy descriptions of QCD would be unified in a common perturbative framework. In particular in [2] the results had the interesting outcome of producing a non vanishing mean value for the relevant quantity $\langle G^2 \rangle$. In addition the effective potential for the condensation parameter in the first order approximation shows a minimum at non vanishing values of the that parameter. Therefore, the procedure is able to reproduce at least some central predictions of the chromomagnetic models and general QCD analysis.

The main objective of the work exposed at the workshop consists in investigating the foundations of the mentioned perturbation theory. The concrete aim is to find a state in the Fock's space of the non interacting theory being able to generate that expansion by also satisfying the physicality condition of the BRST quantization approach.

It follows that it is possible to find the sought for state and it turns out to be an exponential of a product of two gluon and ghost creation operators. That is, it can be interpreted as a coherent superposition of states with many zero momentum gluon and ghost pairs. It is also shown that the state satisfies the linear condition which defines the physical subspace in the BRST quantization for the $\alpha = 1$ value of the gauge parameter. Thus, the indefiniteness in the appropriate value of this parameter to be used which remained in the former work is resolved opening the way for the study of the predictions of the proposed expansion.

This resume of the exposed work is organized as follows: In Section 2 the ansatz for the Fock's space state which generates the wanted form of the perturbative expansion is introduced . The proof that the state satisfies the physical state condition is also given in this section. Then, in Section 3 it is argued that the proposed state can generate the wanted modification of the propagator by a proper selection of the parameters at hand.

THE ALTERNATIVE INITIAL STATE

After applying the Kugo-Ojima operatorial quantization procedure to the non-interacting limit of QCD, some indications were found about that the searched state vector obeying the physical condition in this procedure could have the general structure

$$|\phi\rangle = \exp \sum_a \left(C_1(p) A_{p,1}^{a+} A_{p,1}^{a+} + C_2(p) A_{p,2}^{a+} A_{p,2}^{a+} + C_3(p) \left(B_p^{a+} A_p^{L,a+} + i \bar{c}_p^{a+} c_p^{a+} \right) \right) | 0 \rangle, \tag{1}$$

where $p = (|\vec{p}|, \vec{p})$ is an auxiliary null 4-momentum, $|\vec{p}|$ is chosen as one of the few smallest values of the modulus of the spatial momentum of the quantized theory in a finite volume V. This value will be taken after in the limit $V \to \infty$ for recovering Lorentz invariance. From here on the sum on the color index a will be explicit. The parameters $C_i(p)$ will be fixed below from the condition that the free propagator associated to a state satisfying the BRST physical state condition, coincides with the one proposed in the previous work [2]. The solution of this problem, would then give foundation to the physical implications of the discussion in that work.

It should also be noticed that the states defined by (1) have some similarity with coherent states [3]. However, in the present case, the creation operators appear in squares. Thus, the argument of the exponential creates pairs of physical and non-physical particles. An important property of this function is that its construction in terms of pairs of creation operators determines that the mean value of an odd number of field operator vanishes. This is at variance with the standard coherent state, in that the mean values of the fields are non zero. The vanishing of the mean field is a property in common with the standard perturbative vacuum, which Lorentz invariance could be broken by any non-zero expectation value of the 4-vector of the gauge field. It should be also stressed that this state as formed by the superposition of states of pair of gluons suggests a connection with some recent works in the literature that consider the formation of gluons pairs due to color interactions.

It can be directly shown that the state (1) satisfies the BRST physical conditions

$$Q_B \mid \Phi \rangle = 0, \tag{2}$$
$$Q_C \mid \Phi \rangle = 0. \tag{3}$$

where the expression of the charges in the interaction free limit [4] are

$$Q_B = i \sum_{k,a} \left(c_k^{a+} B_k^a - B_k^{a+} c_k^a \right), \tag{4}$$

$$Q_C = i \sum_{k,a} \left(\overline{c}_k^{a+} c_k^a + c_k^{a+} \overline{c}_k^a \right). \tag{5}$$

Also, the evaluation of the norm gives the result

$$N = \langle \Phi \mid \Phi \rangle = \prod_{\sigma=1,2} \left[\sum_{m=0}^{\infty} |C_\sigma(p)|^{2m} \frac{(2m)!}{(m!)^2} \right]^8. \tag{6}$$

Note that, as should be expected, the norm is not dependent on the $C_3(p)$ parameter which defines the non-physical particle operators entering in the definition of the state.

gluon and ghost propagators

Next, the determination of the form of the main elements of perturbation theory, that is the free particle propagators was considered. It is seen that the propagators associated to the considered state has the same form as proposed in [2] under a proper selection of the parameters. Consider the generating functional of the free particle Green functions as given by

$$Z(J) \equiv \langle \widetilde{\Phi} \mid T \left(\exp \left\{ i \int d^4 x J(x) A^0(x) \right\} \right) \mid \widetilde{\Phi} \rangle. \tag{7}$$

As a consequence of the Wick theorem it can be written in the form [5]

$$Z(J) \equiv \langle \tilde{\Phi} | \exp\left\{i \int d^4 x J(x) A^{0-}(x)\right\} \exp\left\{i \int d^4 y J(y) A^{0+}(x)\right\} | \tilde{\Phi} \rangle$$
$$\times \exp\left\{i \int d^4 y \int d^4 x \theta (y_0 - x_0) J(y) J(x) \left[A^{0-}(x), A^{0+}(y)\right]\right\}. \quad (8)$$

Therefore, the sought for modification to the free propagator is completely determined by the term

$$\prod_{a=1,..,8} \langle \tilde{\Phi} | \exp\left\{i \int d^4 x J^{\mu,a}(x) A_\mu^{a-}(x)\right\} \exp\left\{i \int d^4 x J^{\mu,a}(x) A_\mu^{a+}(x)\right\} | \tilde{\Phi} \rangle \quad (9)$$

where all the color dependent operators are decoupled thanks to the commutation relations.

After computing the above expressions, by also assuming $2C_1(p) = 2C_2(p) = C_3(p)$ (which follows necessarily in order to obtain Lorentz invariance!) and making use of the properties of the basis vectors defined before, the modification to the propagator becomes

$$\exp\left\{\frac{1}{2} \int \frac{d^4 x d^4 y}{2p_0 V} J^{\mu,a}(x) J^{\nu,a}(y) g_{\mu\nu} \left[C_3^*(p) + C_3(p)(C_3^*(p)+1)^2 \frac{1}{\left(1 - |C_3(p)|^2\right)}\right]\right\}. \quad (10)$$

Then, it was necessary to perform the limit process $\vec{p} \to 0$. For its calculation it is considered that each component of the linear momentum p is related with the quantization volume by

$$p_x \sim \frac{1}{a}, \quad p_y \sim \frac{1}{b}, \quad p_z \sim \frac{1}{c}, \quad V = abc \sim \frac{1}{p^3},$$

Since $C_3(p) < 1$ then it follows

$$\lim_{p \to 0} \frac{C_3^*(p)}{4p_0 V} \sim \lim_{p \to 0} \frac{C_3^*(p) p^3}{4p_0} = 0. \quad (11)$$

For the other limit it follows

$$\lim_{p \to 0} \frac{C_3(p)(C_3^*(p)+1)^2 \frac{1}{(1-|C_3(p)|^2)}}{4p_0 V}, \quad (12)$$

Then, after fixing a dependence of the arbitrary constant C_3 of the form $|C_3(p)| \sim (1 - \kappa p^2)$, $\kappa > 0$, y $C_3(0) \neq -1$ the limit(12) becomes

$$\lim_{p \to 0} \frac{C_3(p)(C_3^*(p)+1)^2 p^3 \frac{1}{(1-(1-\kappa p^2)^2)}}{4p_0} = \frac{C}{2(2\pi)^4} \tag{13}$$

where C is an arbitrary constant determined by the also arbitrary factor κ. An analysis of its properties has been done which shows that C can take only real and positive values.

Therefore, the total modification to the propagator including all color values turns to be

$$\prod_{a=1,..,8} \langle \tilde{\Phi} \mid \exp\left\{i \int d^4x J^{\mu,a}(x) A_\mu^{a-}(x)\right\} \exp\left\{i \int d^4x J^{\mu,a}(x) A_\mu^{a+}(x)\right\} \mid \tilde{\Phi} \rangle$$
$$= \exp\left\{\sum_{a=1,..8} \int d^4x d^4y J^{\mu,a}(x) J^{\nu,a}(y) g_{\mu\nu} \frac{C}{2(2\pi)^4}\right\}. \tag{14}$$

Then, it shows that the gluon propagator has the same form proposed in [2], for the selected gauge parameter value $\alpha = 1$ (which corresponds to $\alpha = -1$ in that reference).

ACKNOWLEDGEMENTS

The authors would like to acknowledge the helpful comments and suggestions of A. Gonzalez, F. Guzman, P. Fileviez, D. Bessis, G. Japaridze, C. Handy A. Mueller, E. Weinberg and J. Lowenstein. One of the authors (A.C.M.) is indebted by general support of the Abdus Salam ICTP and its Associate-Membership Scheme during stay (August to September 1999) in which this work was prepared. The support of the Center of Theoretical Studies of Physical Systems of the Clark Atlanta University and the Christopher Reynolds Foundation allowing the visit to U.S. in which the results were commented with various colleagues is also greatly acknowledged.

REFERENCES

1. C. N. Yang and R. Mills, Phys. Rev. 96 (1954) 191.
2. A. Cabo, S. Peñaranda and R. Martinez, Modern Physics A10 (1995) 2413.
3. C. Itzykson and J. -B. Zuber, Quantum Field Theory, New York, McGraw-Hill, 1980.
4. N. Nakanishi and I. Ojima, Covariant Operator Formalism of Gauge Theories and Quantum Gravity, Singapore, Word Scientific, 1990.
5. S. Gasiorowicz, Elementary Particle Physics, New York, Jonh Wiley & Sons, 1966.

FCNC and non-standard soft-breaking terms in weak-scale Supersymmetry [1]

J. Lorenzo Diaz-Cruz [2]

Instituto de Fisica, BUAP,
Ap. Postal J-48, 72500 Puebla, Pue., Mexico

Abstract. We study the inclusion of non-standard soft-breaking terms in the minimal SUSY extension of the SM, considering it as a model of weak-scale SUSY. These terms modify the higgs-sfermion interaction and the sfermion mass matrices, which can induce new sources of flavour violation. Bounds on the new soft parameters can be obtained from current data. The results are then applied to evaluate the FCNC top quark decays $t \to c + h_i$ ($h_i = h^0, H^0, A^0$). Implications of complex soft parameters for CP-violation are also addressed.

1.- Supersymmetric (SUSY) extensions of the Standard model (SM) [1] have been extensively studied, mainly because of the possibilty to solve the hierarchy problem. The minimal SUSY SM (MSSM) [2], has been used as as a framework to search for signals of SUSY. The required breaking of SUSY is incorporated in the model through soft-breaking terms [3], which include gaugino and scalar masses, as well as trilinear interactions. General soft-breaking terms can produce large flavour changing neutral currents (FCNC) [4]. Possible solutions to this problem have been proposed within the main theoretical frameworks of SUSY-breaking [5].

The MSSM reproduces the SM agreement with data, and predicts new signatures associated with the superpartners that are expected to appear in current or future colliders [6]. However, this anaysis usually involves some simplifications about the soft-breaking parameters. For instance, one could work within a particular GUT model and incorporate some specific mechanism of SUSY breaking, then use the structure of the soft-terms to study the mass spectrum of superpartners, evaluate production cross-section and decay rates, and search for their signatures at future colliders. Although this approach makes a certain amount of sense, one could question its generality and whether the future colliders will test weak-scale SUSY or only a particular model of SUSY breaking. In order to study, in a general setting, the possible presence of SUSY in nature, we shall define the MSSM at the weak-scale by considerig the most general structure of soft-breaking terms, whose values

---

[1] Work supported by CONACYT and SNI (México).
[2] email: ldiaz@sirio.ifuap.buap.mx

will be constrained by low-energy phenomenology.

Although it is widely stated that the soft-terms included in the definition of the MSSM are the most general ones, there are extra terms that are not usually considered in the literature [7,8], which should be included in a model-independent analysis of weak-scale SUSY. In this paper we study how the inclusion of non-standard terms in the MSSM, modify the Higgs-sfermion interactions and the sfermion mass matrices, which in turn can induce new sources of flavour violation. We evaluate then the contribution of the trilinear terms to the the FCNC top quark decays $t \to c + h_i$, with h_i denoting the neutral Higgs bosons of the MSSM. We also comment on the implication of complex trilinear terms for CP-violation phenomena.

2.- The usual trilinear terms included in the MSSM correspond to interactions of the sfermions with the Higgs doublets ($H_{1,2}$), of the form

$$\mathcal{L}_3 = \epsilon_{ij}[A^d \tilde{Q}^i H_1^j \tilde{D} - A^u \tilde{Q}^i H_2^j \tilde{U} + A^l \tilde{L}^i H_1^j \tilde{E}], \tag{1}$$

where $\tilde{Q}, \tilde{L}$ represent the squark and slepton doublets, whereas the squark and slepton singlets are denoted by $\tilde{U}, \tilde{D}, \tilde{E}$. Equation (1) resembles the Yukawa Lagrangian of the MSSM, provided that the fermion fields are replaced by their scalar superpatners. However, one could write extra soft-breaking terms that resemble the most general two-Higgs doublet model, known as model III [9], by allowing each sfermion flavour to couple to both Higgs doublets, namely,

$$\mathcal{L}'_3 = \epsilon_{ij}[C^d \tilde{Q}^i H_2^{cj} \tilde{D} - C^u \tilde{Q}^i H_1^{cj} \tilde{U} + C^l \tilde{L}^i H_2^{cj} \tilde{E}], \tag{2}$$

where $H_n^c = i\tau_2 H_n^*$ ($n = 1, 2$); $A^{u,d,l}$ and $C^{u,d,l}$ denote 3×3 matrices in flavour space. These terms are indeed soft, because each of the scalar fields carries $U(1)_Y$ charges that forbidds their appearance in tadpoles graphs, which are the only diagrams that could generate quadratic divergences from these cubic interactions [8]. The resulting squared sfermion mass-matrices (6×6) can be written in terms of 3×3 blocks, as follows

$$M_{\tilde{f}}^2 = \begin{pmatrix} (M_{\tilde{f}}^2)_{LL} & (M_{\tilde{f}}^2)_{LR} \\ (M_{\tilde{f}}^2)_{LR}^\dagger & (M_{\tilde{f}}^2)_{RR} \end{pmatrix} \tag{3}$$

The mass terms $(M_{\tilde{f}}^2)_{LL,RR}$ receive contributions from the F- and D-terms, after the Higgs fields aquire v.e.v.'s $< H_{1,2}^0 > = v_{1,2}$, as well from the chiral-conserving soft-masses. On the other hand, the chirality-changing mass terms $(M_{\tilde{f}}^2)_{LR}$, which receive contributions from F-terms and from the $A-$ and $C-$trilinear interactions, are given by

$$(M_{\tilde{u}}^2)_{LR} = \mu m_u^0 \cot \beta + A^u v \sin \beta + C^u v \cos \beta, \tag{4}$$

$$(M_{\tilde{d}}^2)_{LR} = \mu m_d^0 \tan \beta + A^d v \cos \beta + C^d v \sin \beta, \tag{5}$$

$$(M_{\tilde{l}}^2)_{LR} = \mu m_l^0 \tan \beta + A^u v \cos \beta + C^l v \sin \beta, \tag{6}$$

where $m^0_{u,d,l}$ denote the (non-diagonal) fermionic mass matrices and $v^2 = v_1^2 + v_2^2$, $\tan\beta = v_2/v_1$.

The fermion and sfermion mass matrices must be diagonalized in order to get the mass eigenstates. However, since the general fermion and sfermion mass matrices are not diagonalized by the same rotations, flavour-violating interactions will appear in the MSSM [10]. In our case, since the C^f terms modify the chirality-changing (LR) sfermion mass matrices, they can represent a new source of flavour violation.

3.- To determine the phenomenological predictions of the model, we need to know the values of the parameters A^f and C^f, which requires a complete understanding of the mechanism of SUSY breaking. In supergravity/superstrings [11], these terms are associated to non-holomorphic interactions, whereas in models with horizontal symmetries [12], they will appear as higher-dimensional operators. In gauge-mediated models [13], the non-standard soft-terms will appear as higher-order loops, as the A-terms do (two-loop level). Thus, the C^q parameters appear to be small in the minimal realization of the these SUSY-breaking scheemes. However, their contribution to low-energy processes may not be negligible when compared with the A-terms, for instance when they are proportional to the light fermion masses. Thus, the corresponding $C^{q,l}$ parameters should be included in a model independent analysis of FCNC phenomena.

To discuss FCNC bounds, it is convenient to work in the so-called super-KM basis, where fermion mass matrices and fermion-sfermion gaugino vertices are diagonal; flavour violation arises from the off-diagonal components of the sfermion mass matrices, which are treated as mass-insertions in loop-graphs [14]. The FCNC bounds on M^2_{LR} are expressed in terms of dimensionless parameters:

$$(\delta^{\tilde{q}}_{LR})_{ij} = \frac{1}{m^2_{\tilde{q}}}[V^q_L (M^2_{\tilde{q}})_{LR} V^{q\dagger}_R]_{ij} \qquad (7)$$

where $V^q_{L,R}$ denote the diagonalizing matrices of the fermion masses. Bounds on the off-diagonal elements of $\delta^{\tilde{f}}_{LR}$ could be obtained, for instance, by requiring that the SUSY contribution to the $K-\bar{K}, D-\bar{D}, B-\bar{B}$ mass differences, saturates the observed values. Similarly, the diagonal elements $(\delta^{\tilde{f}}_{LR})_{ii}$ can be bounded using the SUSY correction to the fermion masses. For d-type squarks, the bounds corresponding to $m^2_{\tilde{q}} = m^2_{\tilde{g}} = 500$ GeV, are [14]:

$$(\delta^{\tilde{d}}_{LR}) \simeq \begin{pmatrix} 1.6\times 10^{-3} & 4.4\times 10^{-3} & 3.3\times 10^{-2} \\ 4.4\times 10^{-3} & 2.4\times 10^{-2} & 1.6\times 10^{-2} \\ 3.3\times 10^{-2} & 1.6\times 10^{-2} & 7.3\times 10^{-1} \end{pmatrix} \qquad (8)$$

The C-terms appear in the definition of the δ_{LR} parameter, namely:

$$(\delta^{\tilde{q}}_{LR})_{ij} = \frac{1}{m^2_{\tilde{q}}}(a_q v \bar{A}^q + b_q \mu m_q + c_q v \bar{C}^q) \qquad (9)$$

where $\bar{A}^q = V_L^q A^q V_R^{q\dagger}$, $\bar{C}^q = V_L^q C^q V_R^{q\dagger}$; $m_{\tilde{q}}^2$ denotes an average squark mass, and m_q is the quark mass matrix; a_q, c_q can be read from Eqs. (4-6). However, FCNC data constraints the off-diagonal elements of the combination $A^d \cos\beta + C^d \sin\beta$ and $A^u \sin\beta + C^u \cos\beta$, and the constraints are strong only for A^d and C^d associated with first and second families. Moreover, since the analysis of FCNC constraints is not complete for stop/scharm parameters, one can only estimate A^u and C^u to be in the range $100 - 1000$ GeV, for which the δ_{LR}^u parameters would be one or two orders of magnitude larger than those of the third-family d-type sfermions, still in agreement with present FCNC bounds.

4.- To illustrate the effects of the non-standard soft-breaking terms, we shall consider the FCNC decays of top quark $t \to c + h_i$ [15], including only the contribution arising from the FCNC Higgs-sfermion interaction, with the gluino and squarks circulating in the loop. The resulting expression for the decay width is

$$\Gamma(t \to c + h_i) = \frac{m_t}{16\pi}(1 - \frac{m_h^2}{m_t^2})(|F_L|^2 + |F_R|^2), \qquad (10)$$

where:

$$F_L = \frac{\sqrt{2}\alpha_s}{3\pi} M_{\tilde{g}} r_{h_i} C_0(m_{\tilde{t}L}, m_{\tilde{g}}, m_{\tilde{c}R}, m_t^2, m_c^2, m_h^2), \qquad (11)$$

$$F_R = \frac{\sqrt{2}\alpha_s}{3\pi} M_{\tilde{g}} r_{h_i} C_0(m_{\tilde{t}R}, m_{\tilde{g}}, m_{\tilde{c}L}, m_t^2, m_c^2, m_h^2), \qquad (12)$$

C_0 denotes the scalar Veltman-Passarino scalar function; $m_{\tilde{t}}$, $m_{\tilde{c}}$, $m_{\tilde{g}}$ correspond to the stop, scharm and gluino masses, respectively, with

$$r_{h_i} = \begin{cases} A^u \cos\alpha - C^u \sin\alpha, & \text{for } h^0, \\ A^u \sin\alpha + C^u \cos\alpha, & \text{for } H^0, \\ A^u \cos\beta + C^u \sin\beta, & \text{for } A^0. \end{cases} \qquad (13)$$

Including only the A-term, the resulting branching ration has values of order $10^{-5} - 10^{-6}$. On the other hand, if we include A_{tc}^u and C_{tc}^u terms of similar strenght ($\simeq 500$ GeV), we find that the branching ratio reaches values of order 10^{-4}. If we also

Table. 1 B.R. of top FCNC decay $t \to c + h_i$. Results are shown for $\tan\beta = 2$, $m_{\tilde{q}} = 300$ GeV, $m_{\tilde{g}} = A^u = C^u = 500$ GeV, and the numbers in paranthesis correspond to $\tan\beta = 10$.

m_A GeV	B.R.$(t \to c + h^0)$	B.R.$(t \to c + H^0)$	B.R.$(t \to c + A^0)$
100.	7.1×10^{-4} (4.8×10^{-4})	1.9×10^{-5} (1.1×10^{-5})	5.8×10^{-4} (3.8×10^{-4})
130.	7.0×10^{-4} (5.1×10^{-4})	1.2×10^{-6} (1.7×10^{-7})	3.9×10^{-4} (2.6×10^{-4})
160.	6.8×10^{-4} (3.8×10^{-4})	0 (2.5×10^{-5})	1.4×10^{-4} (9.6×10^{-5})
190.	6.6×10^{-4} (3.3×10^{-4})	0 (0)	0 (0)

include the constrictions from off-diagonal terms in $M^2_{LL,RR}$ it is possible to obtain branching ratios of order 10^{-3}, which could be tested at LHC. Some representative values of B.R. are shown in table 1.

5.- Another interesting aplication of the new soft-breaking terms is in CP-violation phenomena. In a recent paper [16], it has been proposed to use a non-minimal expression for the A-terms, in order to explain the recently observed value of ϵ'/ϵ as having a SUSY origin. Since the C-terms can also be complex, its contribution to the imaginary part of $(\delta^d_{LR})_{12}$ could enhance the amount of CP-violation due to SUSY, and would help to explain the observed effect within the MSSM.

CP-violating Higgs interactions will also receive a contribution from the C^f terms. For instance, the parameter η^l_{CP}, which measures CP-violation in the coupling of Higgs bosons with leptons [17], receives a new contributions from the C-terms, with sleptons and gauginos circulating in the loop, it is given by

$$\eta^l_{CP} = -\frac{6\alpha_{em}}{20\sqrt{2}\cos^2\theta_W y_l} Im[C^l M_1 f(M_1, m_{\tilde{l}})], \tag{14}$$

where y_l denotes the Yukawa coupling of lepton l, $m_{\tilde{l}}$, M_1 corresponds to the slepton and Bino masses, respectively; f is a function that arises from the loop integration. For SUSY masses of order 200 GeV, $\tan\beta = 10$ and $m_A = 100$ GeV, we find that η^μ_{CP} reaches values of order 0.1, which can be detected at a future muon collider [17].

6.- In conclusion, we have studied the effects of non-standard soft-breaking terms in the MSSM, and found that they modify the chirality-changing (LR) components of the squared sfermion mass matrices, which can induce new sources of flavour violation. Given present FCNC data, we can only estimate the A and C parameters. To probe their strength, we evaluate the decays $t \to c + h_i$, and find a B.R. that may be detectable at LHC. The C-terms also give the possibility to explain the newly observed CP-violation phenomena as a SUSY effect, and to measure a CP-violating higgs-lepton coupling at a future muon collider.

Acknowledgment.- Discussions with G. Kane and M.A. Perez are acknowledged. This work was supported by CONACYT and SNI (México).

REFERENCES

1. For a review see: H.P. Nilles, Phys. Rep. 110 (1984) 1; H. Haber and G.L. Kane, Phys. Rep. 117 (1985) 75.
2. H. Haber, "Introductory low-energy superstmmetry", lectures given at TASI-92, U. of Colorado, SCIPP-92/93.
3. L. Girardelo and S. Grisaru, Nucl. Phys. B194 (1986) 65; D.M. Capper, J. Phys. G3 (1977) 731.
4. H. Murayama, Prog. of Theor. Phys. Suppl. 123 (1996) 349.
5. M. Dine, A. Kagan and M. Samuel, Phys. Rev. D48 (1993) 4269.
6. G.L. Kane, hep-ph/9709318, and references therein.

7. Non-standard soft-terms are early mentioned in: K. Inoue et al., Prog. Theor. Phys. 67 (1982) 1889; see also: L. Hall and L. Randall, Phys. Rev. Lett. 65 (1990) 2939; J. Rosiek, Phys. Rev. D41 (1990) 3464.
8. I. Jack and D.R.T. Jones, hep-ph/9903365; see also: F. Borzumati et al., hep-ph/9902443.
9. D. Atwood et al., Phys. Rev. D55 (1997) 3156; J.L. Diaz Cruz and G. Lopez Castro, Phys. Rev. D51 (1995) 5263.
10. A.Masiero and L. Silvestrini, in Perspectives on Supersymmetry, ed. G. Kane, World Scientific, Singapore (1998).
11. R. Arnowitt and P. Nath, hep-ph/9708254, in Perspectives on Supersymmetry, ed. G. Kane, World Scientific, Singapore (1998).
12. Y. Nir and N. Seiberg, Phys. Lett. B309 (1993) 337; M. Leurer, Y. Nir and N. Seiberg, Nucl. Phys. B420 (1994) 468; D. Tomassini and A. Pomarol, Nucl. Phys. B466 (1996) 3.
13. For a review of gauge-mediated models, see: C. Kolda, hep-ph/970825.
14. F. Gabbiani, E. Gabrielli, A.Masiero and L. Silvestrini, Nucl. Phys. B477 (1996) 321.
15. J.M. Yang and C.S. Li, Phys. Rev. D49 (1994) 3412 .
16. A. Masiero and H. Murayama, hep-ph/9903363.
17. K.S. Babu et al., hep-ph/9804355.

Bloch–Wilson Hamiltonian and a Generalization of the Gell-Mann–Low Theorem[1]

Axel Weber[2]

Instituto de Física y Matemáticas, Universidad Michoacana de San Nicolás de Hidalgo, Edificio C-3 Cd. Universitaria, A. P. 2-82, 58040 Morelia, Michoacán, Mexico

Abstract. The effective Hamiltonian introduced many years ago by Bloch and generalized later by Wilson, appears to be the ideal starting point for Hamiltonian perturbation theory in quantum field theory. The present contribution derives the Bloch–Wilson Hamiltonian from a generalization of the Gell-Mann–Low theorem, thereby enabling a diagrammatic analysis of Hamiltonian perturbation theory in this approach.

The presently available techniques for calculations in quantum field theory reflect the dominance of scattering processes for the experimental exploration of the physics of elementary particles. The single most important technique is beyond doubt Lagrangian perturbation theory, the explicit covariance of which has historically played an important rôle in the implementation of the renormalization program. This in turn was the crucial ingredient for converting the formal expressions of Lagrangian perturbation theory into predictions for measurable quantities. On the other hand, the identification of physical states defined as eigenstates of the Hamiltonian and the Hilbert space they span, becomes a complicated task in this approach, which is exemplified by the serious problems arising in the solution of the Bethe–Salpeter equation. In short, Lagrangian perturbation theory is primarily a theory of processes as opposed to a theory of states.

This contribution is concerned with the development of a theory of states, establishing efficient techniques for Hamiltonian perturbation theory. Apart from the possibility of gaining a new perspective on the foundations of quantum field theory, this approach appears to be natural for the description of hadronic structure and of bound state phenomena in general. In a very general setting, consider the problem of solving the Schrödinger equation

$$H|\psi\rangle = E|\psi\rangle \qquad (1)$$

---

[1] This work was supported in part by Conacyt grant 3298P–E9608 and the Coordinación de la Investigación Científica of the Universidad Michoacana de San Nicolás de Hidalgo.
[2] e–mail: axel@io.ifm.umich.mx

for the state $|\psi\rangle$. The Hamiltonian is supposed to be decomposable into a "free" and an "interacting" part, $H = H_0 + H_I$, where the eigenstates of H_0 are explicitly known and span the full Hilbert space (or Fock space) $\mathcal{F}$, which we picture as a direct sum of free n–particle subspaces ($n \geq 0$). The eigenstates of H are expected to be representable as (infinite) linear combinations of the eigenstates of H_0, hence the Schrödinger equation (1) can be written in a Fock space basis, where in general an infinite number of n–particle subspaces are involved. The problem in this generality is obviously too difficult to be solved in practice.

Restricting attention momentarily to the vacuum state, the Gell-Mann–Low theorem [1] states that the free (Fock space) vacuum evolves dynamically into the physical vacuum as H_0 turns adiabatically into H. Explicit expressions can then be given for the physical vacuum state and its energy in terms of the free n–particle states and their energies in the form of a perturbative series. It is natural to ask whether it is possible to generalize the theorem to the case where the perturbative vacuum is replaced by a linear subspace Ω of $\mathcal{F}$ consisting of eigenspaces of H_0, i.e. $H_0 \Omega \subseteq \Omega$, the simplest non–trivial example being the free two–particle subspace of $\mathcal{F}$. When the interaction H_I is switched on adiabatically, one may expect that Ω evolves into the suspace of interacting physical two–particle states, where now different eigenstates of H_0 are allowed to mix during the adiabatic process. If this expectation comes true, the determination of the physical two–particle states may be reduced to a problem within the free two–particle subspace, thus dramatically reducing the number of degrees of freedom to be considered and converting the problem into a (at least numerically) solvable one.

Couched into mathematical jargon, what one is looking for is a map U_{BW} from Ω to a direct sum of eigenspaces of H, i.e. $HU_{BW}\Omega \subseteq U_{BW}\Omega$, where U_{BW} is expected to be related to the adiabatic evolution operator. One would then hope that U_{BW} induces a similarity transformation, so that the problem of diagonalizing H in $U_{BW}\Omega$ is equivalent to diagonalizing $H_{BW} := U_{BW}^{-1} H U_{BW} : \Omega \to \Omega$, which in the example above is equivalent to a relativistic two–particle Schrödinger equation. The simplest (but not unique) choice for $U_{BW}^{-1} : U_{BW}\Omega \to \Omega$ is the orthogonal projector P to Ω,[3] hence we will look for an operator U_{BW} in Ω with

$$PU_{BW} = P = \mathbf{1}_\Omega \,. \qquad (2)$$

Eq. (2) implies in turn the injectivity of U_{BW}, hence also $U_{BW}P = \mathbf{1}$ in $U_{BW}\Omega$. Together with $HU_{BW}\Omega \subseteq U_{BW}\Omega$ one then has that

$$(\mathbf{1} - U_{BW}P)HU_{BW} = 0 \text{ in } \Omega \,. \qquad (3)$$

Eqs. (2) and (3) together in fact characterize U_{BW}: (3) implies $HU_{BW}\Omega = U_{BW}(PHU_{BW}\Omega) \subseteq U_{BW}\Omega$. Consequently, $H|_{U_{BW}\Omega}$ is diagonalizable, and by (2) it is a similarity transform of H_{BW}.

---

[3] That the choice of P for the similarity transformation is not unreasonably simple is suggested by phenomenology: even in the highly non–perturbative situation of low–energy QCD the physical hadrons can be associated with a specific content of constituent quarks (and thus with an element of the free two– or three–particle subspace).

Remarkably, Eqs. (2) and (3) also determine U_{BW} uniquely, at least within the perturbative regime. To see this, rewrite (3) as

$$\begin{aligned} H_I U_{BW} - U_{BW} P H_I U_{BW} &= U_{BW} P H_0 U_{BW} - H_0 U_{BW} \\ &= U_{BW} H_0 - H_0 U_{BW} \,, \end{aligned} \quad (4)$$

where I have used $PH_0 U_{BW} = H_0 P U_{BW} = H_0$. Now consider the matrix element of (4) between $\langle u|$ and $|k\rangle$, where $|k\rangle \in \Omega$ and $|u\rangle \in \Omega^\perp$ (the orthogonal complement of Ω in $\mathcal{F}$) are eigenstates of H_0 with eigenvalues E_k and E_u, respectively,

$$\langle u| H_I U_{BW} - U_{BW} P H_I U_{BW} |k\rangle = (E_k - E_u)\langle u|U_{BW}|k\rangle \,. \quad (5)$$

It then follows that

$$\begin{aligned} U_{BW} &= P + (\mathbf{1} - P) U_{BW} P \\ &= P + \int_\Omega dk \int_{\Omega^\perp} du \, |u\rangle \langle u|U_{BW}|k\rangle \langle k| \\ &= P + \int_\Omega dk \int_{\Omega^\perp} du \, |u\rangle \frac{\langle u|H_I U_{BW} - U_{BW} P H_I U_{BW}|k\rangle}{E_k - E_u} \langle k|\,, \end{aligned} \quad (6)$$

where I have taken k and u to label the eigenstates of H_0 in Ω and $\Omega^\perp$, respectively. Eq. (6) can be solved iteratively to obtain U_{BW} as a power series in H_I. It should be emphasized, however, that the individual terms in the series are not guaranteed to give convergent expressions (let alone the series as a whole). This depends, among other things, on the choice of Ω.

Eqs. (2) and (3) have been used for the characterization of U_{BW} before, first by Bloch [2] in the context of degenerate quantum mechanical perturbation theory, and later by Wilson [3] for the formulation of a non–perturbative renormalization group in Minkowski space. In practical applications, one will calculate U_{BW} to a certain order in the iterative expansion of (6) and solve the Schrödinger equation for the corresponding Hamiltonian $H_{BW} = PHU_{BW}$. Its solution yields an approximation to the eigenvalues of $H|_{U_{BW}\Omega}$ (the eigenvalues are invariant under similarity transformations) and also to the eigenstates via $|\psi\rangle = U_{BW}|\phi\rangle$ where $|\phi\rangle$ are the eigenstates of H_{BW}. The solutions will in general also include bound states (e.g., if Ω is the free two–particle subspace), in contrast to Lagrangian perturbation theory. The reason for this difference is that although in the present formalism H_{BW} is determined perturbatively, the corresponding Schrödinger equation can be solved exactly (at least to arbitrary precision with numerical methods). This is somewhat analogous to the Bethe–Salpeter equation, but avoids the conceptual problems associated with the latter. In this context, it is worth mentioning that the normalizability of the free two–particle component $|\phi\rangle = P|\psi\rangle$ gives a natural criterium for the "boundedness" of the state $|\psi\rangle$, although the latter may not be normalizable in the Hilbert space sense.

The formulation presented so far has two important shortcomings: first, the terms in the perturbative series following from (6) are not well–defined in the case of vanishing energy denominators, and a consistent prescription is at least not obvious from (3) or (6). Second, it is not a priori clear how to translate the terms in the perturbative series into diagrams. A diagrammatic formulation, however, is expected to be at least helpful, if not imperative, for the investigation of such important properties as renormalizability and Lorentz and gauge invariance at finite orders of the expansion, as well as for practical applications of the formalism.

In search of an alternative characterization of U_{BW}, I will now return to the idea of the adiabatic evolution. Consider the adiabatic evolution operator from $t = -\infty$ to $t = 0$,

$$U_\epsilon = T \exp -i \int_{-\infty}^{0} dt\, e^{-\epsilon|t|} H_I(t)$$

$$= \sum_{n=0}^{\infty} \frac{(-i)^n}{n!} \int_{-\infty}^{0} dt_1 \cdots \int_{-\infty}^{0} dt_n\, e^{-\epsilon(|t_1|+\ldots+|t_n|)} T[H_I(t_1)\cdots H_I(t_n)], \quad (7)$$

where

$$H_I(t) = e^{iH_0 t} H_I e^{-iH_0 t} \quad (8)$$

is the usual expression in the interaction picture and T stands for the decreasing time ordering operator. Then the following theorem holds:

Generalized Gell-Mann–Low Theorem. *With the notations introduced before, suppose that the operator $U_{BW} := \lim_{\epsilon \to 0} U_\epsilon (PU_\epsilon P)^{-1}$ exists in Ω. Then it has the properties $PU_{BW} = P$ and $(1 - U_{BW} P) H U_{BW} = 0$ in Ω.*

Remarks. We thus have an explicit expression for U_{BW} in terms of the adiabatic evolution operator. Given that $PU_\epsilon P$ is always formally invertible as a power series in H_I, the implications of the theorem rest on the existence of the limit $\epsilon \to 0$ of $U_\epsilon (PU_\epsilon P)^{-1}$, which in turn depends on the choice of Ω.

Proof. The property $PU_{BW} = P$ follows directly from the definition of U_{BW}. The first part of the proof of $(1 - U_{BW} P) H U_{BW} = 0$ is identical to the original Gell-Mann–Low proof [1] and will not be reproduced here. It establishes by manipulation of the series (7) for U_ϵ that (before taking the limit $\epsilon \to 0$)

$$HU_\epsilon = U_\epsilon H_0 + i\epsilon g \frac{\partial}{\partial g} U_\epsilon, \quad (9)$$

where H_I is assumed to be proportional to some "coupling constant" g.

Now choose any $|\phi\rangle \in \Omega$. Eq. (9) implies

$$HU_\epsilon (PU_\epsilon P)^{-1}|\phi\rangle = U_\epsilon H_0 (PU_\epsilon P)^{-1}|\phi\rangle + i\epsilon \left(g \frac{\partial}{\partial g} U_\epsilon\right) (PU_\epsilon P)^{-1}|\phi\rangle. \quad (10)$$

It follows that

$$HU_\epsilon(PU_\epsilon P)^{-1}|\phi\rangle - i\epsilon g \frac{\partial}{\partial g}\left(U_\epsilon(PU_\epsilon P)^{-1}\right)|\phi\rangle$$

$$= U_\epsilon H_0(PU_\epsilon P)^{-1}|\phi\rangle + i\epsilon U_\epsilon(PU_\epsilon P)^{-1}\left(Pg\frac{\partial}{\partial g}U_\epsilon\right)(PU_\epsilon P)^{-1}|\phi\rangle \quad (11)$$

$$= U_\epsilon(PU_\epsilon P)^{-1} PHU_\epsilon(PU_\epsilon P)^{-1}|\phi\rangle, \quad (12)$$

where in going from (11) to (12) Eq. (10) has been used again, multiplied by $U_\epsilon(PU_\epsilon P)^{-1}P$ from the left, and P has been inserted to the left of H_0, which is possible due to $H_0\Omega \subseteq \Omega$. Taking the limit $\epsilon \to 0$, we have $HU_{BW}|\phi\rangle = U_{BW}PHU_{BW}|\phi\rangle$, which proves the theorem. In taking the limit, the existence of the g–derivative of U_{BW} in Ω has been assumed. Incidentally, this assumption implies that the expression $U_\epsilon(g\,\partial/\partial g)(PU_\epsilon P)^{-1}|\phi\rangle$ is in general *divergent* in the limit $\epsilon \to 0$, since $HU_\epsilon(PU_\epsilon P)^{-1}|\phi\rangle$ cannot be expected to be equal to $U_\epsilon H_0(PU_\epsilon P)^{-1}|\phi\rangle$ in this limit [1]. ∎

The theorem corroborates the expectation detailed at the beginning of this contribution. More importantly, the adiabatic formulation also has the benefit of fixing an $i\epsilon$–prescription for the energy denominators appearing in the series generated by (6). Performing the time integrations in $U_\epsilon(PU_\epsilon P)^{-1}$ yields explicitly to second order in H_I

$$U_{BW} = \int_\Omega dk\,|k\rangle\langle k| + \int_\Omega dk\int_{\Omega^\perp} du\,|u\rangle \frac{\langle u|H_I|k\rangle}{E_k - E_u + i\epsilon}\langle k|$$
$$- \int_\Omega dk\,dk' \int_{\Omega^\perp} du\,|u\rangle \frac{\langle u|H_I|k'\rangle\langle k'|H_I|k\rangle}{(E_k - E_u + 2i\epsilon)(E_{k'} - E_u + i\epsilon)}\langle k|$$
$$+ \int_\Omega dk \int_{\Omega^\perp} du\,du'\,|u\rangle \frac{\langle u|H_I|u'\rangle\langle u'|H_I|k\rangle}{(E_k - E_u + 2i\epsilon)(E_k - E_{u'} + i\epsilon)}\langle k| + \ldots, \quad (13)$$

where the limit $\epsilon \to 0$ is understood. The same expression without the $i\epsilon$–prescription follows from iterating (6).

The second important advantage of the formulation in terms of U_ϵ is the ready translation into diagrams. The diagrams associated with the perturbative expansion of H_{BW} turn out to be similar to Goldstone or time–ordered diagrams, but unlike the latter they do *not* combine into a set of Feynman diagrams. This is essentially due to the fact that the matrix elements of the effective Hamiltonian $\langle k|HU_{BW}|k'\rangle$ in general do not vanish if the energies E_k and $E_{k'}$ are different.

REFERENCES

1. Gell-Mann, M., and Low, F., *Phys. Rev.* **84**, 350 (1951); see also: Fetter, A. L., and Walecka, J. D., *Quantum Theory of Many–Particle Systems*, New York: McGraw-Hill, 1971.
2. Bloch, C., *Nucl. Phys.* **6**, 329 (1958).
3. Wilson, K. G., *Phys. Rev.* **D2**, 1438 (1970); see also: Perry, R. J., *Ann. Phys.* (N.Y.) **232**, 116 (1994).

Breaking of flavor permutational symmetry and the CKM matrix

A. Mondragón and E. Rodríguez-Jáuregui

Instituto de Física, Universidad Nacional Autónoma de México
Apdo. Postal 20-364, 01000 México, D. F., México. [1]

Abstract. The phase equivalence of the theoretical quark mixing matrix $\mathbf{V}^{th}$ derived from the breaking of the flavour permutational symmetry and the standard parametrization $\mathbf{V}^{PDG}$ advocated by the Particle Data Group is explicitly exhibited. From here, we derive exact explicit expressions for the three mixing angles θ_{12}, θ_{13}, θ_{23}, and the CP violating phase δ_{13} in terms of the quark mass ratios $(m_u/m_t, m_c/m_t, m_d/m_b, m_s/m_b)$ and the parameters $Z^{*1/2}$ and Φ^* characterizing the preferred symmetry breaking pattern. The computed values for the CP violating phase and the mixing angles are: $\delta_{13}^* = 75°$, $\sin\theta_{12}^* = 0.221$, $\sin\theta_{13}^* = 0.0034$, and $\sin\theta_{23}^* = 0.040$, which coincide almost exactly with the central values of the experimentally determined quantities.

MASS MATRICES FROM THE BREAKING OF $S_L(3) \otimes S_R(3)$

Under exact $S_L(3) \otimes S_R(3)$ symmetry, the mass spectrum for either up or down quark sectors consists of one massive particle in a singlet irreducible representation and a pair of massless particles in a doublet irreducible representation, the corresponding quark mass matrix is $\mathbf{M}_{3q}$. In order to generate masses for the first and second families, we add the terms $\mathbf{M}_{2q}$ and $\mathbf{M}_{1q}$ to $\mathbf{M}_{3q}$. The term $\mathbf{M}_{2q}$ breaks the permutational symmetry $S_L(3) \otimes S_R(3)$ down to $S_L(2) \otimes S_R(2)$ and mixes the singlet and doublet representation of $S(3)$. $\mathbf{M}_{1q}$ transforms as the mixed symmetry term in the doublet complex tensorial representation of $S_{diag}(2) \subset S_L(2) \otimes S_R(2)$. Putting the first family in a complex representation allows us to have a CP violating phase. Then, in a symmetry adapted basis, $\mathbf{M}_q$ takes the form

---

[1] Presented by E. Rodríguez-Jáuregui

$$M_q = m_{3q} \left[\begin{pmatrix} 0 & A_q e^{-i\phi_q} & 0 \\ A_q e^{i\phi_q} & 0 & 0 \\ 0 & 0 & 0 \end{pmatrix} + \begin{pmatrix} 0 & 0 & 0 \\ 0 & -\triangle_q + \delta_q & B_q \\ 0 & B_q & \triangle_q - \delta_q \end{pmatrix} \right]$$
$$+ m_{3q} \begin{pmatrix} 0 & 0 & 0 \\ 0 & 0 & 0 \\ 0 & 0 & 1 - \triangle_q \end{pmatrix} \quad (1)$$

The entries in the mass matrix may be readily expressed in terms of the mass ratios $\tilde{m}_{1q} = m_{1q}/m_{3q}$ and $\tilde{m}_{2q} = m_{2q}/m_{3q}$: $A_q^2 = \tilde{m}_{1q}\tilde{m}_{2q}(1-\delta_q)^{-1}$, $\triangle_q = \tilde{m}_{2q} - \tilde{m}_{1q}$, $B_q = \delta_q((1 - \tilde{m}_{1q} + \tilde{m}_{2q} - \delta_q) - \tilde{m}_{1q}\tilde{m}_{2q}(1-\delta_q)^{-1})$. If each possible symmetry breaking pattern (SBP) is now characterized by the ratio $Z_q^{1/2} = B_q/(-\triangle_q + \delta_q)$, the small parameter δ_q is obtained as the solution of the cubic equation

$$\delta_q \left[(1 + \tilde{m}_{2q} - \tilde{m}_{1q} - \delta_q)(1-\delta_q) - \tilde{m}_{1q}\tilde{m}_{2q} \right] - Z_q(-\tilde{m}_{2q} + \tilde{m}_{1q} + \delta_q)^2 = 0 \quad (2)$$

which vanishes when Z_q vanishes. In the symmetry adapted basis, the second term $\mathbf{M_{2q}}$ in the right hand side of Eq. (1), is decomposed as a linear combination of two linearly independent numerical matrices, $\mathbf{M_{2A}}$ and $\mathbf{M_{2S}}$, this matrices, are of the same form as $\mathbf{M_{2q}}$ with mixing parameters $Z_A = -\sqrt{8}$ and $Z_S = 1/\sqrt{8}$ respectively. There is a corresponding decomposition of the mixing parameter $Z_q^{1/2}$,

$$Z_q^{1/2} = N_{Aq} Z_A^{1/2} + N_{Sq} Z_S^{1/2} \quad \text{with} \quad 1 = N_{Aq} + N_{Sq}, \quad (3)$$

in this way a unique linear combination of $Z_A^{1/2}$ and $Z_S^{1/2}$ is associated to the SBP. The pair of numbers (N_A, N_S) are a convenient mathematical label of the SBP. The parameter $Z_q^{1/2} = M_{2q23}/M_{2q22}$ is a measure of the amount of mixing of singlet and doublet irreducible representations of $S_L(3) \otimes S_R(3)$. It will be assumed that, the up and down mass matrices are generated following the same SBP: $Z_u^{1/2} = Z_d^{1/2} = Z^{*1/2}$. In a previous paper [1], we found that the $S_L(3) \otimes S_R(3)$ flavour symmetry is broken down to $S_L(2) \otimes S_R(2)$ according to a mixed SBP characterized by $Z^{*1/2} = 1/2 \left(Z_S^{1/2} - Z_A^{1/2} \right) = \sqrt{81/32}$.

A THE MIXING MATRIX

The Hermitian mass matrix $\mathbf{M}_q$ may be written in terms of a real matrix $\bar{\mathbf{M}}_q$ and a diagonal matrix of phases $\mathbf{P}_q$ as $\mathbf{P}_q \bar{\mathbf{M}}_q \mathbf{P}_q^\dagger$. Then, the mixing matrix $\mathbf{V}$ is given by

$$\mathbf{V} = \mathbf{O}_u^T \mathbf{P}^{u-d} \mathbf{O}_d \quad (4)$$

where $\mathbf{P}^{u-d} = diag[1, e^{i(\phi_u - \phi_d)}, e^{i(\phi_u - \phi_d)}]$ is the diagonal matrix of the relative phases and $\mathbf{O}_q$ is the orthogonal matrix that diagonalizes $\bar{\mathbf{M}}_q$

$$\mathbf{O}_q = \begin{pmatrix} (\tilde{m}_{2q}f_1/D_1)^{1/2} & -(\tilde{m}_{1q}f_2/D_2)^{1/2} & (\tilde{m}_{1q}\tilde{m}_{2q}f_3/D_3)^{1/2} \\ ((1-\delta_q)\tilde{m}_{1q}f_1/D_1)^{1/2} & ((1-\delta_q)\tilde{m}_{2q}f_2/D_2)^{1/2} & ((1-\delta_q)f_3/D_3)^{1/2} \\ -(\tilde{m}_{1q}f_2f_3/D_1)^{1/2} & -(\tilde{m}_{2q}f_1f_3/D_2)^{1/2} & (f_1f_2/D_3)^{1/2} \end{pmatrix} \quad (5)$$

where $f_1 = 1-\tilde{m}_1-\delta_q^*$, $f_2 = 1+\tilde{m}_2-\delta_q^*$, $f_3 = \delta_q^*$, $D_1 = (1-\delta_q^*)(1-\tilde{m}_1)(\tilde{m}_{2q}+\tilde{m}_{1q})$, $D_2 = (1-\delta_q^*)(1+\tilde{m}_{2q})(\tilde{m}_{2q}+\tilde{m}_{1q})$ and $D_3 = (1-\delta_q^*)(1+\tilde{m}_{2q})(1-\tilde{m}_{1q})$.

From Eqs. (4) and (5), we derived closed, explicit expressions for all entries in the matrix $\mathbf{V}$ written in terms of four mass ratios $(\tilde{m}_u, \tilde{m}_c, \tilde{m}_d, \tilde{m}_s)$ and two free real parameters $\Phi = \phi_u - \phi_d$ and $Z^{1/2}$ [1]. The CP violating phase Φ measures the mismatch in the $S_L(2) \otimes S_R(2)$ symmetry breaking in the u- and d-sectors. We made a χ^2 fit of the exact expressions for the absolute values of the the entries in the mixing matrix $|V^{th}|$ and the Jarlskog invariant J to the experimentally determined values of $|V^{exp}|$ and J^{exp}. We took the values of the running quark masses evaluated at the scale of m_t from H. Fritzsch [2], and Fusaoka and Koide [3], we left the mass ratios $\tilde{m}_c$, $\tilde{m}_d$ and $\tilde{m}_s$ fixed at their central values $\tilde{m}_c = 0.0044$, $\tilde{m}_d = 0.0015$ and $\tilde{m}_s = 0.034$ but we took the value of $\tilde{m}_u = 0.000032$ close to its upper bound. A detailed account of the computation may be found in Mondragón and Rodríguez-Jáuregui [1]. Therefore, the theoretical expressions for the entries in the mixing matrix $\mathbf{V}$ are functions of the four mass ratios $(\tilde{m}_u, \tilde{m}_c, \tilde{m}_d, \tilde{m}_s)$ with $Z^* = \sqrt{81/32}$ and the CP violating phase $\Phi = 90°$. The quark mixing matrix V^{th} computed from the theoretical expresions is

$$V^{th} = \begin{pmatrix} 0.9753e^{i1°} & 0.221e^{i158°} & 0.0034e^{i84°} \\ 0.220e^{i112°} & 0.9745e^{i89°} & 0.040e^{i90°} \\ 0.0085e^{i270°} & 0.039e^{i270°} & 0.9992e^{i90°} \end{pmatrix} \quad (6)$$

B Phase equivalence of $\mathbf{V}^{th}$ and $\mathbf{V}^{PDG}$

The standard parametrization of the mixing matrix recomended by the Particle Data Group [5] is written in terms of three mixing angles $\theta_{12}, \theta_{23}, \theta_{13}$ and one CP violating phase δ_{13},

$$\mathbf{V}^{PDG} = \begin{pmatrix} c_{12}s_{13} & s_{12}c_{13} & s_{13}e^{-i\delta_{13}} \\ -s_{12}c_{23} - c_{12}s_{23}s_{13}e^{i\delta_{13}} & c_{12}c_{23} - s_{12}s_{23}s_{13}e^{i\delta_{13}} & s_{23}c_{13} \\ s_{12}s_{23} - c_{12}c_{23}s_{13}e^{i\delta_{13}} & -c_{12}s_{23} - s_{12}c_{23}s_{13}e^{i\delta_{13}} & c_{23}c_{13} \end{pmatrix} \quad (7)$$

where $c_{ij} = \cos\theta_{ij}$ and $s_{ij} = \sin\theta_{ij}$. The range of values of the experimentally determined moduli in $|V_{ij}^{exp}|$, as given by Caso et al [3], corresponds to 90% confidence limits on the range of values of the mixing angles of: $0.217 \leq s_{12} \leq 0.222$, $0.036 \leq s_{23} \leq 0.042$, $0.0018 \leq s_{13} \leq 0.0044$. The standard parametrization $\mathbf{V}^{PDG}$ was introduced without taking the possible functional relations between the quark masses and the flavour mixing parameters into account. In contrast, these functional relations are explicitly exhibited in the theoretical expressions, V_{ij}^{th}, derived in the previous sections. Furthermore, we have seen that, when the best values of the parameters $Z^{1/2}$ and Φ are used, the mixing matrix $\mathbf{V}^{th}$ reproduces the central values of all experimentally determined quantities, that is, the moduli $|V_{ij}^{exp.}|$, the Jarlskog invariant $J^{exp.}$ and the three inner angles, α, β and γ, of the unitarity triangle [1]. Since the two parametrizations reproduce the same set of experimental data equally well, we are justified in writing

$$|V_{ij}^{th}| = |V_{ij}^{PDG}| = |V_{ij}^{exp}|. \tag{8}$$

We cannot simply equate $\mathbf{V}^{th}$ and $\mathbf{V}^{PDG}$ because the arguments of corresponding matrix elements in the two parametrizations are not equal: $arg(V_{ij}^{th}) \neq arg(V_{ij}^{PDG})$. This difference is of no physical consequence, it reflects the freedom in choosing the unobservable phases of the quark fields in the mass representation. In the mass basis, the quark charged currents take the form

$$J_c^\mu = \frac{g}{\sqrt{2}} \bar{q}_{Li}^u \gamma^\mu V_{ij}^{th} q_{Lj}^d. \tag{9}$$

A redefinition of the phases of the quark fields which leaves J_c^μ invariant, will change the argument of V_{ij}^{th} but leave the moduli $|V_{ij}^{th}|$ invariant,

$$V_{ij}^{th} \to \tilde{V}_{ij}^{th} = e^{-i\chi_i^u} V_{ij}^{th} e^{i\chi_j^d}. \tag{10}$$

The phases χ_i^u and χ_j^d ocurring in Eq. (10) will be determined from the requirement that corresponding entries in $\tilde{\mathbf{V}}^{th.}$ and $\mathbf{V}^{PDG}$ be equal,

$$|V_{ij}^{th}| e^{i(w_{ij}^{th} - (\chi_i^u - \chi_j^d))} = |V_{ij}^{PDG}| e^{iw_{ij}^{PDG}}, \tag{11}$$

in this expression w_{ij}^{th} and w_{ij}^{PDG} are the arguments of V_{ij}^{th} and V_{ij}^{PDG} respectively. Since the moduli $|V_{ij}^{th}|$ and $|V_{ij}^{PDG}|$ are equal, the arguments of the entries in the two parametrizations are related by the set of nine equations

$$\chi_i^u - \chi_j^d = w_{ij}^{th} - w_{ij}^{PDG}. \tag{12}$$

The set of Eqs. (12) relate the differences of the unobservable quark field phases to the differences of the arguments of corresponding entries in $\mathbf{V}^{th}$ and $\mathbf{V}^{PDG}$. Using an elimination procedure for all possible combinations $\left(\chi_i^{(u)} - \chi_j^{(d)}\right) - \left(\chi_i^{(u)} - \chi_{j'}^{(d)}\right)$ we derive a set of nine equations, only four of which are linearly independent. Since, in $\mathbf{V}^{PDG}$ there are five entries with non-vanishing arguments, namely, $w_{13}^{PDG} = -\delta_{13}, w_{21}^{PDG}, w_{22}^{PDG}, w_{31}^{PDG}$ and w_{32}^{PDG}, we require still one more equation relating the arguments of the entries of the two parametrizations. This is obtained from the phase relations between the determinants of the two matrices, $\mathbf{V}^{th}$ and $\mathbf{V}^{PDG}$. From Eqs. (10) and (11), it follows that

$$\det \mathbf{V}^{th} = \det \left[\mathbf{X}_u^\dagger \mathbf{V}^{PDG} \mathbf{X}_d \right], \qquad (13)$$

in this expression $\mathbf{X}_u$ and $\mathbf{X}_d$ are the diagonal unitary matrices of phases ocurring in Eq. (10). The quark field phases themselves are determined only up to a common additive constant. Since the quark field phases are unobservable, without loss of generality, we may fix one of them, and solve for the others. In this way, if we set $\chi_1^d = 0$, we get the diagonal matrices of phases required to compute the phase transformed $\tilde{\mathbf{V}}^{th}$ are

$$\mathbf{X}_u = diag[e^{iw_{11}^{th}}, e^{i(-w_{12}^{th} - w_{33}^{th} + 2\Phi^*)}, e^{i(-w_{23}^{th} - w_{12}^{th} + 2\Phi^*)}] \qquad (14)$$

and

$$\mathbf{X}_d = diag[1, e^{i(w_{11}^{th} - w_{12}^{th})}, e^{i(w_{12}^{th} - w_{23}^{th} - w_{33}^{th} + 2\Phi^*)}]. \qquad (15)$$

Hence, with the help of Eqs. (14)-(15), we verify that $\mathbf{X}_u^\dagger \mathbf{V}^{th} \mathbf{X}_d = \mathbf{V}^{PDG}$, is satisfied as an identity, provided that $|V_{ij}^{th}| = |V_{ij}^{PDG}|$. The computed values for $\sin\theta_{12}^* = 0.221, \sin\theta_{23}^* = 0.040, \sin\theta_{13}^* = 0.0034$ and the CP violating phase $\delta_{13}^* = 75°$. For the three inner angles α, β and γ of the unitarity triangle, we get, $\alpha = 83°, \beta = 221°, \gamma = 75°$ in good agreement with current data on CP violation in the $K^o - \bar{K}^o$ mixing system [5], [6] and oscillations in the B_s°-$\bar{B}_s^\circ$ system [5], [7]. A detailed account of this work may be found in [8].

ACKNOWLEDGMENT(S)

This work was partially supported by DGAPA-UNAM under contract No. PAPIIT-IN125298 and by CONACYT (México) under contract 3909P-E9607.

REFERENCES

1. A. Mondragón, E. Rodríguez-Jáuregui *Phys. Rev.* **D 59**, 093009, (1999). see also A. Mondragón and E. Rodríguez-Jáuregui, *Rev. Mex. Fis.* **44(S1)**, 33 (1998), hep-ph/9804267
2. H. Fritzsch, *Mass hierarchies, Hidden Symmetry and Maximal CP-violation*", hep-ph/9807551 See also H. Fritzsch, *The symmetry and the Problem of Mass Generation. Proceedings of the XXI International Colloquium on Group Theoretical Methods in Physics (Group 21), Goslar, Germany*, (1996), edited by H.-D. Doebner, W. Scherer, and C. Schutte (World Scientific, Singapore, 1997), Vol. II, p. 543.
3. H. Fusaoka and Y. Koide *Phys. Rev.* **D 57**, 3986 (1998).
4. C. Jarlskog, *Phys. Rev. Lett.* **55**, 1039 (1985).
5. Particle Data Group, C. Caso et al., *Eur. Phys. J.* **C3**, 1 (1998).
6. S. Mele *Phys. Rev.* **D 59**, 113011, (1999).
7. A. Ali and D. London, *Eur. Phys. J* **C9**, 687-703, (1999).
8. A. Mondragón, and E. Rodríguez-Jáuregui, "*The CP violating phase δ_{13} and the quark mixing angles θ_{13}, θ_{23} and θ_{12} from flavour permutational symmetry breaking*", hep-ph/9906429, (1999).

The Possibility of Discovering New Boson in e^-e^-, $\mu^-\mu^-$, $e^-\mu^-$ Colliders [1]

J. C. Montero, V. Pleitez and M. C. Rodriguez

Instituto de Física Teórica
Universidade Estadual Paulista
Rua Pamplona, 145
01405-900– São Paulo, SP
Brazil

Abstract. Several left-right asymmetries in Møller (electron-electron), muon-muon and electron-muon scattering are considered in the context of the electroweak standard model and in a model with $SU(3)_C \otimes SU(3)_L \otimes U(1)_Y$ gauge symmetry at tree level in collider experiments. We show that these asymmetries are very sensitive to a doubly charged vector bilepton in the case of ee and $\mu\mu$ colliders and to an extra Z' neutral vector boson contribution in $e^-\mu^-$ collider.

In collider experiments the cross sections are small ($\sim 10^{-3}$ nb) but the left-right asymmetries are large, on the other hand we have opposite situation on the case of fixed target experiments, those two facts were showed on [1]. In that paper together with [2,3] were first showed the relevance of these asymmetries as a tool in searching new vectors bosons, in those works were show that a doubly charged vector is more sensitive to electron-electron or muon-muon scatering while a neutral vector is sensitive in electron-muon scattering and that those asymmetries are insensitive to the scalars. On those papers one problem is that the energies used on the calculations are below the engergies available on the futures linear collider or same in the first muon collider.

This work is a continuation of the papers [2,3], because now we are considerating the energy of those accelerators, and worried in show the difference between the diferents asymetries calculated.

This work is organized as following. First we show the lagrangian that we are using for the differents vector bosons. Then we show our results and some comments about them.

In the Electroweak Standard Model (ESM) at tree level, the relevant part of the lagrangian of interactions between the leptons and the bosons is

---

[1] Presented by M. C. Rodriguez at VII Taller de Partículas y Campos, Mérida, Yucatan, México, November 10–17, 1999.

$$\mathcal{L}_F = -\sum_i \frac{g\, m_i}{2 M_W} \bar{\psi}_i \psi_i H^0 - e \sum_i q_i \bar{\psi}_i \gamma^\mu \psi_i A_\mu$$
$$- \frac{g}{2\cos\theta_W} \sum_i \bar{\psi}_i \gamma^\mu (g_V^i - g_A^i \gamma^5) \psi_i Z_\mu, \tag{1}$$

where $\theta_W \equiv \tan^{-1}(g'/g)$ is the weak mixing angle, $e = g \sin\theta_W$ is the positron electric charge and

$$g^2 = \frac{8 G_F M_W^2}{\sqrt{2}},$$
$$g_V^i \equiv t_{3L}(i) - 2 q_i \sin^2\theta_W,$$
$$g_A^i \equiv t_{3L}(i),$$

the $t_{3L}(i)$ is the weak isospin of the fermion and q_i is the charge of the fermion.

Together with the ESM we have also considered a model with $SU(3)_C \otimes SU(3)_L \otimes U(1)_N$ gauge symmetry [4]. In this model appear a Doubly Charged Bilepton, here denoted by U^{--}. In this model its interaction with the leptons are given by the following lagrangian

$$\mathcal{L}^U = \frac{g}{\sqrt{2}} \bar{l}_L^c E_R^{lT} E_L^l l_L U_\mu^{++} + H.c, \tag{2}$$

where the mixing matrix is $\mathcal{K} = E_R^{lT} E_L^\nu$ and in this work we make the following consideration $\mathcal{K}_{ee} = \mathcal{K}_{\mu\mu} = \mathcal{K}_{\tau\tau} \approx 1$ [2]. In this model we have an extra Z' and its interactions with the charged leptons have this form

$$\mathcal{L}_{NC}^{Z'} = -\frac{g}{2\cos\theta_W}[\bar{l}_{aL}\gamma^\mu L_l l_{aL} + \bar{l}_{aR}\gamma^\mu R_l l_{aR}] Z'_\mu, \tag{3}$$

where the couplings are

$$L_l = -\sqrt{1 - 4\sin^2\theta_W}/\sqrt{3},$$
$$R_l = 2 L_l.$$

Calculating the assymetries defined in the reference [1] for the case of electron-muon scattering, considerating $E_\mu = 1$ TeV and the scattering angle given by $\theta = 0.5$ rad, for more detail see [5], we got the following results

$$A_{RL}^{CO,ESM}(e\mu \to e\mu) = -0.19,$$
$$A_{R;RL}^{CO,ESM}(e\mu \to e\mu) = 0.29. \tag{4}$$

In the electron-electron and muon-muon case, we will discuss only the asymmetry with both beams are polarized using the same parametrs used in the electron-muon case we have

$$A_{R;RL}^{CO,ESM}(ee \to ee) \approx 1,$$
$$A_{R;RL}^{CO,ESM}(\mu\mu \to \mu\mu) \approx 1. \quad (5)$$

Using the Eq.(3) and calculation the same assymetries, for the case electron-muon scatering, and considerating that the neutral vector boson has $M_{Z'} = 500$ GeV and using the same parameters considered in the case of ESM we got

$$A_{RL}^{CO,ESM+Z'}(e\mu \to e\mu) = -0.030,$$
$$A_{R;RL}^{CO,ESM+Z'}(e\mu \to e\mu) = 0.29. \quad (6)$$

In this case the A_{RL} is considerably enhanced, compare Eq.(4) with Eq.(6). In the figure 1 we see that this assymetry will be appropiate in searching for extra neutral vector bosons although the $A_{R;RL}$ is almost insensitive to the contribuitions of this new boson, look the reference [5].

For the case electron-electron and $\mu\mu$, and considerating the doubly charged vector with $M_U = 500$ GeV, and the others parameters are the same as in the case of the ESM, we found the following results

$$A_{R;RL}^{CO,ESM+U}(ee \to ee) = -1,$$
$$A_{R;RL}^{CO,ESM+U}(\mu\mu \to \mu\mu) = -1. \quad (7)$$

Compare Eq.(5) with Eq.(7) we see that the $A_{R;RL}$ change the signal, and it is a very good signal. In the reality $A_{R;RL}$ is different of the value of the ESM in a wide energy range, to see this result look at [5]. In the figure 2 we show the results for the A_{RL}.

All the asymmetries that you have discussed above depend of the scattering angle, if we want to have a result that don't depend of the angle, we can define an integrate asymmetry, see [1]. Using the ESM lagrangian we got the following results to this integrate asymmetries

$$\bar{A}_{RL}^{CO,ESM}(ee \to ee) = -0.025,$$
$$\bar{A}_{RL}^{CO,ESM}(e\mu \to e\mu) = -0.006. \quad (8)$$

When we considerate the new bosons the results, considering the same mass as showed above, we had

$$\bar{A}_{RL}^{CO,ESM+U}(ee \to ee) = -0.199,$$
$$\bar{A}_{RL}^{CO,ESM+Z'}(e\mu \to e\mu) = -0.022, \quad (9)$$

and we see that this asymetries are, again, differents of the ESM to a wide range of the energy.

REFERENCES

1. Montero, J. C., Pleitez, V., Rodriguez, in Particles and Fields Eighth Mexican School, eds Juan Carlos D'Olivo, Gabriel López Castro and Myriam Mondragón, AIP New York 1999, page 397.
2. Montero, J. C., Pleitez, V., Rodriguez, M. C., *Phys. Rev. D*, 094026 (1998); hep-ph/9802313.
3. Montero, J. C., Pleitez, V., Rodriguez, M. C., *Phys. Rev. D*, 097505 (1998); hep-ph/9803450.
4. F. Pisano and V. Pleitez, *Phys. Rev. D* **46**, 410 (1992).
5. Montero, J. C., Pleitez, V., Rodriguez, M. C., in preparation.

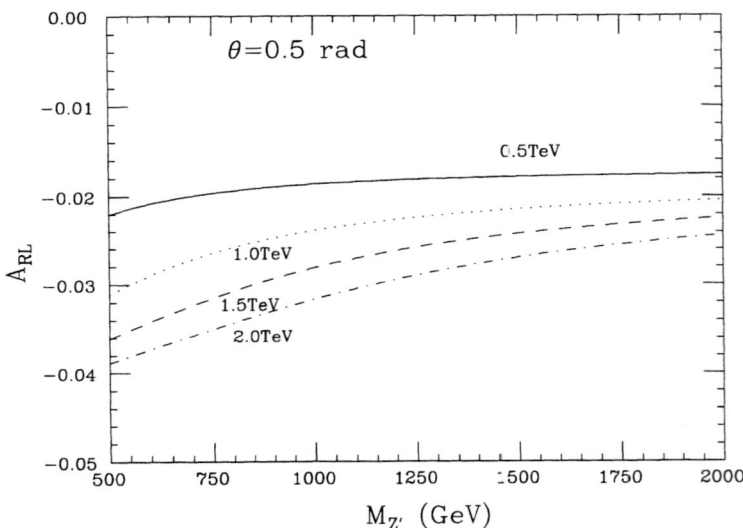

FIGURE 1. The A_{RL} asymmetry for a fixed scattering angle, $\theta = 0.5$ rad, and several values of $\sqrt{s}$ of $\mu^- e^-$ colliders for ESM+U as a function of $M_{Z'}$.

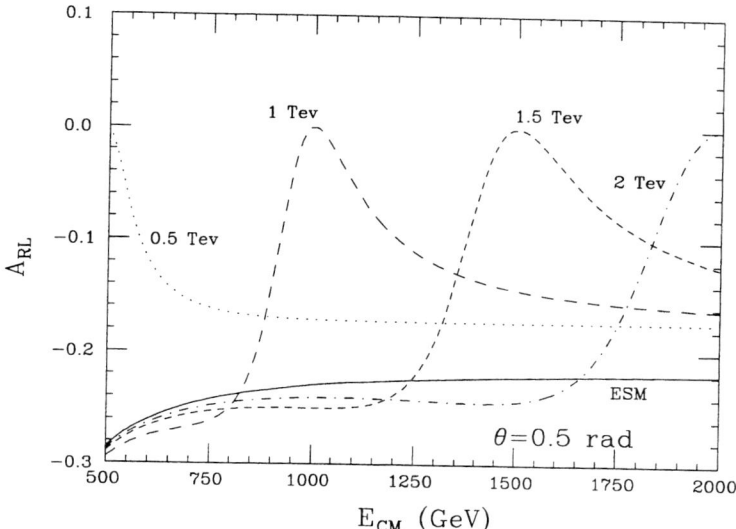

FIGURE 2. The A_{RL} asymmetry for a fixed scattering angle $\theta = 0.5$ rad for the ESM (solid line) and for the ESM+U for several U-masses as function of E_{CM}.

The Cosmological Constant and Quintessence

A. de la Macorra* and G. Piccinelli[†]

*Instituto de Física, UNAM
Apdo. Postal 20-364, 01000 México D.F., México
[†] Centro Tecnológico ENEP Aragón, UNAM
Av. Rancho Seco s/n, Col. Impulsora, Ciudad Nezahualcoyotl, México

Abstract. We study the possibility of parametrizing the cosmological constant as the contribution of a scalar field. Scalar field potential are not uniquely determined by symmetries so we analize what kind of scalar potentials leads to an interpretation of the scalar energy density as quintessence, i.e. a slow varying cosmological constant. All model dependence is given in terms of $-V'/V$ and we study all possible asymptotic limits. Finally, we show that during the accelerating period the expansion of the universe is model independent.

Recent observations show that the universe has entered an accelerating expansion regime [1]. If these observations are confirmed it would be the first experimental evidence for an energy density different from the radiation or matter and with negative pressure, i.e. for a "cosmological constant". The cosmological constant can be given by constant energy density with equation of state $p_\Lambda = -\rho_\Lambda$, where p_Λ is the pressure and ρ_Λ the energy density, or it could be parameterised in terms of a slow varying scalar field ϕ, quintessence, with a potential $V(\phi)$ and a time varying equation of state $p_\phi = (\gamma_\phi - 1)\rho_\phi$, with $p_\phi = \frac{1}{2}\dot\phi^2 - V(\phi)$ and $\rho_\phi = \frac{1}{2}\dot\phi^2 + V(\phi)$. The equation of state has $0 \leq \gamma_\phi \leq 2$, it is model dependent and it may vary with time. For an accelerating universe we have the condition $(\gamma_\phi - 1)\Omega_\phi < -1/3$, where Ω_ϕ is the ratio between the energy density of ϕ and the critical energy density.

Different models have been proposed to give an accelerating universe and a general analisys can be found in [6], [7]. Scalar field with negative pressure and a time-varying, spatially fluctuating energy density, received the name of quintessence [2]. Constraints on the equation of state of a quintessence-like component have been placed from observational data [3], [8]. Some examples are the exponential potential, $V \sim e^{-\lambda\phi}$ with $\lambda < 3$ [9], [13], the inverse power potentials $V \sim 1/\phi^n$, $n > 0$ [10], [4], $V \sim (e^{1/\phi}-1)$, $V \sim e^{\phi^2}/\phi^n$ [11] or $V \sim (c+(\phi-\phi_0)^n)^a e^{-\lambda\phi}$ [12] and mixtures of any of these potentials. All these potentials have a finite $\lambda = -V_\phi/V < O(1)$, with $V_\phi \equiv \partial V/\partial \phi$, when V approaches its minimum and this is indeed required for any model to give an accelerating universe since in this limit the potential energy dominates the energy density of ϕ. Some of these potentials may arise from non-perturbative effect or form string theory [14], [15]. As we have discussed above, the behaviour of scalar fields is fundamental in understanding the evolution of the universe. In this paper we are interested in giving an overview of the cosmological evolution of scalar field and to determine what kind of potential leads to a possible interpretation of the scalar field as quintessence and to a dominating energy density.

Another important question is the coincidence problem [4],i.e. why the energy density of the cosmological constant is of the same order of magnitude as matter even though the expansion rate of both energies is quite different today. A possible answer is by setting the initial conditions such that the ϕ field start to dominate only recently. However, this solution introduces a fine tuning problem on the initial conditions. To avoid such a fine tuning problem *tracker* solutions have been proposed [4]. In this models the energy density of the scalar field ϕ redshifts mimics the dominant energy density (radiation) at the beginning with $\gamma_\phi > 1$ and when matter dominates the equation of state for ϕ changes to $\gamma_\phi < 1$. With this kind of fields one avoids the initial fine tuning problem but they have difficulties to explain (without any fine tuning of the potential) the central values of $\Omega_\phi = 2/3 \pm 0.05$, $\gamma_\phi \leq 0.35 \pm 0.07$ [5] since they have $\gamma_\phi > 0.3$ [4]. However, they remain a very interesting possibility. Other models with more the one term in the potential $V = V_1 + V_2$ where one term dominates during radiation and scales as radiation or matter while the other dominates at a recent time giving the acceleration of the universe can be studied. In particular if during the radiation dominated epoch on has a potential $V_1 = a\,\phi^4$, it will redshift as radiation and the ratio ρ_r/ρ_ϕ remains constant. On the other for $V_1 = a\,\phi^2$ the ratio ρ_m/ρ_ϕ is constant. In either case the initial value can be set to $\rho_r/\rho_\phi = O(1)$ or $\rho_r/\rho_\phi = O(1)$ avoiding an initial fine tuning problem but not the coincidence problem.

Our starting point is a universe field with a barotropic energy density, which can be either matter or radiation, and the energy density of the moduli field. The barotropic fluid is described by an energy density ρ_γ and a pression p_γ with a standard equation of state $p_\gamma = (\gamma_\gamma - 1)\rho_\gamma$, where $\gamma_\gamma = 1$ for matter and $\gamma_\gamma = 4/3$ for radiation. We do not make any hypothesis on which energy density dominates, that of the barotrpoic fluid or that of the moduli field. For a canonically normalized scalar field ϕ with a selfinteraction given in terms of the scalar potential $V(\phi)$ but with gravitational interaction with all other fields (as moduli fields) the equations to be solved, in a spatially flat Friedmann–Robertson–Walker (FRW) Universe, are given by $\dot H = -\frac{1}{2}(\rho_\gamma + p_\gamma + \dot\phi^2)$ $\dot\rho = -3H(\rho+p)$ and $\ddot\phi = -3H\dot\phi - \frac{dV(\phi)}{d\phi}$ where H is the Hubble parameter, $V(\phi)$ is the scalar field potential and we have taken $8\pi G = 1$. It is useful to make a change of variables [13] $x \equiv \dot\phi/\sqrt{6}H$, $y \equiv \sqrt{V}/\sqrt{3}H$ and the dynamical equations become

$$x_N = -3x + \sqrt{\frac{3}{2}}\lambda y^2 + \frac{3}{2}x[2x^2 + \gamma_\gamma(1 - x^2 - y^2)]$$

$$y_N = -\sqrt{\frac{3}{2}}\lambda\, x\, y + \frac{3}{2}y[2x^2 + \gamma_\gamma(1 - x^2 - y^2)] \tag{1}$$

$$H_N = -\frac{3}{2}H[\gamma_\gamma(1 - x^2 - y^2) + 2x^2]$$

where N is the logarithm of the scale factor a, $N \equiv ln(a)$, $f_N \equiv df/dN$ for $f = x, y, H$ and $\lambda(N) \equiv -V_\phi/V$. Notice that all model dependence in eqs.(1) is through the quantities $\lambda(N)$ and the constant parameter γ_γ. Eqs.(1) must be supplemented by the Friedmann or constraint equation for a flat universe $\frac{\rho_\gamma}{3H^2} + x^2 + y^2 = 1$ and they are valid for any scalar potential as long as the interaction between the scalar field and matter or radiation is gravitational only. This means that it is possible to separate the energy and pression densities into contributions from each component, i.e. $\rho = \rho_\gamma + \rho_\phi$ and $p = p_\gamma + p_\phi$, where ρ_ϕ (p_ϕ) is the energy density (pression) of the scalar field.

As a result of the dynamics, the scalar field will evolve to its minimum and if we do not wish to introduce any kind of unnatural constant or fine tuning problem, the minimum of the potential must have zero energy, i.e. $V|_{min} = V'|_{min} = 0$ at ϕ_{min}. We will consider here only these kind of potentials. For finite ϕ_{min} the scalar field will naturally oscillate around its vacuum expectation value (v.e.v.). If the scalar field has a non zero mass or if the potential V admits a Taylor expansion around ϕ_{min} than, using the Hôpital rule, one has $\lim_{t\to\infty}|\lambda| = \infty$ and it will oscillate. On the other hand, if $\phi_{min} = \infty$ then ϕ will not oscillate and λ will approach either zero, a finite constant or infinity. The oscillating behaviour of ϕ or λ is important in determining the cosmological evolution of x, y and $\Omega_\phi \equiv \rho_\phi/\rho = x^2 + y^2$ and we will show that any scalar field with a non-vanishing mass redshifts as matter field.

A complete analysis of the solutions of eqs.(1) has been obtained [7] and the values of x, y at late times depend on the asymptotic behaviour of λ. If $\lambda \to 0$ then $x \to 0$, $y \to 1$ and $H \to cte$ becoming the scalar field a "true" cosmological constant. If $\lambda \to \infty$ then depending on whether λ oscillates or not we will have different asymptotic behaviour. If the v.e.v. is at $\varphi \to \infty$ then λ will not oscillate and $x \propto y \propto \lambda^{-1} \to 0$ and $\Omega_\phi \to 0$. If, on the other hand, the v.e.v. of ϕ is finite, λ will oscillate and approach infinity. In this class of potentials we can expand around the minimum with a leading term $V(\phi) = V_0(\phi - \phi_0)^n$ giving an equation of state with $\gamma_\phi = 2n/(2+n)$ [6]- [7]. If $\gamma_\phi < \gamma_\gamma$ ($\gamma_\phi > \gamma_\gamma$) then $\Omega_\phi \to 1$ ($\Omega_\phi \to 0$) and for $\gamma_\phi = \gamma_\gamma$, Ω_ϕ goes to a finite ($\neq 0, 1$) constant value. We have therefore seen that asymptotic limit of the quantity λ determines the cosmological evolution of the scalar field and the redshift of the energy density.

To determine the cosmological relevant potentials, it is useful to use the cosmological acceleration parameter α and expansion rate parameter Γ. The acceleration parameter is defined as

$$\alpha \equiv \frac{\rho + 3p}{(3\gamma_\gamma - 2)\rho} = \frac{3\gamma - 2}{3\gamma_\gamma - 2} \tag{2}$$

with $\gamma = (\rho + p)/\rho$. If $\alpha = 1$ then the acceleration of the universe is the same as that of the barotropic fluid and any deviation of α from one implies a different cosmological behaviour of the universe due to the contribution of the scalar field. A positive accelerating universe requires a negative α while for $0 < \alpha < 1$ the acceleration of the universe is negative (deceleration) but smaller than that of the barotropic fluid. For $\alpha > 1$ the deceleration is larger than for the barotropic fluid. In terms of the standard deceleration parameter $q \equiv -\frac{\ddot{a}a}{\dot{a}^2}$ one has $\alpha = \frac{2q}{3\gamma_\gamma - 2}$ or in terms of x, y one finds $\alpha = 1 - \frac{3\gamma_\gamma}{3\gamma_\gamma - 2}(y^2 - x^2 \frac{2-\gamma_\gamma}{\gamma_\gamma}) = 1 - 3\Omega_\phi \frac{\gamma_\gamma - \gamma_\phi}{3\gamma_\gamma - 2}$. It is also useful to define the normalized equation of state parameter $\Gamma = \frac{\gamma}{\gamma_\gamma}$ which gives the relative expansion rate of the universe with respect to the barotropic fluid. A Γ smaller than one means that the universe expands slower than the barotropic fluid and a Γ larger than one says that the universe expands faster due to the contribution of the scalar field. In our case α and Γ are not independent since $\Gamma = 1 - (1-\alpha)(3\gamma_\gamma - 2)/3\gamma_\gamma = 1 - \Omega_\phi \frac{\gamma_\gamma - \gamma_\phi}{\gamma_\gamma}$.

In table 1 we give a summary of the asympotic limits of the cosmological relevant quantities for all different asymptotic limits of λ.

$\lambda(\phi) = -V'/V$	$\Omega_\phi = \rho_\phi/\rho$	γ_ϕ	$\alpha(\phi)$	$\Gamma(\phi)$	e.g. $V(\phi)$
$c =$cte $(> \sqrt{3\gamma_\gamma})$	$\frac{3\gamma_\gamma}{c^2}$	γ_γ	1	1	$V_0 \, e^{-c\phi}$
$c =$cte $(< \sqrt{6})$	1	$\frac{c^2}{3}$	$\frac{c^2-2}{3\gamma_\gamma - 2}$	$\frac{c^2}{3\gamma_\gamma}$	$V_0 \, e^{-c\phi}$
∞ (no oscil.)	0	γ_γ	1	1	$V_0 \, e^{-ce^\phi}$
∞ (oscil.)	0	$\frac{2n}{2+n} (> \gamma_\gamma)$	1	1	
	cte	$\frac{2n}{2+n} (= \gamma_\gamma)$	1	1	$V_0 \, \phi^n, \; n > 0$ even
	1	$\frac{2n}{2+n} (< \gamma_\gamma)$	$\frac{3\gamma_\phi - 2}{3\gamma_\gamma - 2}$	$\frac{\gamma_\phi}{\gamma_\gamma}$	
0	1	0	$-\frac{2}{3\gamma_\gamma - 2}$	0	$V_0 \, \phi^{-n}, \; n > 0$

Table 1. In this table we show the asymptotic behaviour of Ω_ϕ, γ_ϕ the acceleration parameter $\alpha(\phi) = \frac{3\gamma - 2}{3\gamma_\gamma - 2}$ and the expansion rate parameter $\Gamma = \frac{\gamma}{\gamma_\gamma}$ for different limiting cases of $\lambda(\phi)$. In the last column we give an example of potential $V(\phi)$ which satisfies this limit.

As mentioned in the introduction a late time acceleration period (as the one observed by SN1a results) introduces the coincidence problem. Form eqs.(1) we can derive [16]

$$\Omega_\phi(N) = \frac{\Omega_{o\phi} e^{-3\gamma_\gamma(N_o - N) + 3\int \gamma_\phi dN}}{1 - \Omega_{o\phi} + \Omega_{o\phi} e^{-3\gamma_\gamma(N_o - N) + 3\int \gamma_\phi dN}} \qquad (3)$$

where $\Omega_\phi(N_o) = \Omega_{o\phi}$ and we have $N < N_0$ for a time $t < t_0$. In the limit of $\Omega_{o\phi} \ll 1$ and $\gamma_\phi \simeq$ cte eq.(3) reduces to $\Omega_\phi \simeq e^{3N(\gamma_\gamma - \gamma_\phi)}$. However we are not interested in this region since the initial condition has $\Omega_{o\phi} \simeq 2/3$. During the accelerating period we have $x^2 < y^2$ and $\gamma_\phi < \gamma_\gamma$ and we can expand eq.(3) as a function of $R \equiv 3\int \gamma_\phi dN$. Eq.(3) becomes

$$\Omega_\phi = \tilde{\Omega}_\phi \left(1 + \frac{1 - \Omega_{o\phi}}{1 - \Omega_{o\phi} + \Omega_{o\phi} e^{-3\gamma_\gamma(N_o - N)}} R + O(R^2)\right) \qquad (4)$$

with

$$\tilde{\Omega}_\phi = \frac{\Omega_{o\phi} e^{-3\gamma_\gamma(N_o-N)}}{1 - \Omega_{o\phi} + \Omega_{o\phi} e^{-3\gamma_\gamma(N_o-N)}}. \qquad (5)$$

Notice that as long as R is constant or $R \ll 1$, eq.(4) is model independent and it is certainly valid during the acceleration period of the universe. We see, therefore, that the accelerating expansion of the universe is model independent and the energy density of quintessence evolves as $\tilde{\Omega}_\phi$. Furthermore, $\tilde{\Omega}_\phi$ remains a good approximation to eq.(3) as long as γ_ϕ is small. We can establish a relationship between the amount of discrepancy that we wish to accept between Ω_ϕ and $\tilde{\Omega}_\phi$. For a r percentage of difference, i.e. $\Omega_\phi/\tilde{\Omega}_\phi = 1+r$ we have $r = \frac{1-\Omega_{o\phi}}{1-\Omega_{o\phi}+\Omega_{o\phi}e^{-3\gamma_\gamma(N_o-N)}} R \simeq R$ (at large $N_0 - N$). Since $(\gamma_\phi)_N$ is proportional to γ_ϕ it varies rapidly when γ_ϕ is not small [16]. R becomes large when $\gamma_{\phi N} \simeq -6\gamma_\phi$ (as a first approximation) and we find $R = 3 \int \gamma_\phi dN = (\Delta\gamma_\phi)/2$. If we allow an $r = 20\%$ discrepancy between the exact and the model independent $\tilde{\Omega}_\phi$, this will happen at $\Delta\gamma_\phi = 2r = 0.4$. Since before the accelerating epoch one has $x \ll y$ and $\gamma_\phi \ll 1$, the model independent energy density stops being a good approximation only until $\gamma_\phi \simeq 2r$. When will this happen is model and initial condition depended, however, for a large number of cases $N_0 - N_i$ can be quite large (N_i is the beginning of the model independent evolution of ϕ). In fact, if, for example, we set initial conditions such that $x < 2y$, i.e. $\gamma_\phi < 0.4$, at N_i (regardless of its value) the evolution of Ω_ϕ is model independent (within a maximum of 20% discrepancy).

To conclude, we have shown that scalar fields may parametrize the cosmological constant giving an accelarating epoach at late times. However, a convincing model still needs to be obtained.

A.M. research was supported in part by CONACYT project 32415-E and by DGAPA, UNAM, project IN-103997.

REFERENCES

1. A.G. Riess et al., Astron. J. 116 (1998) 1009; S. Perlmutter et al, ApJ 517 (1999) 565; P.M. Garnavich et al, Ap.J 509 (1998) 74.
2. R.R. Caldwell, R. Dave and P.J. Steinhardt, Phys. Rev. Lett. 80 (1998) 1582.
3. M.S. Turner and M. White Phys. Rev. D 56 (1997) 4439; G. Efstathiou, astro-ph/9904356; L. Wang et al, astro-ph/9901388.
4. I. Zlatev, L. Wand and P.J. Steinhardt, Phys. Rev. Lett.82 (1999) 8960; Phys. Rev. D59 (1999)123504
5. G. Efstathiou, S.L. Bridle, A.N. Lasenby, M.P. Hobson and R.S. Ellis, MNRAS 303L (1999) 47; I. Zlatev, L. Wand and P.J. Steinhardt, astro-ph/9901388; M. Roos and S.M. Harun-or Rashid, astro-ph/9901234
6. A.R. Liddle and R.J. Scherrer,Phys. Rev. D59, (1999)023509;
7. A. de la Macorra and G. Piccinelli hep-ph/9909459

8. K. Freese, F.C. Adams, J.A. Frieman and E. Mottola, Nucl. Phys. B 287 (1987) 797; M. Birkel and S. Sarkar, Astropart. Phys. 6 (1997) 197.
9. C. Wetterich, Nucl. Phys. B302 (1998) 668 P. Ferreira, M. Joyce, Phys. Rev. D 58 (1998) 503; E.J. Copeland, A. Liddle and D. Wands, Ann. N.Y. Acad. Sci. 688 (1993) 647.
10. P.J.E. Peebles and B. Ratra, ApJ 325 (1988) L17;B. Ratra and P.J.E. Peebles, Phys. Rev. D37 (1988) 3406
11. P. Brax and J. Martin, astro-ph/9905040
12. A. Albrecht and C. Skordis, astro-ph/9908085, V. Sahni and L.Wang, astro-ph/9910087
13. E.J. Copeland, A. Liddle and D. Wands, Phys. Rev. D57 (1998) 4686
14. A. de la Macorra hep-ph/9910330
15. M.C. Bento and O. Bertolami, gr-qc/9905075, O. Bertolami and R. Schiappa, Class.Quant.Grav.16 (1999) 2545
16. A. de la Macorra astro-ph/9911079

Z physics effects of an additional non-sequential bottom quark

R. Martinez, J-Alexis Rodriguez, M. Vargas and I.D. Zuluaga

Dpto. de Fisica, Universidad Nacional

Bogota, Colombia

Abstract. We analyze the Zff vertex in the framework of models that add a new bottom quark in a non-sequential way and we evaluate the tree level contribution to the LEP/SLC observables Γ_Z, R_b and R_l. We obtain bounds for the mixing angles from the experimentally allowed contour regions of the parameters $\Lambda_{L,R}$ introduced here. In order to get a more restrictive region, we consider the experimental results for $B \to \nu\nu X$ as well.

INTRODUCTION

The SM contains three generations of quarks in irreducible representations of the gauge symmetry group $SU(3)_C \times SU(2)_L \times U(1)_Y$. The possibility of extending them has been studied in different frameworks [1]-[5] which are based either on a fourth generation sequential family, or on non-sequential fermions, regularly called exotic representations because they are different from those of the SM.

These unusual representations emerge in other theories, like the E_6 model where a singlet bottom type quark appears in the fundamental representation [2]; also, top-like singlets have been suggested in Supersymmetric gauge theories [3].

The possibility of indirect consequences of singlet quark mixing for FCNC and CP violation has been used to get bounds on the flavor changing couplings. Heavy meson decays like B^0 and $D^0 \to \mu^+\mu^-$ [1], [5], rare decays $b \to s\gamma$ [1], [2], [5], measurements like $K_L \to \mu\mu$, $B \to \mu\mu X$, $B \to \nu\nu X$, K meson physics [1], [2], [4], or even $Z \to ll$, $l \to lll$ [6] have been considered for this purpose.

In the last years, the LEP and SLC colliders have brought to completion a remarkable experimental program by collecting an enormous amount of electroweak precision data on the Z resonance. This activity, together with the theoretical efforts to provide accurate SM predictions have formed the apparatus of electroweak precision tests [7]. We are interested in using the electroweak precision test quantities in order to get bounds on the mixing angles for additional fermions in exotic representations. Specifically, we want to consider models that include a new quark with charge $-1/3$ which is mixed with the SM b quark. This kind of new physics was taken into account by Barmert, et. al. [8] during the discrepancy between experiment and SM theory in the R_b ratio. They analyzed a broad class of models in order to explain the discrepancy, and they considered those models in which new Zbb couplings arise at tree level through Z or b quark mixing with new particles. The parametrization of the vertex in a general way has been reviewed by Barger et. al. [1], [4] as well as Cotti and Zepeda [6]. The LEP precision test parameters that we use are the total Z width Γ_Z, the fractions R_l and R_b.

The procedure to get bounds on the mixing angles is the following. First, we analyze the Zff vertex as obtained after a rotation of a general quark multiplet (common charge) into mass eigenstates. In particular, we write down the neutral current terms for the bottom quarks, which are assumed to be mixed.

With these expressions we can evaluate the tree level contribution to the process $Z \to bb$; we enclose this new contribution within the coupling constants g_V and g_A. We then write down the total Z width Γ_Z, the fractions R_l and R_b including the new contributions, and we obtain bounds on the new parameters by using the experimental values from LEP and SLC [9]. We also use the result obtained by Grossman et. al. [4], involving $B \to \nu\nu X_s$, in order to narrow down the bounds.

THE MODEL

Following closely the notation of ref. [6], if we have a multiplet $\Psi^O_{a=L,R}$ with n_a ordinary fermions and m_a exotic fermions with the same electric charge q:

$$\Psi^O_a = U_a \Psi_a \qquad (1)$$

where U_a is the unitary matrix that rotates the mass eigenstate Ψ_a into the interaction eigenstate Ψ^O_a. U_a can be further the composed as follows [6]:

$$U_a = \begin{pmatrix} A & E \\ F & G \end{pmatrix}_a \qquad (2)$$

where

$$\left(U^+ U\right)_a = \begin{pmatrix} A^+A + F^+F & A^+E + F^+G \\ E^+A + G^+F & E^+E + G^+G \end{pmatrix}_a = \begin{pmatrix} 1 & 0 \\ 0 & 1 \end{pmatrix} . \qquad (3)$$

If we suppose that the up quark sector of the SM is diagonal and that there are no exotic quarks, then A_L corresponds to the classical Kobayashi-Maskawa matrix. In the SM this matrix is unitarity, whereas in our model it is not:

$$\left(A^+A\right)_L = I - \left(F^+F\right)_L . \qquad (4)$$

F_L corresponds to the mixing of the ordinary-exotic quarks. As mentioned, A_L is not quite unitary and the factor $(F^+F)_L$ indicates Flavor Changing transitions in the light-light sector. The neutral current Lagrangian for the multiplet Ψ is given by

$$-\mathcal{L}^{NC} = \frac{e}{c_w s_w} \sum_{a=L,R} \overline{\Psi^O}_a \gamma^\mu D_a \Psi^O_a Z^0_\mu,$$

$$= \frac{e}{c_w s_w} \sum_{a=L,R} \overline{\Psi}_a \gamma^\mu U^+_a D_a U_a \Psi_a Z^0_\mu \quad (5)$$

where $s_w = \sin\theta_w$ and D_a are diagonal matrices which contain the couplings of the neutral gauge bosons to the matter fields; they have the form:

$$D_a = \left(\mathbf{T}_3 - \mathbf{Q} s_w^2\right)_a,$$

$$= \begin{pmatrix} t_{30} - q s_w^2 & 0 \\ 0 & t_{3E} - q s_w^2 \end{pmatrix}_a \quad (6)$$

where t_{30} and t_{3E} are the standard and exotic weak isospin 3rd componend of the multiplets. The product $(U^+ D U)_{a=L,R}$ in eq. (5) can be written as:

$$\left(U^+ D U\right)_L = \begin{pmatrix} F^+ F & -A^+ E \\ -E^+ A & -E^+ E \end{pmatrix}_L (t_{3E} - t_{30})_L + \mathbf{T}_{3L} - \mathbf{Q} s_w^2,$$

$$\left(U^+ D U\right)_R = \begin{pmatrix} F^+ F & F^+ G \\ G^+ F & G^+ G \end{pmatrix}_R t_{3ER} - \mathbf{Q} s_w^2 \quad (7)$$

and the neutral current Lagrangian in the light-light sector can be written as [6]:

$$-\mathcal{L}^{NC} = \frac{e}{c_w s_w} \sum_{a=L,R} \overline{\Psi}_{l,a} \gamma^\mu K_a \Psi_{l,a} Z^0_\mu \quad (8)$$

where

$$K_L = \left(F^+ F\right)_L (t_{3EL} - t_{30L}) + I_{3\times 3} \left(t_{30L} - q s_w^2\right),$$

$$K_R = \left(F^+ F\right)_R t_{3ER} - I_{3\times 3} \, q s_w^2. \quad (9)$$

In this work, we only consider one *bottom* exotic quark (i.e. not mixing with d and s). Then, U_a and the $(F^\dagger F)_a$ product become:

$$U_a = \begin{pmatrix} & & 0 \\ & A_a & 0 \\ & & -s_a \\ 0 & 0 & s_a & c_a \end{pmatrix}, \quad (F^\dagger F)_a = \begin{pmatrix} 0 & 0 & 0 \\ 0 & 0 & 0 \\ 0 & 0 & \sin^2\theta_a \end{pmatrix}. \quad (10)$$

Therefore, the coupling bbZ gets modified by the $\Lambda_{L,R}$ factors:

$$K^b_L = \Lambda_L + t_{30L} - q s_w^2,$$

$$= \sin^2\theta_L \left(t_{3EL} + \frac{1}{2}\right) + \left(-\frac{1}{2} + \frac{1}{3} s_w^2\right),$$

$$K^b_R = \Lambda_R - q s_w^2,$$

$$= \sin^2\theta_R \, t_{3ER} + \left(\frac{1}{3} s_w^2\right). \quad (11)$$

PRECISION TEST PARAMETERS

To restrict new physics, we will use parameters measured at the Z pole. These parameters are the total decay width of the Z boson Γ_Z, the fractions $R_b = \Gamma\left(Z \to b\bar{b}\right)/\Gamma\left(Z \to hadrons\right)$ and $R_l = \Gamma\left(Z \to hadrons\right)/\Gamma\left(Z \to l\bar{l}\right)$ [9], [7]. Considering the new physics (NP) and the SM couplings, we can write

$$\Gamma\left(Z \to b\bar{b}\right) = \frac{G_F m_Z^3}{2\sqrt{2}\pi}\left[\frac{3\beta - \beta^3}{2}\left(g_V^{SM} + g_V^{NP}\right)^2 + \beta^3\left(g_A^{SM} + g_A^{NP}\right)^2\right]R \quad (12)$$

where R are the QCD and QED corrections, and $\beta = \sqrt{1 - 4\frac{m_b^2}{m_Z^2}}$ is the kinematic factor [7] with $m_b = 4.7$ GeV. It is convenient to separate the SM and NP contributions as follows:

$$\Gamma\left(Z \to b\bar{b}\right) = \Gamma_b^{SM}\left(1 + \delta_b^{NP}\right). \quad (13)$$

The symbol δ_b^{NP} is given by:

$$\delta_b^{NP} = \frac{(3-\beta^2)\left[\left(g_V^{NP}\right)^2 + 2g_V^{NP}g_V^{SM}\right] + 2\beta^2\left[\left(g_A^{NP}\right)^2 + 2g_A^{NP}g_A^{SM}\right]}{(3-\beta^2)\left(g_V^{SM}\right)^2 + 2\beta^2\left(g_A^{SM}\right)^2}. \quad (14)$$

Similarly,

$$\Gamma\left(Z \to hadrons\right) = 2\Gamma_u^{SM} + 2\Gamma_d^{SM} + \Gamma_b ,$$
$$= \Gamma_{had}^{SM}\left(1 + \frac{\Gamma_b^{SM}}{\Gamma_{had}^{SM}}\delta_b^{NP}\right). \quad (15)$$

Here, only Γ_b gets NP corrections because only the SM bottom mixes with the exotic quark. The Z partial decay into d and s quarks remains unchanged.

On the other hand, Γ_Z is equal to

$$\Gamma_Z = 3\Gamma\left(Z \to \nu\bar{\nu}\right) + 3\Gamma\left(Z \to l\bar{l}\right) + \Gamma\left(Z \to hadrons\right) \quad (16)$$

which again is re-written as follows:

$$\Gamma_Z = \Gamma_Z^{SM}\left(1 + \frac{\Gamma_b^{SM}}{\Gamma_Z^{SM}}\delta_b^{NP}\right). \quad (17)$$

Similarly, for R_l and R_b we obtain the following expressions:

$$R_l = R_l^{SM}\left(1 + R_b^{SM}\delta_b^{NP}\right),$$
$$R_b = R_b^{SM}\left[1 + \delta_b^{NP}\left(1 - R_b^{SM}\right)\right]. \quad (18)$$

In a general way, R_b is mainly a measure of $\left|g_L^b\right|^2 + \left|g_R^b\right|^2$; therefore, the fraction R_b is very sensitive to anomalous couplings of the b quark.

RESULTS

With the expressions for Γ_Z, R_l and R_b in terms of the new physics contribution, and with the experimental data from LEP we get bounds on the parameters $\Lambda_{L,R}$ introduced in eq.(11). We consider only mixing between an exotic bottom quark with the third SM family. The experimental data that we used for the LEP parameters, as well as their SM values are in Table 1 [7], [9].

TABLE 1. SM predictions and experimental values measured at LEP for the Γ_Z, R_l and R_b

	Experimentals	Standard Model
Γ_Z	2.4939 ± 0.0024	2.49582
R_l	20.765 ± 0.026	20.7468
R_b	0.21656 ± 0.00074	0.215894

In order to get a more restrictive region we use the bound $|\Lambda_{L,R}| < 0.0018$ obtained by Grossman et. al. [4]. The allowed region is the intersection of the regions for Γ_Z, the fractions R_l and R_b with the bound obtained by Grossman et. al. The region is delimited in each case, respectively, by

$$-5.9 \times 10^{-4} \leq \Lambda_L \leq 1.8 \times 10^{-3} ,$$
$$-1.8 \times 10^{-3} \leq \Lambda_L \leq 7.0 \times 10^{-4} ,$$
$$-1.8 \times 10^{-3} \leq \Lambda_L \leq 4.1 \times 10^{-4}. \quad (19)$$

For Λ_R we obtain for the three cases the same bound given by

$$-1.8 \times 10^{-3} \leq \Lambda_R \leq 1.8 \times 10^{-3}. \quad (20)$$

The intersection between the three regions is given by:

$$-5.9 \times 10^{-4} \leq \Lambda_L \leq 4.1 \times 10^{-4} ,$$
$$-1.8 \times 10^{-3} \leq \Lambda_R \leq 1.8 \times 10^{-3} . \quad (21)$$

We note that for Λ_L the region is more restrictive than the one obtained by Grossman et. al. [4], while the Λ_R parameter is not modified. We can use these bounds in order to get constraints for the left and right mixing angles of each model. The couplings for several $SU(2)_L$ representations are given in table 2, and the bounds are shown in table 3.

Summarizing, we have used the total decay width Γ_Z, the fractions R_l and R_b to obtain bounds on the mixing angles of new quark bottom-type representations with the SM bottom quark. Taking into account the results of Grossman et. al. [4] and the intersection obtained from the experimental parameters measured at LEP, we have gotten the allowed intervals $-5.9 \times 10^{-4} \leq \Lambda_L \leq 4.1 \times 10^{-4}$ and $-1.8 \times 10^{-3} \leq \Lambda_R \leq 1.8 \times 10^{-3}$. Our results reduce the allowed region for the parameter Λ_L while the parameter Λ_R is not modified with respect to the results

TABLE 2. The parameters Λ_a for different representations according with the quantum numbers in eq.(8)

(t^3_{EL}, t^3_{ER})	Λ_L	Λ_R	Model
$(0,0)$	$\frac{1}{2}\sin^2\theta_L$	0	Vector singlets
$(-\frac{1}{2}, -\frac{1}{2})$	0	$-\frac{1}{2}\sin^2\theta_R$	Vector Doublets
$(0, -\frac{1}{2})$	$\frac{1}{2}\sin^2\theta_L$	$-\frac{1}{2}\sin^2\theta_R$	Mirror fermions
$(-1, -1)$	$-\frac{1}{2}\sin^2\theta_L$	$-\sin^2\theta_R$	Self-conjugated triplets

TABLE 3. Bounds on the mixing angles for different representations of the exotic quarks

Model	$|\sin\theta_L| \leq$	$|\sin\theta_R| \leq$
Vector Singlets	2.02×10^{-2}	0
Vector doublets	0	6×10^{-2}
Mirror fermion	2.86×10^{-2}	6×10^{-2}
Self-conjugated triplets	3.43×10^{-2}	4.24×10^{-2}

obtained by Grossman, et. al. [4]. The lower bound for Λ_L is coming from the measurement of the total width Γ_Z while the upper bound is from the fraction R_b.

This work was partially supported by COLCIENCIAS. One of us (M. V.) acknowledges the scholarship by Fundación MAZDA para el Arte y la Ciencia.

REFERENCES

1. V. Barger, M. S. Berger and R. Phillips, *Phys. Rev.* **D52**, 1663 (1995); F. del Aguila, et. al., *Phys. Rev. Lett.* **82**, 1628 (1999); J. Diaz Cruz et. al., *Phys. Rev.* **D41**, 1981 (1990).
2. Y. Nir and D. Silverman, *Phys. Rev.* **D42**, 1477 (1990); *Nucl. Phys.* **B345**, 301 (1990); D. Silverman, *Phys. Rev.* **D45**, 1800 (1992); *Int. J. Mod. Phys.* **A11**, 2253 (1996); *Phys. Rev.* **D58**, 095006 (1998).
3. G. C. Branco, et. al. *Phys. Rev.* **D48**, 1167 (1993).
4. Y. Grossman, Z. Ligeti and E. Nardi, *Nucl. Phys.* **B465**, 369 (1996).
5. C. O. Dib, D. London and Y. Nir, *Int. J. Mod. Phys.* **A6**, 1253 (1991); J. Silva and L. Wolfenstein, *Phys. Rev.* **D55**, 5331 (1997).
6. U. Cotti and A. Zepeda, *Phys. Rev.* **D55**, 2998 (1997).
7. G. Altarelli, hep-ph/9811456; Hollik W.,hep-ph/9811313; and references therein.
8. P. Bamert, C. P. Burguess, J. M. Cline, D. London and E. Nardi, *Phys. Rev.* **D56**, 1 (1998).
9. *The Eur. Phys. J.* **C3**, 1 (1998). F. del Aguila and M. Bowick, *Nucl. Phys.* **B224**, 107 (1983); P. Fishbane, R. Norton and M. Rivard, *Phys. Rev.* **D33**, 2632 (1986); W. Buchmuller and M. Gronau, *Phys. Lett.* **B220**, 641 (1989).

Horizontal interactions and gauge theories

William A. Ponce

*Departamento de Física, Universidad de Antioquia,
Medellín, Colombia. A.A. 1226*

Abstract. We analyze the experimental and theoretical facts that constraint an $SU(3)_c \otimes SU(2)_L \otimes U(1)_Y \otimes U(1)_H$ local gauge symmetry, where $U(1)_H$ is an abelian horizontal symmetry which can be anomalous supersymmetric, non-anomalous supersymmetric, and non-anomalous non-supersymmetric.

INTRODUCTION

Even though the standard model (SM) local gauge group $SU(3)_c \otimes SU(2)_L \otimes U(1)_Y \equiv G_{SM}$ with $SU(2)_L \otimes U(1)_Y$ hidden and $SU(3)_c$ confined, has been very sucessfull in explaining a huge amount of experimental data related with particle interactions, it has failed in providing a clue in the explanation of the fermion mass spectrum (masses and mixing angles).

In this note we attempt to gain insight in the predictions of the masses and mixing angles of the elementary quarks and leptons by introducing a gauge group $G_{SM} \otimes U(1)_H$, where $U(1)_H$ is an abelian horizontal symmetry which forbids all the fermion yukawa couplings at tree level, but the one of the top quark [1,2]. Then we look for theoretical and experimental constraints on such $U(1)_H$ which are imposed by consistence, and by the mass spectrum of the known particles, which is generated by a set of effective operators that couple the SM fermions to the electroweak Higgs fields.

In what follows we analyze first the $U(1)_H$ supersymmetric (SUSY) case and next the non-SUSY local gauge models.

$U(1)_H$ SUPERSYMMETRIC

For a SUSY $U(1)_H$ with a tree level yukawa coupling involving only the top quark, we expect that the effective theory provides a Yukawa lagrangian of the form

$$\mathcal{L}^Y = Y_{ij}^u \bar{q}_i u_j \phi_u + Y_{ij}^d \bar{q}_i d_j \phi_d + Y_{ij}^l \bar{L}_i e_j \phi_d \tag{1}$$

with

$$Y_{ij}^u = \begin{cases} A_{ij}^u \lambda^{[h(q_i)+h(u_j)+h(\phi_u)]} & h(q_i)+h(u_j)+h(\phi_u) \geq 0 \\ 0 & h(q_i)+h(u_j)+h(\phi_u) < 0; \end{cases} \qquad (2)$$

$$Y_{ij}^d = \begin{cases} A_{ij}^d \lambda^{[h(q_i)+h(d_j)+h(\phi_d)]} & h(q_i)+h(d_j)+h(\phi_d) \geq 0 \\ 0 & h(q_i)+h(d_j)+h(\phi_d) < 0; \end{cases} \qquad (3)$$

$$Y_{ij}^L = \begin{cases} A_{ij}^L \lambda^{[h(L_i)+h(e_j)+h(\phi_d)]} & h(L_i)+h(e_j)+h(\phi_d) \geq 0 \\ 0 & h(L_i)+h(e_j)+h(\phi_d) < 0. \end{cases} \qquad (4)$$

Where $Y_{33} = 1$ $(h(q_3)+h(u_3)+h(\phi_u) = 0)$, $Det(A^u) \sim Det(A^d) \sim Det(A^L) \sim 1$, and the zeroes, corresponding to negative charges, are a result of the holomorphy of the superpotential [3].

In the former expressions $\bar{L}_i = (\bar{\nu}_i, \bar{e}_i^-)_L$, $\bar{q}_i = (\bar{u}_i, \bar{d}_i)_L$, $i=1,2,3$ is a generational index, u_i, d_i, ν_i and e_i stand for the fields of the up quark, down quark, neutrino and electron of the ith family respectively, and λ is a small Cabibbo-like expansion parameter which is a model dependent quotient between two mass scales (as specified in what follows). The $U(1)_H$ charges are defined as in the following Table (where we have simultaneos quoted the family independent SM $U(1)_Y$ hypercharges for each multiplet):

	q_i	u_i^c	d_i^c	L_i	e_i^+	ϕ_u	ϕ_d
Y	1/3	−4/3	2/3	−1	2	−1	1
H	$h(q_i)$	$h(u_i)$	$h(d_i)$	$h(L_i)$	$h(e_i)$	$h(\phi_u)$	$h(\phi_d)$

With ϕ_u and ϕ_d in the table standing for the two scalar fields in the minimal supersymmetric standard model (MSSM).

THE ANOMALOUS CASE

From the equations in the previous section we may writte:

$$Det(M^u) \sim <\phi_u>^3 \lambda^{\sum_i [h(q_i)+h(u_i)]+3h(\phi_u)}$$
$$Det(M^d) \sim <\phi_d>^3 \lambda^{\sum_i [h(q_i)+h(d_i)]+3h(\phi_d)}$$
$$Det(M^L) \sim <\phi_d>^3 \lambda^{\sum_i [h(L_i)+h(e_i)]+3h(\phi_d)}$$

Now, if the only fields that are in chiral representations of $U(1)_H$ and transform non-trivially under the SM gauge group are the MSSM supermultiplets, we can then calculate the mixed anomalies which are:

$$[SU(3)]^2 U(1)_H : C_3 = \sum_i [2h(q_i) + h(u_i) + h(d_i)]$$

$$[SU(2)]^2 U(1)_H : C_2 = \sum_i [3h(q_i) + h(L_i)] + h(\phi)$$

$$[U(1)_Y]^2 U(1)_H : C_1 = \sum_i [\frac{h(q_i)}{3} + \frac{8h(u_i)}{3} + \frac{2h(u_i)}{3} + h(L_i) + 2h(e_i)] + h(\phi)$$

$$[Grav]^2 U(1)_H : C_{Gr} = \sum_i [3(2h(q_i) + h(u_i) + h(d_i)) + 2h(L_i) + h(e_i)] + 2h(\phi)$$

$$[U(1)_H]^2 U(1)_Y : C^{(2)} = \sum_i [h(q_i)^2 - 2h(u_i)^2 + h(d_i)^2 - h(L_i)^2 + h(e_i)^2] + h^2(\phi)$$

$$[U(1)_H]^3 : C^{(3)} = \sum_i [6h(q_i)^3 + 3h(u_i)^3 + 3h(d_i)^3 + 2h(L_i)^3 + h(e_i)^3]$$
$$+ 2h^3(\phi).$$

Where $h(\phi) = h(\phi_u) + h(\phi_d)$, $h^2(\phi) = h^2(\phi_d) - h^2(\phi_u)$ and $h^3(\phi) = h^3(\phi_u) + h^3(\phi_d)$, and the hypercharges $h(\phi_u)$ and $h(\phi_d)$ are included in the anomalies due to the existence of the spin 1/2 shiggsses in the MSSM. Now if $m_{up} \neq 0$, the symmetries in $\mathcal{L}$ allow us to combine the former expressions in order to writte [2]

$$Det(M^L).Det(M^u)^{1/3}.Det(M^d)^{-2/3} \sim <\phi_u><\phi_d>\lambda^{(C_1+C_2-2C_3)/2}. \quad (5)$$

Assuming approximate geometrical hierarchies and $Det(M^L) \sim Det(M^d)$, then the left hand side of the former equation is $\sim m_s m_c$, while the right hand side is $\sim m_b m_t \lambda^{(C_1+C_2-2C_3)/2}$ (for $\tan\beta \leq m_t/m_b$). The conclusion is then that $C_1 + C_2 - 2C_3 > 10$ for $\lambda \sim 0.22$, and the mixed anomalies can not vanish simultaneously (conclusion independent of the value for $h(\phi)$). Then $U(1)_H$ must be anomalous and broken at a large scale by an induced Fayet-Iliopoulos term [4] and the cancellation of the anomalies must be implemented with a Green-Schwartz mechanism [5]. The consistent condition for this anomaly cancellation reads [6]:

$$\frac{C_1}{\kappa_1} = \frac{C_2}{\kappa_2} = \frac{C_3}{\kappa_3} = \delta_{GS}$$

where κ_1, κ_2 and κ_3 are the Kac-Moody levels of $U(1)_Y, SU(2)_L$ and $SU(3)_c$ respectively, and δ_{GS} is a constant that gives the transformation of an axion field S under $U(1)_H$. For ilustrations of this anomalous case we reffer the reader to the extensive litterature [2,3,7] (where $\lambda \sim M_{string}/M_{Planck}$, with $M_{string} \sim 10^{17.4} GeV$ is the perturbative SUSY string mass scale).

THE NON-ANOMALOUS CASE

If $m_u = 0$, the analysis leading to Eq. (5) is not longer valid, and then we do not have any constraint on the values for the mixed anomalies. It is then possible to look for a suitable horizontal charge assignment such that [8] $C_\alpha = C^2 = C^3 =$

0 ($\alpha = 1, 2, 3, Gr$), which provides us with an anomaly-free horizontal abelian gauge supersymmetry which may reproduce the observed pattern of fermion masses and mixing angles. Some implications of this symmetry are analysed in Ref. [8], where λ can take the natural value $\lambda \sim M_{SM}/M_{SUSY}$ (where $M_{SM} \sim 246\ GeV \sim \sqrt{2}\langle\phi_u\rangle$ is the electroweak mass scale).

Then, it may be possible that a non anomalous horizontal $U(1)_H$ gauge supersymmetry, can be the responsible for the fermion mass hierarchies of the MSSM. So far does not exist a decisive model of this kind in the literature, but the road is open for the model builders.

$U(1)_H$ NON-SUSY

In the non-SUSY SM case there does not exist holomorphy zeroes, and the texture of the mass matrices must be implemented in a nontrivial way as shown anon. There is also only one higgs scalar ϕ_d without its SUSY spin 1/2 partner. It is also imposible to implement a Green-Scwartz mechanism, so a consistent non-SUSY $U(1)_H$ must be anomaly free. That is, it must satisfy $C_\alpha = C^2 = C^3 = 0$, ($\alpha = 1, 2, 3, Gr$) with $h(\phi) = h^2(\phi) = h^3(\phi) = 0$ in C_α, C^2 and C^3 (scalars do not contribute to triangle anomalies and shiggsses do not exist in the SM).

Now the Yukawa coefficients in $\mathcal{L}$ in Eq. (1) are given by:

$$Y_{ij}^u = A_{ij}^u \lambda^{|h(q_i)+h(u_j)-h(\phi_d)|}$$
$$Y_{ij}^d = A_{ij}^d \lambda^{|h(q_i)+h(d_j)+h(\phi_d)|} \quad (6)$$
$$Y_{ij}^L = A_{ij}^L \lambda^{|h(L_i)+h(e_j)+h(\phi_d)|}$$

Where the top quark mass exists at tree level as far as $h(q_3)+h(u_3) = h(\phi_d)$, which in turn implies $Y_{33}^u = A_{33}^u \simeq 1$ (with a top quark mass of $m_t \simeq \langle\phi_d\rangle = 174$ GeV.).

Now, if for all $i, j = 1, 2, 3$, $h(q_i)+h(u_j)-h(\phi_d) > 0$ and $h(q_i)+h(d_j)+h(\phi_d) > 0$ then we get the relationship $Det(M^u).Det(M^d) \equiv \langle\phi_d\rangle^6 \lambda^{C_3}$ which in turn implies $C_3 \neq 0$ producing an inconsistent $U(1)_H$ non-SUSY local gauge theory. So, in order to get a consistent mass spectrum we must demand the existence of exponents with opposite signs in Y_{ij}^u or either in Y_{ij}^d or in both.

The texture zeroes of the mass matrices can now be generated by demanding that $h(q_i) + h(u_j) - h(\phi_d) =$ an integer number, modulus n, for $n = 2, 3, 4, ...$ [9].

ACKNOWLEDGMENTS

This work was partially supported by Colciencias in Colombia and BID.

REFERENCES

1. William A. Ponce *et al*, Phys. Rev. **D44** (1991) 2166.

2. Y.Nir, Phys. Lett. **B354** (1995) 107
3. M.Leurer, Y.Nir and N.Seiberg, Nucl. Phys. **B398** (1993) 319; **B420** (1994) 468.
4. M.Dine, N.Seiberg and E.Witten, Nucl. Phys. **B289** (1987) 589; J.Atick, L.Dixon and A.Sen, Nucl. Phys. **B292** (1987) 109.
5. M.Green and J.Schwartz, Phys.Lett. **B149** (1984) 117.
6. L.E.Ibañez, Phys. Lett. **B303** (1993) 55.
7. P.Binétruy and P.Ramond, Phys.Lett **B350** (1995) 49; V.Jain and R.Shrock , Phys.Lett. **B352** (1995) 83; P.Binétruy, S.Lavignac and P.Ramond, Nucl. Phys. **B477** (1996) 353; E.J.Chun and A. Lukas, Phys. Lett. **B387** (1996) 99; E.Dudas, C.Grojean, S.Pokorski and C.A.Savoy, Nucl. Phys. **B354** (1996) 107.
8. E.Nardi, J.M.Mira and D.R.Aristizabal, [hep-ph/9911212].
9. W.A.Ponce, L.E.Epele and C.G.Canal, Phys. Lett. **B411** (1997) 159.

Bounds on neutrino mixing angles within the context of $SU(6)_L \otimes U(1)_Y$ model

R. Gaitán†, E. García‡, A. Hernández-Galeana*[1] and J. M. Rivera-Rebolledo*[1]

* Departamento de Física, Escuela Superior de Física y Matemáticas, I.P.N.,
U.P. Adolfo L. Mateos, México D.F., 07738, México.
† Centro de Investigaciones Teóricas, FES, UNAM,
Apartado Postal 142, Cuautitlán-Izcalli, Estado de México,
Código postal 54700, México.
‡ Departamento de Física, CINVESTAV-IPN
México D.F., 07000, México.

Abstract. We obtain limits on the mixing angles between ν_e, ν_μ, ν_τ and the heavy neutral fermions in the model $SU(6)_L \otimes U(1)_Y$. We also give values for g_e^2/g_μ^2, for the first row of CKM matrix and for the invisible decay rate of the Z boson; its close agreement with the experimental data is shown.

I INTRODUCTION

We first review briefly some kind of unified models before dealing with the $SU(6)_L \otimes U(1)_Y$ model. We are interested in generic models with gauge group $G \supset G_{SM} \otimes G_H$, where $G_{SM} = SU(2)_L \otimes U(1)_Y \otimes SU(3)_C$ is the gauge group of the "standard model" (SM), while G_H represents the gauge group of a horizontal symmetry. The spontaneous symmetry breaking (SSB) of G is performed at least in the following steps:

$$G \xrightarrow{<\phi_1>} G_{SM} \otimes G_H \xrightarrow{<\phi_2>} G_{SM} \xrightarrow{<\phi_{SM}>} U(1)_Q \otimes SU(3)_C \quad (1)$$

where $<\phi_1>$, $<\phi_2>$, and $<\phi_{SM}>$ are the vacuum expectation values of the scalars. Some examples of such G models are $SU(6)_L \otimes U(1)_Y \otimes SU(3)_C$ [1] and $[SU(6)]^3 \otimes Z_3$ [2]. In general, extended models introduce new particles (scalars, gauge bosons and fermions), which can mix with ordinary particles. In this work we obtain bounds on the mixing angles between the new neutral fermions and the known neutrinos in the $SU(6)_L \otimes U(1)_Y \otimes SU(3)_C$ model. We study the consequences of these mixing angles on experimental constraints on universality,

---

[1] With support from COFAA, EDD and SNI

the unitarity of the CKM matrix, the invisible width of the Z boson, and leptonic decays of the π mesons and the τ lepton [3]

II THE MODEL

Besides the ordinary fermions, in order to cancel anomalies, the model $SU(6)_L \otimes U(1)_Y$ introduce a $\{\overline{15}(0)\}_L = \psi^{(0)}_{[\alpha\beta]L}$ multiplet of exotic leptons, where the number in parenthesis stands for the hypercharge, the symbol $[\alpha\beta]$ means antisymmetric ordering and the label L refers to the left handed Weyl spinors. Further,

$$\psi^L_{[\alpha\beta]}(0) = \begin{pmatrix} 0 & N_1 & E_1^- & N_4 & E_2^- & N_6 \\ & 0 & N_5 & E_1^+ & N_7 & E_2^+ \\ & & 0 & N_2 & E_3^- & N_8 \\ & & & 0 & N_9 & E_3^+ \\ & & & & 0 & N_3 \\ & & & & & 0 \end{pmatrix}_L \tag{2}$$

is the multiplet of exotic leptons which are classified according to $SU(2)_L$ into 3 triplets,

$$\begin{pmatrix} E_1^+ \\ (N_4+N_5)/\sqrt{2} \\ E_1^- \end{pmatrix}, \begin{pmatrix} E_2^+ \\ (N_6+N_7)/\sqrt{2} \\ E_2^- \end{pmatrix}, \begin{pmatrix} E_3^+ \\ (N_8+N_9)/\sqrt{2} \\ E_3^- \end{pmatrix},$$

and six neutral singlets,

$$N_1, \quad N_2, \quad N_3, \quad (N_4-N_5)/\sqrt{2}, \quad (N_6-N_7)/\sqrt{2}, \quad (N_8-N_9)/\sqrt{2}.$$

The symmetry breaking is realized in three stages: at the scale M_1

$$SU(6)_L \otimes U(1)_Y \longrightarrow SU(2)_L \otimes SU(2)_H \otimes U(1)_Y, \tag{3}$$

and the six $SU(2)_L$ exotic neutral singlets acquires mass of order M_1.

At the scale M_2, $SU(2)_L \otimes SU(2)_H \otimes U(1)_Y \longrightarrow SU(2)_L \otimes U(1)_Y$ and the exotic leptons which transform as triplets of $SU(2)_L$ now have a mass of order M_2. The final stage of the symmetry breaking chain, $SU(2)_L \otimes U(1)_Y \longrightarrow U(1)_{EM}$ is achieved using a Higgs $\phi_3 = \phi_{3a}(1) = \{\overline{6}(1)\}$,

We arrange all the independent neutral fermionic degrees of freedom as the vector of left-handed spinors

$$\begin{aligned} n_L^{0T} = & (N_1, N_2, N_3, (N_4-N_5)/\sqrt{2}, (N_6-N_7)/\sqrt{2}, (N_8-N_9)/\sqrt{2}, \\ & \nu_e, \nu_\mu, \nu_\tau, (N_4+N_5)/\sqrt{2}, (N_6+N_7)/\sqrt{2}, (N_8+N_9)/\sqrt{2})^T, \end{aligned} \tag{4}$$

where ν_e, ν_μ, ν_τ are the ordinary neutrinos and the rest the new heavy neutral fermions. The fields in n_L^{0T} are gauge eigenstates. The mass terms for these neutral states are given by [1] $L_{mass} = n_L^{0T} C M n_L^0$ where the charge conjugation matrix C completes a Lorentz invariant form. The structure of this symmetric mass matrix M and the scales M_1, M_2, and v give us the model dependence. Note that M has the form

$$M = \begin{pmatrix} A_{6x6} & B_{6x3} & 0_{6x3} \\ B_{3x6}^T & 0_{3x3} & 0_{3x3} \\ 0_{3x6} & 0_{3x3} & C_{3x3} \end{pmatrix}, \qquad (5)$$

that is, M decouple in the matrix C_{3x3} and a M'_{9x9} see-saw type mass matrix. All the nonzero entries of C are of the order of M_2, and this is the order of the mass eigenvalues corresponding to the triplets of the neutral exotic leptons. The mass matrix we are interested in is M'_{9x9}, because this matrix contains the mixing of the ordinary neutrinos with the six neutral exotic singlets, which get mass in the first step of SSB of order M_1.

The diagonalization of M'_{9x9} is achieved by the matrix U, which satisfies $U^T M' U = M_{diagonal}$, where

$$U = \begin{pmatrix} A & G \\ F & H \end{pmatrix}, \qquad (6)$$

with the unitarity conditions $A^\dagger A + F^\dagger F = AA^\dagger + GG^\dagger = I$. The light eigenvalues in the limit $M_1 >> M_2$, with $\bar{m} = \sqrt{5} \frac{v^2}{M_1}$, are

$$m_{\nu_e} = \frac{3}{2}\bar{m}, \quad m_{\nu_\mu} = \frac{17}{8}\bar{m}, \quad m_{\nu_\tau} = \frac{129}{16}\bar{m}. \qquad (7)$$

III PHENOMENOLOGICAL IMPLICATIONS.

a Leptonic Universality. As a consequence of diagonalization of the mass matrix of the neutral fermions in the $SU(6)_L \otimes U(1)_Y$ model [4] one has

$$H_L H_L^+ \simeq diag(c_{\nu_e}^2, c_{\nu_\mu}^2, c_{\nu_\tau}^2)_L, \qquad (8)$$

with

$$c_{\nu_e L}^2 = \frac{1}{1+\frac{45}{2}\delta^2}, \quad c_{\nu_\mu L}^2 = \frac{1}{1+\frac{285}{16}\delta^2}, \quad c_{\nu_\tau L}^2 = \frac{1}{1+\frac{17055}{128}\delta^2}, \qquad (9)$$

where $\delta = \frac{v}{M_1}$, $v \sim 250 GeV$, and $M_1 > 2.9947 \times 10^{10} GeV$. From eqs. (9) we find $s_{\nu_e}^2 < 1.57 \times 10^{-15}$, $s_{\nu_\mu}^2 < 1.24 \times 10^{-15}$ and $s_{\nu_\tau}^2 < 0.928 \times 10^{-14}$.

$$\frac{g_e^2}{g_\mu^2} = \frac{c_{\nu_e L}^2}{c_{\nu_\mu L}^2} = 1 - 0.327 \times 10^{-15}, \quad \frac{g_\tau^2}{g_\mu^2} = \frac{c_{\nu_\tau L}^2}{c_{\nu_\mu L}^2} = 1 - 1.026 \times 10^{-14} \times 10^{-7} \qquad (10)$$

whereas the experiment [5] gives $\frac{g_e^2}{g^2} \approx 1 - 0.327 \times 10^{-15}$, and $\frac{g_\tau}{g_\mu} = 1.0025 \pm 0.0002$ In this way we test universality.

Quantity	Model	Ref. ([3])	Experiment
$s^2_{\nu_e}$	1.57×10^{-15}	7.1×10^{-3}	
$s^2_{\nu_\mu}$	1.24×10^{-15}	1.4×10^{-3}	
$s^2_{\nu_\tau}$	9.28×10^{-15}	3.3×10^{-2}	
$\frac{\Gamma(Z \to inv)}{\Gamma^{SM}(Z \to inv)}$	$1 - 2.625 \times 10^{-15}$		0.9919 ± 0.0077
$\sum_{i=1}^{3} |V_{ui}|^2$	$1 + 1.24 \times 10^{-15}$		0.9969175 ± 0.0029608
g_e^2/g_μ^2	$1 - 0.327 \times 10^{-15}$		0.9585 ± 0.0031
g_τ^2/g_μ^2	$1 - 1.026 \times 10^{-14}$		1.00025 ± 0.0002

TABLE 1. Summary of results

b **The invisible width of the Z boson.** Following [3] the $SU(6)_L \otimes U(1)_Y$ predicts for 'the invisible width of the Z boson with $t_3 = 0$ for singlets and real triplets,

$$\frac{\Gamma(Z \to inv)}{\Gamma^{SM}(Z \to inv)} = 1 - 2.625 \times 10^{-15} \qquad (11)$$

which is very close to experimental value [5] ($m_t = (173.3 \pm 5.6)$ GeV):

$$\frac{\Gamma(Z \to inv)}{\Gamma^{SM}(Z \to inv)} = 0.9919 \pm 0.0077. \qquad (12)$$

c **CKM unitarity.** Finally the CKM unitarity in our model is expressed as

$$\sum_{i=1}^{3} |V_{ui}|^2 = 1 + 1.24 \times 10^{-15} \qquad (13)$$

and the experimental result is [3], $\sum_{i=1}^{3} |V_{ui}|^2 = 0.9969 \pm 0.0029$.
All the above results are collected in the Table 1.

IV CONCLUSIONS

We have found that the limits for the mixing angles between the standard neutrinos and neutral heavy fermions in the context of the $SU(6)_L \otimes U(1)_Y$ model are in close agreement with that reported in Ref.([3]). The same good agreement is exhibited in Table 1 for the lepton universality, CKM unitarity and the invisible width of the Z boson, when compared with the experiment.

V ACKNOWLEDGMENTS

R.G. acknowledges financial support from DGAPA (Programa de Estimulos de Iniciación a la Investigación). We thank M. Pérez Angón for fruitful conversations.

REFERENCES

1. A. Hernández, W. A. Ponce, and A. Zepeda, Z Phys.C 55, 423(1992).
2. A. Hernández Galeana, R. E. Martínez, W. A. Ponce and A. Zepeda, Phys. Rev. D44, 2166(1991).
3. E. Nardi, E. Roulet, and D. Tommasini, Phys. Lett.B 327, 319(1994).
4. R. Gaitán, A. Hernández, S. Tomás, W. A. Ponce, and A. Zepeda, Phys. Rev. D 51, 6474(1995).
5. Particle Data Group Collaboration, D. Haidt et. al., Review of particle physics, Eur. Phys. J. C3, 1-174(1998).

Flavor changing neutral current decays of the top quark

M. A. Pérez [a,b] and G. Tavares-Velasco [a]

[a] *Departamento de Física, CINVESTAV, Apartado Postal 14-740, 07000, México D. F., México*
[b] *Facultad de Ciencias Físico Matemáticas, Universidad Autónoma de Puebla, Apartado Postal 1152, Puebla, México*

Abstract.
A review of the top quark rare decay modes $t \to cV_i$ and $t \to cV_iV_j$ ($V_i = \gamma, Z, g$) is presented within the framework of the Standard Model (SM) and some of its extensions. Since these decays are induced by flavor changing neutral current interactions (FCNC), they constitute a strong test for the SM. We found that these transitions open up the possibility of looking for signals from new physics.

NEW PHYSICS EFFECTS

More than three decades after its raise, a plethora of experimental evidences support the standard model (SM). Although the Higgs boson remains as its only missing piece, the SM lacks of answers to some fundamental questions. Many extensions have been conjectured as it is believed that the electroweak gauge group is a low energy representation embedded into a larger gauge group. Then, it is important to test some SM issues in order to find any effect which could arise from any of its extensions. The phenomenology of the top quark turns out to be very interesting since this fermion is unexpectedly heavier than the remaining fundamental particles, so it is likely that some new physics effects might be more apparent in processes involving the top quark than in those involving any other light sector of the theory. As the top quark mass is of order of the Fermi scale $\nu = (\sqrt{2}G_F)^{-1/2} = 246$ GeV, which in turn characterizes the electroweak symmetry-breaking scale, it might be a useful tool to probe the SM scalar sector. According to the planned projects for the Tevatron and the large hadron collider (LHC), it will be possible an adequate production of top quark pairs $t\bar{t}$ to allow higher precision measurements of the observable parameters m_t, $\Gamma(t \to bW)$ and $B(t \to bW)$, and also to perform electroweak precision studies. The aim of this presentation is to bring the attention to some FCNC top quark transitions which would be useful to test the SM and some of its extensions at the above mentioned colliders.

RARE DECAYS OF THE TOP QUARK

The following rare decays of the top quark are kinematically allowed: $t \to c\gamma$ [1], $t \to cg, cZ$ [2], $t \to cW^+W^-$ [3,4], $t \to c\gamma\gamma$, $c\gamma Z$, and cgg [4]. All these transitions are negligible in the SM due to the Glashow-Illiopoulos-Maiani (GIM) mechanism; except for the decay $t \to cW^+W^-$, the remaining ones are induced at one-loop level. Therefore, these decay modes represent an interesting playground to search for new physics effects. In table 1 we show some estimates for the respective branching ratios, within the SM and some of its extensions. For completeness we also include the rare decay $t \to bW^+Z$ [3], though it is not suppressed by the GIM mechanism but instead by phase space. The values shown in the table suggest that the FCNC decays $t \to cV_iV_j$ might serve as a test not only for the SM but for the model III, where certain branching fractions reach their most significant values.

TABLE 1. Branching ratios for some rare decays of the top quark, in the SM, supersymmetry (SUSY), Model III, technicolor models (TC), and model independent estimates (MI).

	SM	SUSY	Model III	TC	MI
$B(t \to c\gamma)$	10^{-12} [1]	10^{-8} [5]	$10^{-8} - 10^{-10}$ [6]	10^{-8} [7]	$< 10^{-3}$ [8]
$B(t \to cZ)$	10^{-12} [3]	10^{-8} [5]	10^{-6} [6]	10^{-7} [7]	
$B(t \to cg)$	10^{-10} [3]	10^{-7} [5]	10^{-5} [6]	10^{-6} [7]	< 0.4 [9]
$B(t \to bW^+Z)$	10^{-6} [3]		10^{-2} [10]		
$B(t \to cW^+W^-)$	10^{-10} [4]		10^{-5} [11,4]		
$B(t \to c\gamma\gamma)$	10^{-16} [4]		10^{-5} [4]		
$B(t \to c\gamma Z)$	10^{-16} [4]		10^{-5} [4]		
$B(t \to cgg)$	10^{-15} [4]		10^{-5} [4]		

Unlike earlier versions of the two-higgs doublet model (2HDM), in Model III no *ad hoc* symmetries are invoked to eliminate tree-level scalar FCNC couplings but instead a more realistic pattern for the Yukawa matrices is imposed and constraints on the scalar FCNC are derived from phenomenology [6]. The tree-level scalar FCNC interactions are given by

$$\mathcal{L}_{Y,FCNC}^{III} = \xi_{ij} \sin\alpha \bar{f}_i f_j h^0 + \xi_{ij} \cos\alpha \bar{f}_i f_j H^0 + \xi_{ij} \cos\alpha \bar{f}_i \gamma^5 f_j A^0 + h.c., \qquad (1)$$

where we are using the Higgs boson mass-eigenstate basis with the light and heavy CP-even Higgs bosons h^0 and H^0, respectively, and the CP-odd one A^0, α denotes the mixing angle, and ξ_{ij} corresponds to the off-diagonal Yukawa couplings. It is usual to use the parametrization $\xi_{ij} = \lambda_{ij}\sqrt{m_i m_j}/v$ [12], where the mass factor gives the strength of the interaction while the dimensionless parameters λ_{ij} are fixed by low energy experiments. There are strong bounds on couplings involving light quarks but no stringent bounds exist for λ_{tc}, so it is feasible a less suppressed strength for the interaction $tc\phi_k^0$, with ϕ_k^0 any of the three physical higgs bosons of Model III.

In the SM and Model III the FCNC decays $t \to cV_iV_j$ may proceed through a Higgs boson mediated resonant Feynman diagram. The decays $t \to c\gamma\gamma$, $t \to c\gamma Z$, $t \to cgg$ and $t \to cWW$ have been studied in the framework of the SM [4], the results are shown in Table 1. As far as the Model III is concerned, the branching ratios become proportional to the factor $f_\lambda = \lambda_{tc}\sin\alpha\cos\alpha$, chosen as real for simplicity. The respective values are shown as a function of the Higgs boson mass m_{h^0} in Fig. 1. It can be observed that the most significant enhancement is reached in the resonance region $(m_{V_i}+m_{V_j}) \leq m_{h^0} \leq (m_t-m_c)$, where the channels $t \to c\gamma\gamma$ and $t \to c\gamma Z$ may have branching ratios as high as 10^{-5}-10^{-4}, while those of the modes $t \to cWW$ and $t \to cgg$ may reach values up to 10^{-4}.

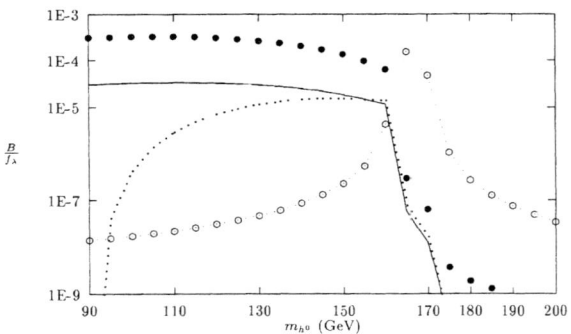

FIGURE 1. Scaled branching ratios for $t \to cV_iV_j$ in Model III: $t \to c\gamma\gamma$ (solid line), $t \to c\gamma Z$ (points), $t \to cW^+W^-$ (hollow circles), and $t \to cgg$ (full circles). A value of 750 GeV is used for the masses m_{A^0}, m_{H^0} and $m_{H^\pm}$.

We will briefly discuss the mode $t \to c + \gamma\gamma$ as a representative case of new signatures associated with FCNC top quark decays. This reaction turn out to be very clean when it is attempted to separate the signal from background. Let us consider the situation at the Tevatron, where it will be possible a production of top quark pairs through quark and gluon fusion with a cross-section of 5×10^3 fb [13]. If the top quark decays into $c + \gamma\gamma$ with a branching ratio of order 10^{-5}, an integrated luminosity of 10^2 fb^{-1} would be needed in order to produce at least one event, which exceeds the planned luminosity (of order 1 fb^{-1}). Then it seems hopeless to detect the FCNC two-photon top quark decay at the Tevatron. The situation looks more promising at the CERN Large Hadron Collider (LHC), where the cross-section to produce top quark pairs will be 4.3×10^5 fb, and a branching ratio of order 10^{-5} will reduce this cross-section down to 4.3 fb. We will use a conservative efficiency (0.05) for the production of the $t + c + \gamma + \gamma$ events at the LHC [9]. An integrated luminosity of 100 fb^{-1} will produce about 21 events per year at the LHC. The main background to this signal comes from the production of $qb \to tc + \gamma + \gamma$ and $qq' \to t + b + \gamma + \gamma$ events, where the final b may be misidentified as c. After the inclusion of the suppression factor $(\frac{4\alpha}{9\pi})^2$ to account for the two emitted photons, these background processes will have less than one event at the LHC and thus it will be feasible the detection of the top quark decay.

FINAL REMARKS

Together with some unwanted features, the scalar sector of the SM with its elusive Higgs boson remains as a puzzle. Supersymmetric Grand Unified Theories allow for a relatively light Higgs boson in the range $100-200$ GeV [14], and it is possible that in the next years it will occur the discovery of such a light boson. Meanwhile there is an incessant search for any evidence of new physics that could give any clue to the achievement of a more comprehensive theory. The results presented in here open up the possibility of looking for signals from physics beyond the SM through the decays $t \to c\gamma\gamma$, $t \to c\gamma Z$, $t \to cWW$, and $t \to cgg$.

ACKNOWLEDGEMENTS

This work was partially supported by CONACYT (México).

REFERENCES

1. J. L. Diaz-Cruz, R. Martinez, M. A. Pérez, A. Rosado, *Phys. Rev. D* **41**, 891 (1990).
2. B. Grzadkowski, J. F. Gunion, P. Krawczyk, *Phys. Lett.* **B 268**, 106 (1991); G. Eilam, J. L. Hewett, A. Soni, *Phys. Rev. D* **44**,1473 (1997); M. Luke, M. J. Savage, *Phys Lett.* **B 307**, 387 (1993); G. Couture, C. Hamzauni, K. König, *Phys. Rev. D* **52**, 1713 (1995).
3. E. Jenkins, *Phys. Rev. D* **52**, 1713 (1995).
4. J.L Diaz-Cruz, M.A. Pérez, G. Tavares-Velasco, J.J Toscano, *Phys. Rev. D* **60**, 115014 (1999).
5. C.-S Li *et al. Phys. Rev. D* **49**, 293 (1994).
6. D. Atwood, L. Reina, A. Soni, *Phys. Rev. D* **55**, 3156 (1997).
7. X. Wang *et al.*, *Phys. Rev. D* **50** (1994) 5781.
8. R. Martinez, M.A. Pérez, J.J. Toscano, *Phys. Lett.* **B 340**, 91 (1994).
9. T. Han, K. Whisnant, B. -L. Young, and X. Zhang, *Phys. Rev. D* **55**, 7241 (1997).
10. J.L Diaz-Cruz, D.A. Lopez-Falcon, *Phys. Rev. D* **61**, 051704 (2000).
11. S. Bar-Shalom, G. Eilam, A. Soni, J. Wudka, Phys. Rev. Lett. **19**,1217 (1997); *Phys. Rev. D* **57**, 2957 (1998).
12. T.P. Cheng, M. Sher, *Phys. Rev. D* **35**,3484 (1987); A. Antaramian, L. Hall, A. Rasin, *Phys. Rev. Lett.* **69**, 1871 (1992).
13. C.P. Yuan in *Lectures on top quark physics*, Procc. of the VI Mexican School of Particles and Fields, edited by J. C. D'Olivo, M. Moreno, M.A. Pérez, World Scientific, Singapore, 1995.
14. M. Carena, *et. al.* in *Higgs Physics at LEP2*, CERN 96-01, edited by G. Altarelli, T. Sjöstrand, F. Zwiner.

Itemization of Trilinear Couplings for Neutral Gauge Bosons

F. Larios[a], M. A. Pérez[b,c], G. Tavares-Velasco[b] and J.J. Toscano[c]

[a] *Departmento de Física Aplicada, CINVESTAV-Mérida,*
A.P. 73, 97310 Mérida, Yucatán, México.
[b] *Departmento de Física, CINVESTAV,*
Apdo. Postal 14-740, 07000 México.
[c] *Facultad de Ciencias Físico Matemáticas, Universidad Autónoma de Puebla,*
Apdo. Postal 1152, 72000 Puebla, Pue., México.

Abstract. We list all possible operators up to dimension 8 which induce the triple neutral gauge boson couplings $ZZ\gamma$, $Z\gamma\gamma$, and ZZZ within the decoupling effective Lagrangian approach.

INTRODUCTION

There has been recently increased interest in testing the gauge boson couplings predicted by the Standard Model (SM) [1]. While the experimental limits on possible anomalous $W^+W^-\gamma(Z)$ couplings have reached an accuracy of the few percent level in both hadronic [2] and leptonic [3] colliders, the situation is not that good for the case of triple neutral gauge boson (TNGB) couplings [4], which are absent in the SM at the tree level. Previous studies of the TNGB couplings have used a parametrization which is only $U(1)_{EM}$ gauge invariant [5]. Despite the generality of such studies, few vertex functions (twelve indeed) are required to describe all possible ZZZ, $ZZ\gamma$ and $Z\gamma\gamma$ couplings. This is because of the *a priori* requirement that two of the boson fields be on-shell. Our study allows for all three of the fields to be off-shell, this will result in a greater set of possible vertices. The purpose of the present paper is to classify all possible $SU(2) \times U(1)$ gauge invariant operators in the context of the linearly realized effective Lagrangian approach. We find that these operators arise at least at dimension 8.

I CHARACTERIZATION OF OPERATORS

We will find out all dimension 8 C-odd operators that generate one or more of the vertices ZZZ, $ZZ\gamma$ or $Z\gamma\gamma$. As it turns out, no dimension 6 operator gives rise to TNGB couplings.

Any $SU(2) \times U(1)$ gauge invariant operator is constructed out of the appropriate combinations of the following building blocks:

Number	Operator	SU(2)	Lorentz
T.1	$W^i_{\mu\nu}$	vector	tensor
T.2	$B_{\mu\nu}$	scalar	tensor
T.3	$\Phi^\dagger\Phi$	scalar	scalar
T.4	$\Phi^\dagger\sigma^i\Phi$	vector	scalar
T.5	$\Phi^\dagger D_\mu\Phi$	scalar	vector
T.6	$\Phi^\dagger\sigma^i D_\mu\Phi$	vector	vector
T.7	$\Phi^\dagger(D_\mu D_\nu + D_\nu D_\mu)\Phi$	scalar	tensor
T.8	$\Phi^\dagger\sigma^i(D_\mu D_\nu + D_\nu D_\mu)\Phi$	vector	tensor

In order to simplify the complete characterization of the operators of interest we divide them into two classes:

I) Operators involving no terms T.5-8, and

II) Operators involving at least one of the terms T.5-8.

Let's start with class I. Another division is convenient, a TNGB coupling can come from either the product of two tensor fields:

I.A.1) $W^i_{\mu\nu}W^j_{\alpha\beta}$, and I.A.2) $W^i_{\mu\nu}B_{\alpha\beta}$,

or, it can come from the product of three tensor fields:

I.B.1) $W^i_{\mu\nu}W^j_{\alpha\beta}W^k_{\lambda\rho}$, I.B.2) $B_{\mu\nu}B_{\alpha\beta}B_{\lambda\rho}$, I.B.3) $B_{\mu\nu}B_{\alpha\beta}W^k_{\lambda\rho}$, and I.B.4) $B_{\mu\nu}W^j_{\alpha\beta}W^k_{\lambda\rho}$.

Cases I.A.1 and 2 can not give rise to a TNGB vertex, this is because two of the fields will necessarily come from the nonabelian part of the W tensor and then at least one of the two charged weak boson components W^1 or W^2 will be there. Then we are left with operators I.B.1-4. The first type does not generate any TNGB vertex; I.B.1 will at least contain one charged W boson component. For the other three cases, I.B.2-4, unless we add two derivatives, they will vanish after Lorentz contraction. However, by adding two derivatives we can generate dimension 8 operators. The ones coming from I.B.2 are listed below. Next, I.B.3 necessarily requires an $SU(2)$ contraction with a term like T.4 which automatically adds two more dimensions; upon lorentz contraction it will vanish except if one of the B fields is the dual field. For I.B.4 we obtain (1+7) operators.

$$\mathcal{O}_{BBB} = \partial^\rho B_{\mu\nu}\partial^\lambda B_{\mu\rho}B_{\lambda\nu}, \tag{1}$$

$$\mathcal{O}_{BWW1} = \partial^\rho B_{\mu\nu}\partial^\lambda \left(W^i_{\mu\rho}W^i_{\lambda\nu}\right), \tag{2}$$

$$\mathcal{O}_{WB\widetilde{B}} = \Phi^\dagger\sigma^i W^i_{\mu\nu}\Phi\widetilde{B}_{\mu\rho}B_{\nu\rho}, \tag{3}$$

$$\mathcal{O}_{\widetilde{B}BB1} = \partial^\rho \widetilde{B}_{\mu\nu} \partial^\lambda B^\mu_\rho B^\nu_\lambda, \tag{4}$$

$$\mathcal{O}_{\widetilde{B}BB2} = \partial^\rho \widetilde{B}_{\mu\nu} \partial^\rho B^{\mu\lambda} B^\nu_\lambda, \tag{5}$$

$$\mathcal{O}_{\widetilde{B}BB3} = \partial^\rho \widetilde{B}_{\mu\nu} \partial^\nu B_{\lambda\rho} B^{\mu\lambda}, \tag{6}$$

$$\mathcal{O}_{\widetilde{B}BB4} = \partial^\rho B_{\mu\nu} \partial^\rho B^\nu_\lambda \widetilde{B}^{\mu\lambda}, \tag{7}$$

$$\mathcal{O}_{B\widetilde{W}W1} = \partial_\alpha B_{\mu\nu} \partial^\alpha (\widetilde{W}^{i\mu\rho} W^{i\nu}_\rho), \tag{8}$$

$$\mathcal{O}_{B\widetilde{W}W2} = \partial_\rho B_{\mu\nu} \partial_\lambda (\widetilde{W}^{i\mu\rho} W^{i\nu\lambda}), \tag{9}$$

$$\mathcal{O}_{B\widetilde{W}W3} = \partial_\rho B_{\mu\nu} \partial_\lambda (\widetilde{W}^{i\lambda\nu} W^{i\mu\rho}), \tag{10}$$

$$\mathcal{O}_{B\widetilde{W}W4} = \partial^\rho B_{\mu\nu} \partial^\nu (\widetilde{W}^i_{\lambda\rho} W^{i\mu}_\lambda), \tag{11}$$

$$\mathcal{O}_{B\widetilde{W}W5} = \partial^\rho B_{\mu\nu} \partial^\nu (\widetilde{W}^{i\mu\lambda} W^i_{\lambda\rho}). \tag{12}$$

Let us discuss the operators of class II, those that involve at least one of the terms T.5-8; here it is also convenient to separate the list into three classes, whether

II.A) two fields come from T.1-2 and one from T.5-8, or
II.B) one field from T.1-2 and two from T.5-8, or
II.C) all three fields from T.5-8.

No TGNB coupling can come from the nonabelian part of $W^i_{\mu\nu}$, the two fields must then come from the product of abelian parts of the following field tensors:

II.A.1) $W^i_{\mu\nu} W^j_{\alpha\beta}$,
II.A.2) $W^i_{\mu\nu} B_{\alpha\beta}$, and
II.A.3) $B_{\mu\nu} B_{\alpha\beta}$.

Let us write down all the possible operators for each case.

II.A.1) Here, T.7 and T.8 do not fit in any dimension 8 operator. Using T.5 we obtain

$$\mathcal{O}_{WW1} = \partial^\lambda (T5)_\lambda W^i_{\mu\nu} W^{i\mu\nu}, \tag{13}$$

$$\mathcal{O}_{WW2} = \partial_\lambda (T5)^\mu W^i_{\mu\nu} W^{i\lambda\nu}, \tag{14}$$

$$\mathcal{O}_{\widetilde{W}W1} = \partial^\lambda (T5)_\lambda \widetilde{W}^i_{\mu\nu} W^{i\mu\nu}, \tag{15}$$

$$\mathcal{O}_{\widetilde{W}W2} = \partial_\lambda (T5)^\mu \widetilde{W}^i_{\mu\nu} W^{i\lambda\nu}, \tag{16}$$

$$\mathcal{O}_{W\widetilde{W}3} = \partial_\lambda (T5)^\mu W^i_{\mu\nu} \widetilde{W}^{i\lambda\nu}. \tag{17}$$

A similar operator could be written with $T.6^i \epsilon^{ijk}$ instead of T.5 but this one would not generate a TNGB coupling.

II.A.2) To build an $SU(2)$ invariant we must make use of T.6:

$$\mathcal{O}_{WB1} = (T6)^i_\lambda W^i_{\mu\nu} \partial^\lambda B^{\mu\nu}, \tag{18}$$

$$\mathcal{O}_{WB2} = (T6)^i_\lambda W^i_{\mu\nu} \partial^\mu B^{\lambda\nu}, \tag{19}$$

$$\mathcal{O}_{\widetilde{W}B1} = (T6)_\lambda \widetilde{W}^i_{\mu\nu} \partial^\lambda B^{\mu\nu} \tag{20}$$

$$\mathcal{O}_{\widetilde{W}B2} = (T6)_\lambda \widetilde{W}^i_{\mu\nu} \partial^\mu B^{\lambda\nu}, \tag{21}$$

$$\mathcal{O}_{W\tilde{B}3} = (T6)_\lambda W^i_{\mu\nu}\partial^\lambda \tilde{B}^{\mu\nu}, \tag{22}$$

$$\mathcal{O}_{W\tilde{B}4} = (T6)_\lambda W^i_{\mu\nu}\partial^\mu \tilde{B}^{\lambda\nu}. \tag{23}$$

II.A.3) Using T.5 we obtain

$$\mathcal{O}_{BB1} = (T5)_\lambda B_{\mu\nu}\partial^\lambda B^{\mu\nu}, \tag{24}$$

$$\mathcal{O}_{BB2} = (T5)_\mu B_{\lambda\nu}\partial^\lambda B^{\mu\nu}, \tag{25}$$

$$\mathcal{O}_{\tilde{B}B1} = (T5)_\lambda \tilde{B}_{\mu\nu}\partial^\lambda B^{\mu\nu}, \tag{26}$$

$$\mathcal{O}_{B\tilde{B}2} = (T5)_\lambda B_{\mu\nu}\partial^\lambda \tilde{B}^{\mu\nu}, \tag{27}$$

$$\mathcal{O}_{\tilde{B}B3} = (T5)_\mu \tilde{B}_{\lambda\nu}\partial^\lambda B^{\mu\nu}, \tag{28}$$

$$\mathcal{O}_{B\tilde{B}4} = (T5)_\mu B_{\lambda\nu}\partial^\lambda \tilde{B}^{\mu\nu}. \tag{29}$$

Let us now turn to the case where only one term of T.1-2 is used.

II.B) Here, the only gauge invariant, non-zero Lorentz invariant operator that can generate a TNGB coupling is the following

$$\Phi^\dagger(D_\mu D_\rho + D_\rho D_\mu)\Phi \partial_\rho \partial_\nu B^{\mu\nu}.$$

However, we can use the equations of motion for the B field and relate this operator to others that involve fermion fields.

II.C) There is one operator that generates a ZZZ coupling:

$$\Phi^\dagger(D_\mu D_\nu + D_\nu D_\mu)\Phi \partial^\mu \Phi^\dagger D^\nu \Phi. \tag{30}$$

Similar operators, with T.6 and T.8, could be written but turn out to be equivalent to this one.

As we can see, this is a rather extensive list of $SU(2) \times U(1)$ operators that generate ZZZ, $ZZ\gamma$ and $Z\gamma\gamma$ couplings. Some remarks are now in order:

- Equations of motion for Z_μ and A_μ cannot relate these operators to $ZZf\bar{f}$ couplings [5,6].

- However, in the case where two of the bosons are on-shell, the vertex functions produced by these operators become equivalent to the ones used in the literature [1,5].

- Some operators can generate all three ZZZ, $ZZ\gamma$ and $Z\gamma\gamma$ couplings; some do not generate ZZZ; some generate charged boson couplings as well. Further analysis is due in order to set up a consistent and adequate characterization of their possible effects [7].

In conclusion, we have constructed a list of 30 independent dimension 8 $SU(2) \times U(1)$ invariant operators that generate TNGB couplings. This itemization can now be used to perform a systematic study, in the context of a linearly realized effective Lagrangian, that includes effects on the low energy observables and LEP/SLC measurements as is done for the charged boson couplings [1].

ACKNOWLEDGMENTS

We want to thank Conacyt for support.

REFERENCES

1. J. Ellison and J. Wudka, hep-ph/9804322, submitted to Ann. Rev. of Nucl. and Particle Science.
   S. Alam, S. Dawson and R. Szalapski, Phys. Rev. **D57**, (1998) 1577.
2. B. Abbot *et al.*, Phys. Rev. Lett. **79**, 1441 (1997).
3. B. Abcd *et al.*, Phys. Rev. Lett. **00**, 441 (1997).
4. B. Abbot *et al.* Phys. Rev. **D57**, R3817 (1998).
5. G.J. Gounaris, J. Layssac and F.M. Renard, hep-ph/9910395.
   S.Y. Choi, Z. Phys. C **68**, 163 (1995).
6. M.B. Einhorn and J. Wudka, Phys. Rev. **D55**, 3219 (1997).
7. F. Larios, M.A. Pérez, G. Tavares-Velasco and J. Toscano, in preparation.

The neutrino telescope ANTARES

A. Rostovtsev [1]

Institute f. Theoretical and Experimental Physics, Moscow, Russia

Abstract. ANTARES is a deep-underwater neutrino telescope project. The planned apparatus consists of an array of photomultipliers, arranged in a lattice near the sea bed at a depth of 2400 m, to detect the Cherenkov light from muons produced by neutrino in the seawater and rock beneath. A good water transparency allows an angular resolution of few tenth of degree necessary for the search of point-like cosmic neutrino sources. Detection of the muon neutrino with energy below 50 GeV would provide sensitivity to neutrino oscillation phenomenon recently reported by Super-Kamiokande experiment.

In order to observe the inner structure of the most luminous astrophysical objects over large energy range, one needs a probe which is a) electrically neutral, so that its trajectory is not affected by magnetic fields, b) stable, so that it can reach us over large distances, and c) weakly interacting, so that it penetrates regions opaque to photons and hadrons. The only candidate presenting these characteristics is the neutrino. The neutrinos as well as high energy gamma rays result from the decay of pions produced by accelerated protons. If so, the candidate sources of neutrinos are the mysterious cosmic accelerators of high energy particles. Presently we believe those include supernova remnants and X-ray binaries in our galaxy, and extragalactic sources like active galactic nuclei and gamma ray bursters.

Another hypothetical cosmological source of high energy neutrino is a decay of massive particles. The most popular models in this context are based on the annihilation of topological defects such as cosmic strings or monopoles formed in the early universe. Dark matter could also be detected with the high energy neutrino telescope if heavy super-symmetric neutralinos are responsible for the missing mass. Primordial neutralinos might get accumulated at the core of heavy bodies like the earth, the sun of the galaxy center. Neutralino annihilation products include neutrinos with energies around a half of the neutralino mass. Obviously, no other decay products than neutrino would escape the heavy bodies. Whatever the source, the neutrino telescope with an excellent angular resolution will be a new instrument to cover unexplored domain in astrophysics.

Neutrino oscillations can be studied in ANTARES using atmospheric neutrino

---

[1] On behalf of the ANTARES collaboration

and a long baseline of the order of the diameter of the earth. Atmospheric neutrinos result from decay of hadrons produced by the interaction of cosmic rays in the atmosphere. A measurement of neutrino energy and direction gives a ratio L/E, where L is a distance between the detector and neutrino production point, and E is the neutrino energy. A variation of atmospheric neutrino flux as function of L/E has a clear oscillation pattern. Assuming the Super-Kamiokande oscillation parameters ($\Delta m^2 = 0.0035 eV^2$, $sin(2\Theta) = 1$), the first dip in the oscillation pattern occurs at $350 km/GeV$, which turns out to be in the region of ANTARES sensitivity for low energy muon neutrino.

The drawback of neutrino detection is the weak interactions of neutrinos, which imply a very massive detector with extremely good background rejection. The pioneering idea to use the seawater as active detector media for neutrino was proposed by Russian physicist M.A. Markov [1] in 1960. Charged-current interaction of muon neutrino with water or rock produces a muon in the detector volume. The direction and energy of the muon can be inferred from the measurement of Cherenkov light produced by the relativistic muon travelling through the water. The background of atmospheric muons is strongly suppressed by absorption in the sea. Since the earth acts as a shield against all other particles except of neutrinos with energy below $1 PeV$, the detection of upward-going neutrinos is practically background free.

The ANTARES project [2] is aimed at the construction of an array of about 1000 photomultiplier tubes arranged in a volume of base area of $0.1 km^2$ with active height of $0.3 km$. A site of the detector location is about 20 nautical miles from Toulon (France) in the Mediterranean at the depth of $2400 m$. The 10-inch Hamamatsu photomultiplier tubes are housed in pressure resistant glass spheres to build the basic element of the detector called Optical Module (OM). The OM also contains a high voltage converter to operate the photomutiplier tube and electronic circuit, which stores the signals into the buffer and provides a trigger. Three optical modules are grouped together in a storey such, that the axes of the photomultiplier tubes point 45° down from horizontal. Each storey is fixed to a flexible string anchored at the sea bottom and strung up by buoyancy. At present design the detector consists of 13 strings. Each string contains 30 stories spaced vertically by 12 meters. The strings are placed in a spiral with a minimal distance between strings of 60-80 meters. Each string is connected to common junction box, which is connected to the on-shore counting room by the electro-optical cable. A schematic view of the ANTARES detector is shown in Fig.1.

A software package was developed to simulate neutrino interactions in the medium surrounding the detector, neutrino propagation, Cherenkov light emission, and the detector response. This package was used to optimize the detector and estimate its performance in terms of efficiency, energy and angular resolutions. The angular resolution improves with increase of the muon energy and is estimated to be about 0.2^o for muon energy above 10TeV. At this energy the angular resolution is not dominated by the reconstruction angle and not by the physical angle between neutrino and muon. The amount of light produced by the muon track allows an estimation of the muon energy. For energies above 1TeV, catastrophic bremsstrahlung and pair production dominate the muon energy loss, which increases with the energy. The error on the energy estimate is a factor 3 for energy below 10TeV and a factor 2 for energy above 10TeV. Below 100GeV the muon energy is estimated by measuring its range. For energy above 10TeV the cosmic neutrinos dominate over the atmospheric neutrino background. Several models predict more than 100 detected cosmic neutrinos per year in the detector above this energy threshold.

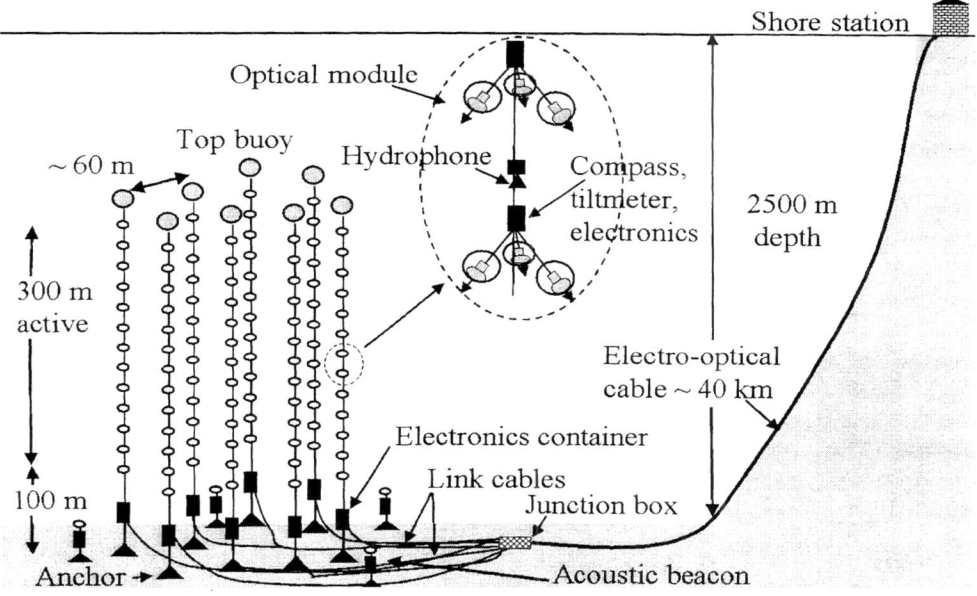

Fig.1 Schematic view of the ANTARES detector.

Three years of data taking for neutrino energy below 100 GeV will allow to confirm or exclude at 90% confidence level the muon neutrino oscillations over the parameter space restricted by the recent Super-Kamiokande observation.

Since 1996 the ANTARES collaboration makes the extensive survey of the site which has been selected for the detector deployment. The results of this survey include a climatological studies of the area, a survey of the sea bed, measurement of the currents, and detailed study of the optical properties at the site [3]. The latter plays a crucial role for the detector performance. The sets of measurements have been performed to define water transparency, amount of background light and fouling of the glass spheres. The fouling of the modules is mainly due to sedimentation process, which affects only the upper half of the glass spheres. At the equator of the glass spheres the fouling produces a loss of the light transmission of only 1.5% after eight months of immersion. Background light measured by the optical module in situ displays three components:

- about 20 kHz contribution from β-decay of ^{40}K,
- a component of unknown origin which slowly varies on typical time scale of few hours reaching 20 kHz rate,
- short (about 1 second) bursts with counting rate reaching 1MHz of one photoelectron signals. These bursts are probably due to the light emission of living organisms in the deep sea.

The background light affects strongly the final trigger rates. It is important to note that most of the background signals have the amplitude equivalent to the one photoelectron signal and are local in space. The bioluminescence bursts introduce a dead time up to 5% for each optical module. These sources of the background light are included in the simulation programs used to estimate the detector performance. An effective attenuation length of light with wavelength of 466 nm was measured at the ANTARES site to be between 40 and 60 m. This value seems to show seasonal variations. The light attenuation is an integrated effect of absorption and diffusion. Further measurements of light arrival time demonstrated that the light diffusion is practically negligible for ANTARES allowing the good angular resolution of the detector.

The project started in 1996 is taking shape now and plans to take first data in 2003. We are looking forward to make exciting discoveries in the future.

REFERENCES

1. Markov M.A., in Proc. 1960 Annual International Conference on High Energy Physics in Rochester, 1960.
2. ANTARES collab, astro-ph/990743, 1999.
3. ANTARES collab, astro-ph/9910170, 1999.

Supernova neutrino oscillation in the presence of a random magnetic field

Sarira Sahu

*Instituto de Ciencias Nucleares, Universidad Nacional Autonoma de Mexico,
Circuito Exterior C.U., A. Postal 70-543, 04510 Mexico D.F.*

Abstract. The active-sterile neutrino conversion is studied for neutrino propagating in the axial potential generated by magnetised electron plasma in the supernova medium. We consider the effect of random magnetic field B_{rms} on the average neutrino conversion probability and obtained the constraint on Δm^2 and $sin^2 2\theta$ for different strength of the random magnetic fields, inside the supernova core. Our calculation shows that, to have $\Delta m^2 \lesssim keV^2$, the ramdom magnetic field $B_{rms} \lesssim 0.63 \times 10^{14}$ Gauss is preferable.

INTRODUCTION

Neutrino propagation in the magnetised medium has interesting consequences in the astrophysical and cosmological scenarios. Large scale magnetic fields in the early universe hot plasma and in the core of the supernova can effect the neutrino conversion [1]. It has been shown that random magnetic fields can strongly influence neutrino conversion rates and this could have important implications, especially in the case of conversion involving a light sterile neutrino [2]. In the magnetised medium neutrino acquires an axial potential which is proportional to the scalar product of the neutrino momentum and the magnetic field vector (**k.B**). The effect of axial potential on neutrino propagation in media with regular and/or random magnetic fields are considered in the literature [2,3].

Here we have considered the effect of axial potential on neutrino propagation in the supernova medium in the presence of random magnetic fields. Using the positive definiteness condition of the average conversion probability inside a supernova core, we have shown that $B_{rms} \lesssim 0.63 \times 10^{14}$ Gauss is preferred, because large values of B_{rms} exclude small values of Δm^2, which are not supported by present experiments.

NEUTRINO PROPAGATION IN THE MAGNETISED MEDIUM

The evolution equation for a system of two neutrinos ν_a and ν_b, where ν_a is the active one and ν_b is active/sterile one is given by

$$i\frac{d}{dt}\begin{pmatrix}\nu_a\\\nu_b\end{pmatrix} = \begin{pmatrix}H_{aa}(t) & H_{ab}(t)\\H_{ba}(t) & H_{bb}(t)\end{pmatrix}\begin{pmatrix}\nu_a\\\nu_b\end{pmatrix}, \quad (1)$$

where the quantity H is in general the potential for the neutrino in the medium. Let us define the functions $R = Re(\langle \nu_a^* \nu_b \rangle)$ and $I = Im(\langle \nu_a^* \nu_b \rangle)$. Then using these in Eq.(1) we obtain

$$\dot{R}(t) = -H_d(t)I(t), \dot{I}(t) = H_{ab}(t)\left(2P(t)-1\right) + H_d R(t), \dot{P}(t) = -2H_{ab}(t)I(t), \quad (2)$$

where the function $P(t)$ is the neutrino conversion probability $P_{\nu_a \to \nu_b}(t)$ and $H_d = H_{aa}(t) - H_{bb}(t)$ and dot on the top corresponds to derivative with respect to t. For fluctuation in the magnetic field we can write $B(t) = B_0 + B'(t)$, where B_0 is the constant background field and $B'(t)$ is the random fluctuation over it and similar decomposition to H_d and H_{ab}. For the neutrino conversion length greater than the domain size i.e. $l_{conv} \gg L_0$ (where $l_{conv} \sim 1/\mathcal{P}\Gamma_W$ and Γ_W is the weak interaction rate), a neutrino will cross many magnetic field domains before it flips its helicity. Thus the neutrino will experience an average field before it flips its helicity. So one can average the propagation equation over the random magnetic field distribution [2,4]. The magnetic field in different domains is randomly oriented with respect to the neutrino propagation direction. So the neutrino conversion probability depends on the root mean square (rms) value of the random magnetic field. We have used the delta correlation among the magnetic field domains [3,4], and the rms value of the averaged magnetic field is given as $B_{rms} = \sqrt{\langle B^2 \rangle}$.

Let us consider the propagation of a system of active (doublet) and light sterile (singlet) neutrinos ($\nu_e \to \nu_s$), with masses m_1 and m_2, mixing angle θ, and no transition magnetic moments, in the presence of a magnetised plasma. Then the Hamiltonian in the evaluation equation Eq.(1) will be

$$\begin{pmatrix} V - \Delta\cos 2\theta + \mu_{eff}\mathbf{k}.\mathbf{B}/k & \Delta\sin 2\theta/2 \\ \Delta\sin 2\theta/2 & 0 \end{pmatrix}, \quad (3)$$

where $\Delta = (m_2^2 - m_1^2)/2E$. For active-sterile neutrino conversion the resultant vector potential experienced by ν_e is given by $V = \sqrt{2}G_F n_e(3Y_e + 4Y_{\nu_e} - 1)$, where G_F is the Fermi coupling constant, n_e is the electron density in the medium and Y_e and Y_{ν_e} are the electron and ν_e abundances respectively in the medium. For neutrino propagating along the z axis the $V_{axial} = \mu_{eff} B_z \frac{k_z}{k}$. The quantity μ_{eff} for $\nu_e \to \nu_s$ is given by $\mu_{eff} = \frac{eG_F P_F}{\sqrt{2}} \frac{P_F}{2\pi^2}$ and P_F is the Fermi momentum of electron.

Let us define the average of the functions $\langle P(t) \rangle = \mathcal{P}(t)$, $\langle R(t) \rangle = \mathcal{R}(t)$ and $\langle I(t) \rangle = \mathcal{I}(t)$. Because of the averaging the average probability $\mathcal{P}(t)$ will only depend on the even powers of the magnetic field correlation. Using the average functions we obtain

$$\dot{\mathcal{I}}(t) = H_{ab}(0)\langle(2\mathcal{P}(t)-1)\rangle + \langle H'_{ab}(2\mathcal{P}(t)-1)\rangle + H_{aa}(0)\langle \mathcal{R}(t)\rangle + \langle H'_{aa}(t)\mathcal{R}(t)\rangle, \quad (4)$$

and $\dot{\mathcal{R}}(t) = -H_{aa}(0)\langle \mathcal{I}(t)\rangle - \langle H'_{aa}(t)\mathcal{I}(t)\rangle$ respectively. Using the delta correlations for the magnetic fields we obtain $\langle H'_d(t)\mathcal{R}(t)\rangle \simeq -2\Gamma_\parallel \mathcal{I}(t)$, $\langle H'_{ab}(t)\mathcal{P}(t)\rangle \simeq -\Gamma_\perp \mathcal{I}(t)$, and $\langle H'_{ab}(t)\mathcal{I}(t)\rangle \simeq \frac{\Gamma_\perp}{2}(2\mathcal{P}(t)-1)$, where we have defined $\Gamma_\perp = \frac{4}{3}\mu^2\langle B^2\rangle L_0$ and $\Gamma_\parallel = \frac{1}{6}\mu^2_{eff}\langle B^2\rangle L_0$. We obtain the master equation for the average conversion probability as,

$$\dddot{y}(t) + A_0\ddot{y}(t) + B_0\dot{y}(t) + 2C_0 y(t) = 0, \quad (5)$$

where

$$A_0 = 4\left(\Gamma_\perp + \Gamma_\parallel\right), \quad B_0 = 4\left(3\Gamma_\perp\Gamma_\parallel + \Gamma_\perp^2 + \Gamma_\parallel^2 + \frac{\omega_s^2}{4}\right), \quad (6)$$

and

$$C_0 = 4\left(\Gamma_\perp^2\Gamma_\parallel + \Gamma_\perp\Gamma_\parallel^2 + H_{ab}^2(0)\Gamma_\parallel + \frac{H_d^2(0)\Gamma_\perp}{4}\right) \quad \text{and} \quad \mathcal{P}(t) = \frac{1}{2} + y(t) \quad (7)$$

respectively and the boundary conditions are $\mathcal{P}(0) = 0$, $\dot{\mathcal{P}}(0) = \Gamma_\perp$ and $\ddot{\mathcal{P}}(0) = 2H_{ab}^2(0) - 2\Gamma_\perp^2$. The qunatity $\omega_s^2 = 4H_{ab}^2(0) + H_d^2(0)$. The solution to the above third order differential equation is given by

$$y(t) = e^{-(\frac{Z_1}{2} + \frac{A_0}{3})t}\left[A_1\left\{e^{\frac{3}{2}Z_1 t} - \cos(Z_4 t) - \frac{3}{2}\frac{Z_1}{Z_4}\sin(Z_4 t)\right\} \right.$$
$$\left. -\frac{\cos(Z_4 t)}{2} + \frac{(\Gamma_\perp - \frac{A_0}{6} - \frac{Z_1}{4})}{Z_4}\sin(Z_4 t)\right]. \quad (8)$$

The coefficient A_1 is a function of A_0, B_0 and C_0 and Z_1 and Z_4 are given by

$$Z_1 = \left[X - \frac{p}{3X}\right], \quad Z_4 = \frac{\sqrt{3}}{2}\left[\frac{p}{3X} + X\right], \quad (9)$$

respectively. The qunatities

$$X = \left(-\frac{q}{2} + \sqrt{\left(\frac{p}{3}\right)^3 + \left(\frac{q}{2}\right)^2}\right)^{\frac{1}{3}}, p = (B_0 - \frac{A_0^2}{3}), q = (2C_0 - \frac{A_0 B_0}{3} + \frac{2}{27}A_0^3). \quad (10)$$

The solution $y(t)$ is very complicated and it is difficult to conclude any thing from the solution. On the other hand the interesting part of the solution is that, the positive definiteness of the average conversion probability ($0 \leq \mathcal{P} \leq 1$) depends on the quantity X. If the quantity $[(p/3)^3 + (q/2)^2] \geq 0$ then $\mathcal{P}$ is positive definite otherwise for $[(p/3)^3 + (q/2)^2] < 0$, $\mathcal{P}$ will be complex and does not make any sense. The sign of this quantity is determined by p, because q^2 is always positive. There are two situations, for X to be real, either (i) $p \geq 0$ (ii) or $p < 0$, but $[(p/3)^3 + (q/2)^2] \geq 0$.

RESULTS AND DISCUSSIONS

We consider the active sterile neutrino conversion in the magnetised medium, with neutrino mixing and $\mu = 0$ ($\Gamma_\perp = 0$) as shown in (3). Only the fluctuation in the parallel component of the magnetic field will contribute. For $p > 0$, the condition in $[(p/3)^3 + (q/2)^2] > 0$ will be simplified to

$$\omega_f^2 > \frac{4\Gamma_\parallel^2}{3}, \text{ where } \omega_f = \sqrt{(V - \frac{\Delta m^2}{2E}\cos 2\theta)^2 + \left(\frac{\Delta m^2}{2E}\right)^2 \sin^2 2\theta}. \quad (11)$$

is square of the flavor conversion frequency of the neutrino in the medium. Let us estimate the range of Δm^2 and $sin^2 2\theta$ from the the above inequality in Eq.(11) in a supernova environment. Inside the supernova the vector potential for the process $\nu_e \to \nu_s$ is $V \simeq 4.48$ eV and $\mu_{eff} \simeq 4.3 \times 10^{-13} \mu_B \left(\frac{P_F}{MeV}\right)$, where μ_B is the Bohr magneton. The quantity μ_{eff} has the same magnetic moment as it does not change the helicity of the particle. The Fermi momentum of the electron inside the core is $P_F \simeq 428$ MeV. Inside the supernova core, neutrinos have energy in the range 30 to 100 MeV. Thomson and Dunkan [5] have argued that very strong magnetic fields might be generated inside the supernova core due to small scale dynamo mechanism. If these fields are generated after core collapse, then it could be viewed as random superposition of many small dipoles of size $L_0 \sim 1$ Km. Then $\Gamma_\parallel \simeq 9.6 B_{14}^2$ eV, where B_{rms} is expressed in units of 10^{14} Gauss. Putting ω_f and $\Gamma_\parallel$ in Eq.(11), we can find the ranges of Δm^2 and $sin^2 2\theta$ for which the condition in Eq.(11) is satisfied. These are shown in the contour plots for different values of the magnetic fields in Figure 1. Figure 1 (a) shows that, a very narrow range of $\Delta m^2/eV^2$ ($3.16 \times 10^8 - 1.6 \times 10^9$) is excluded for $0 \lesssim sin^2 2\theta \lesssim 0.63 \times 10^{-3}$ for the random magnetic field strength $B_{rms} = 0.1 \times 10^{14}$ Gauss. By increasing the strength of the random magnetic field to 0.5×10^{14} and 0.63×10^{14} Gauss, we see in Figure 1 (b) and (c) that the width of the excluded range of Δm^2 has increased. The spread of Δm^2 towards smaller values is large compared to the large values. Also the large values of $sin^2 2\theta$ are excluded. In Figure 1 (b) we see that $3.5 \times 10^8 \lesssim \Delta m^2/eV^2 \lesssim 1.6 \times 10^9$ is excluded and for this the excluded range of $sin^2 2\theta$ is $0 \lesssim sin^2 2\theta \lesssim 0.4$ (for $B_{rms} = 0.5 \times 10^{14}$ Gauss). In Figure 1 (c) it is shown that for $B_{rms} = 0.63 \times 10^{14}$ Gauss, the excluded region of $\Delta m^2/eV^2$ is $1.6 \times 10^7 - 1.6 \times 10^9$ and in this range all the values of $sin^2 2\theta$ are excluded. It shows that, two distinct allowed regions for Δm^2 are there in both sides of the excluded region. In these allowed regions all ranges of $sin^2 2\theta$ are allowed. But so far as Δm^2 value is concerned, it is interesting to consider only the region left to the excluded curve, because it corresponds to smaller Δm^2 values ($\Delta m^2 \lesssim (keV)^2$). Going from Figure 1 (c) to (d) (for $B_{rms} = 10^{14}$ Gauss) we observe that, the left arm of the curve vanishes and the right arm spreads towards the higher values of Δm^2 for all ranges of $sin^2 2\theta$. This implies that small values of Δm^2 are excluded. By further increasing the strength of the random magnetic field, we have observed that all values of $\Delta m^2/eV^2 \lesssim 1.6 \times 10^{13}$ are excluded for all ranges of $sin^2 2\theta$ values.

This range of Δm^2 are obviously not at all interesting, because it corresponds

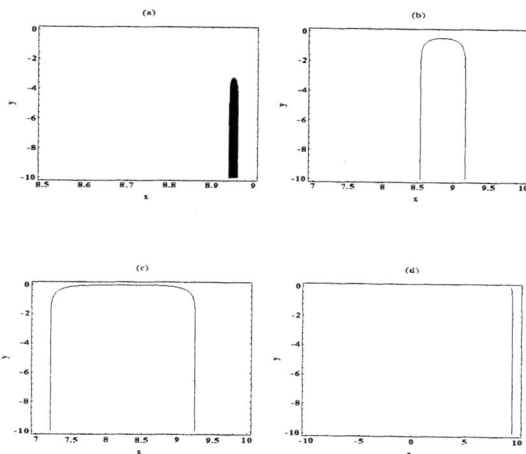

FIGURE 1. We have defined $\mathbf{x} = log[\Delta m^2/eV^2]$ and $\mathbf{y} = log[sin^2 2\theta]$ (a) for $B_{rms} = 10^{13}$ Gauss and the shaded region is excluded, (b) and (c) for $B_{rms} = 0.5 \times 10^{14}$ and 0.63×10^{14} Gauss, region inside the curve is excluded, (d) for $B_{rms} = 10^{14}$ Gauss and left side of the curve is excluded.

to very high values of Δm^2, and no astrophysical, cosmological and laboratory observations favour this. Thus if we consider only the effect of random magnetic field in a magnetised electron plasma along the neutrino propagation direction, then magnetic field should satisfy $B_{rms} \lesssim 10^{14}$ Gauss so that small Δm^2 ranges should not be excluded. On the other hand for $B_{rms} < 10^{13}$ Gauss we found that, all the parameter ranges are allowed. Thus we found that for magnetic field $B_{rms} \lesssim .63 \times 10^{14}$ Gauss, all the interesting ranges of parameters are allowed. Also this implies that the maximum value of Δm^2 can be in the keV^2 range.

REFERENCES

1. Ya. Zeldovich, A. A. Ruzmaikin and D. D. Sokoloff, Magnetic Fields in Astrophysics, Mc Graw Hill, 1983; E. Parker, Cosmological Magnetic Fields, Oxford Univ. Press, 1979; Peter Meszaros, High Energy Radiation from Magnetised Neutron Stars, Chicago Univ. Press, 1992.
2. S. Pastor, V. Semikoz and J.W.F. Valle, Phys. Lett. **B369**, 301 (1996).
3. S. Sahu, Phys. Rev. **D 56**, 4378 (1997); S. Sahu and V. M. Bannur, to appear in Phys. Rev. D.
4. K. Enqvist, A. Rez and V. Semikoz, Nucl. Phys. **B436**, 49 (1995).
5. C. Thomson and R. C. Dunkan, Astrophys. J. **408**, 194 (1993).

Silicon Drift Detectors in the ALICE Experiment

V. Bonvicini[b], P. Cerello[a], E. Crescio[a], P. Giubellino[a],
R. Hernandez-Montoya[a,1], A. Kolojvari[a,2], G. Mazza[a],
L. Montano[a,1], J. Nissinen[c], D. Nouais[a], A. Rashevsky[b],
A. Rivetti[d,a,3], F. Tosello[a], A. Vacchi[b]
for the ALICE Collaboration

[a] INFN Sezione di Torino, Italy
[b] INFN Sezione di Trieste, Italy
[c] Jyväskylä University, Finland
[d] Politecnico di Torino, Italy
[1] CINVESTAV Mexico City, Mexico
[2] St. Petersburg University, St. Petersburg, Russia
[3] CERN, Geneva, Switzerland

Abstract. Silicon Drift Detectors (SDDs) are well suited to high-energy physics experiments with relatively low event rates. In particular SDDs will be used for the two intermediate layers of the Inner Tracking System of the ALICE experiment. Beam test results of linear SDD prototypes have shown a resolution of 40x30 μm^2 and a cluster finding efficiency of essentially 100% with E=600 V/cm.

INTRODUCTION

Silicon Drift Detectors will equip the two middle layers of the Inner Tracking System (ITS) of ALICE, an heavy-ion experiment at the Large Hadron Collider (LHC). The aim of ALICE is the study of the behaviour of nuclear matter at extreme energy densities and temperatures, where the formation of a new phase of matter, the quark-gluon plasma, is expected. For this purpose heavy-ion collisions at a centre-of-mass energy ~ 5.5 TeV per nucleon will be studied.
The extremely high track density expected (8000 charged particles per unit of rapidity at midrapidity) led to the choice of silicon detectors for the ITS [1], due to their high granularity and excellent spatial resolution.
In particular, the SDD, which couple a very good multi-track capability with dE/dx information, will equip the two intermediate of the six cylindrical layers of the ITS. The two innermost layers will be made of pixel detectors, while the outer two layers,

where a lower occupancy is expected, will be equipped with double-sided microstrip detectors.

A description of the principle of operation of the SDDs will be given and some results from beam tests, concerning the efficiency, the resolution, and the calibration of the drift velocity will be discussed.

PRINCIPLE OF OPERATION

The principle of operation of the SDDs (Figure 1), firstly proposed in [2], is based on the measurement of the drift time of the electron cloud produced by the crossing particle from its impact point to collecting anodes under the effect of a built-in electric field. This allows to determine the coordinate along the drift direction; by using an array of separate anodes (as shown in Figure 1) to measure the signal it is possible to determine also the perpendicular one. The SDDs, therefore, represent an unambiguous 2-D position sensitive detector.

The electrostatic field necessary to transport the electrons towards the collection region is built by polarizing parallel, implanted p^+ strips (drift cathodes), which have also the function to fully deplete the n-type silicon wafer.

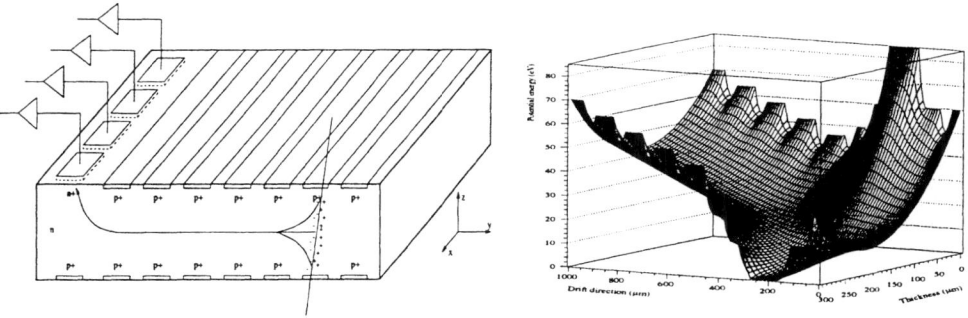

FIGURE 1. Left: the operation principle of a SDD. Right:Potential energy of electrons (negative electric potential) in a SDD.

In Figure 1 on the right the potential energy of the electrons is shown. In the drift region, the parabolic potential in the thickness direction focuses the electrons into the middle plane of the detector. They then drift, at a constant velocity $v=\mu_e E$, towards the collection region, where the cathodes opposite to the anodes force the drifting charge towards the anode array.

Owing to the diffusion and mutual Coulombian repulsion during the drift, the electron cloud can be collected by more than one anode. The coordinate along the direction perpendicular to the drift can be localized with a precision much better than the anode pitch by calculating the centroid of the charge deposited on the anodes. Typically a 200 μm pitch allow to determine this coordinate with a precision better than 30-40 μm. The signal measured on each anode is amplified and sampled with a typical frequency of a few tens of MHz; the coordinate along the drift direction is measured by the centroid of the sampled signal with a precision of the same order of the perpendicular one.

I SDDS IN ALICE

The middle layers of the ITS will be equipped with 260 SDDs, mounted on linear structures called ladders.
The detectors are produced on NTD 5-inch silicon wafers with a resistivity of 3 kΩcm. Their active area, 7.25×7.53 cm^2, is divided in two halves with opposite drift directions, from the central strip towards two arrays of 256 anodes with a pitch of $\sim$300 μm. The drift length of each half-detector is 35.0 mm.
The drift cathodes are biased through a voltage divider integrated on the substrate itself.
The detectors will be operated at a drift field E$\sim$600 V/cm in order to limit the maximal drift time to $\sim$5 μs.
Each detector is equipped with MOS point-like charge injectors, placed at known distances from the anodes, dedicated to the drift velocity monitoring.
The front-end electronics consist of two integrated circuits with a low power consumption. The first one, PASCAL, contains an array of 64 preamplifiers connected to a 64×256 cell analog memory and performs the preamplification of the signals, the analog storage and the analog-to-digital conversion at a sampling frequency of about 40 MHz. The second one, named AMBRA, is a digital two-event buffer which allows data derandomization and transmission. A detailed description of the SDD and its front-end electronics is given in [1].

II BEAM TEST RESULTS

Several drift detector prototypes have been tested on beams both at the PS (3 GeV/c) and at the SPS (370 GeV/c) accelerators at CERN [3]. For the readout during the tests the OLA [4] integrated amplifiers and a Flash ADC system have been used. During the beam tests the SDDs were placed on the beam line between two 50 μm-pitch microstrip telescopes, used for the track reconstruction [5].
For the prototypes tested, a cluster finding efficiency of 100% up to 25 mm drift path have been observed at a drift field of 460 V/cm, and of 100% for all distances at a field of 650 V/cm. This can be easily explained by the fact that the signal amplitude depends on the charge deposited by the particle and is inversely proportional to the drift time, owing to the diffusion. In Figure 2 the distribution of cluster amplitudes vs. drift time for different electric field is shown. The separation between signal and noise (represented by the band at low amplitudes) depends on the drift field.

The field planned for ALICE is 600 V/cm, for which a full efficiency for all drift distances is expected, with a preamplifier noise of 250 e r.m.s.

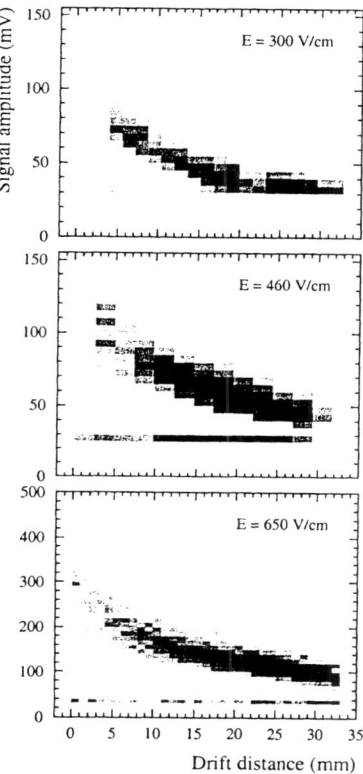

FIGURE 2. Signal amplitude vs. drift distance for different electric fields

The linearity curve along the drift region for three different values of the drift field is shown in Figure 3. The systematic deviations from the linearity are less than ±15 μm peak to peak.

The detector spatial resolution has been evaluated as the difference between the cluster position measured by the SDD and the impact point projected by the microstrip telescope. For the prototype with an anode pitch of 200 μm a resolution of 30 μm in the anode direction and of 40 μm in the drift direction have been calculated.

During the beam tests, the performance of the drift velocity monitoring system has been studied on a 16 mm maximum drift path prototype operated at low drift field, E=345 V/cm, in order to have the maximum drift time comparable to the large area prototype [6]. In fact, since the drift velocity depends on the temperature through the mobility ($\mu_e \approx T^{-2.4}$), the temperature needs to be controlled to

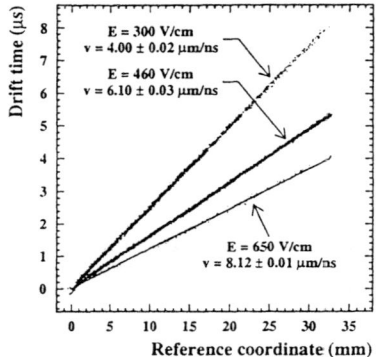

FIGURE 3. Drift time vs. corresponding coordinate of the track point on the SDD as reconstructed by microstrip telescopes.

better than 0.1°C in order to avoid the deterioration of the spatial resolution (the drift velocity variations at room temperature are about 1%/K). In Figure 4 the improvement of the resolution obtained using the charge injectors is shown. The effect of the velocity calibration is more evident at higher drift times as expected.

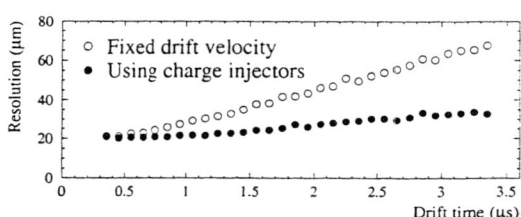

FIGURE 4. Resolution as a function of drift time in the two cases, using the charge injectors and without the velocity calibration

REFERENCES

1. ALICE Collaboration, CERN/LHCC 99-12.
2. E. Gatti and P. Rehak, *Nucl. Instr. and Meth.* **A225**, 608 (1984).
3. V. Bonvicini et al., ALICE INT-98-24.
4. W. Dąbrowski et al., *Nuclear Physics B (Proc. Suppl.)* **44**, 637 (1995)
5. D. Nouais et al., *Nuclear Physics B (Proc. Suppl.)* **78**, 252 (1999).
6. Drift velocity monitoring of SDDs using MOS charge injectors, D. Nouais et al., submitted to *Nucl. Instr. and Meth.*.

On the quark structure of the $f_0(980)$ meson and the VEPP-2M experimental results for the $\phi \to \pi^0\pi^0\gamma$ decay

M. Napsuciale [1] and J.L. Lucio M.[2]

Instituto de Física, Universidad de Guanajuato
Loma del Bosque # 103, Lomas del Campestre, 37150 León, Guanajuato; México

Abstract. It is shown that VEPP-2M results for the $\phi \to \pi^0\pi^0\gamma$ decay **are** consistent with a (mostly) $\bar{s}s$ content for the $f_0(980)$, provided we take into account the effects of the $U_A(1)$ breaking in the scalar sector.

INTRODUCTION

The Particle Data Group (PDG) [1] candidates for the ground state $\bar{q}q$ scalar nonet are : the $f_0(980)$, $f_0(1370)$ and the recently resurrected $f_0(400-1250)$ (or σ) meson for two sites in the $I = 0$ sector; the $a_0(980)$ and $a_0(1450)$ for the isovector scalar meson, and the $K_0^*(1430)$ for the isospinor scalar meson. However, over the past years experimental evidence has accumulated for the existence of light scalar mesons [2], and different proposals exist for the $\bar{q}q$ lowest lying scalar meson nonet.

The main argument against the identification of the $a_0(980)$ and $f_0(980)$ as $\bar{q}q$ states is their tiny coupling to two photons which can not be reproduced by quark model calculations [3]. Molecule [4] and a $\bar{q}q\bar{q}q$ [5] structures for these mesons have been explored and found to be consistent with their tiny coupling to two photons. More recently, phenomenology involving the $a_0(980)$ and $f_0(980)$ has been argued to be also consistent with a four-quark picture for these mesons [6].

The second most important property of QCD in this energy region is the physical effect preventing the η' from being a pseudo-goldstone boson (breaking of the $U_A(1)$ symmetry). It seems natural to formulate a chiral theory incorporating the physical degrees of freedom appearing at this energy and below, and which takes into account both, spontaneous symmetry breaking and the $U_A(1)$ breaking.

A model which tkes into account both effects has been recently reconsidered and it was pointed out that the $U_A(1)$ breaking combined with the spontaneous

---

[1] mauro@ifug2.ugto.mx
[2] lucio@ifug.ugto.mx

breaking of chiral symmetry has important effects in the scalar sector [8]. In this theory, the $f_0(980)$ turns out to be a mostly $\bar{s}s$ meson whereas the $a_0(980)$ meson is the chiral partner of the pion, thus interpreted as a $\bar{q}q$ meson with $q = u, d$. A calculation for the $a_0(980) \to \gamma\gamma$ and $f_0(980) \to \gamma\gamma$ decays in this framework [9] turn out to be consistent with the experimental results.

Recently, the CMD2 and SND groups working with the VEPP-2M e^+e^- collider at Novosibirsk, reported new measurements on $\phi \to \pi^0\pi^0(\eta)\gamma$ and $\phi \to \pi^+\pi^-\gamma$ [7].

On the theoretical side, calculations have been performed for the $\phi \to \pi^0\pi^0\gamma$ decay within the framework of χPT [10]. These calculations yield a branching ratio too small as compared with the CMD2 an SND experimental results. Contributions from intermediate vector mesons were also estimated using Vector Meson Dominance and turned out to be negligible even compared with χPT results [11].

In this contribution we report calculations for the $\phi \to \pi^0\pi^0\gamma$ decay within the $U(3) \times U(3)$ LSM with $U_A(1)$-breaking [8]. This is a clean process since there exist no final state radiation (as in the $\phi \to \pi^+\pi^-\gamma$ case) and the pseudoscalar mixing angle is not involved (as in the $\phi \to \pi^0\eta\gamma$ case).

SCALAR MESON CONTRIBUTIONS TO $\phi \to \pi^0\pi^0\gamma$

This decay is generated at one loop level in the model. The diagrams contributing to this process are depicted in Fig. 1.

A straightforward calculation yield

$$M(\phi(Q,\eta) \to \pi^0(p)\pi^0(p')\gamma(k,\epsilon)) = e\, G(m_{\pi\pi}^2)\, T_{\mu\nu}\eta^\mu\epsilon^\nu \qquad (1)$$

where

$$T_{\mu\nu} = Q.k\, g_{\mu\nu} - k_\mu Q_\nu \qquad (2)$$

and

$$G(m_{\pi\pi}^2) = \frac{g_{\phi K^+K^-}}{2\pi^2 M_\phi^2} FL(m_{\pi\pi}^2) \qquad (3)$$

with the loop function

$$L(m_{\pi\pi}^2) = \frac{1}{2(a-b)} - \frac{2}{(a-b)^2}[f(\frac{1}{b}) - f(\frac{1}{a})] + \frac{a}{(a-b)^2}[g(\frac{1}{b}) - g(\frac{1}{a})]. \qquad (4)$$

where

$$f(z) = \begin{cases} -\left(arcsin[\frac{1}{2\sqrt{z}}]\right)^2 & z > \frac{1}{4} \\ \frac{1}{4}\left(ln\frac{\eta_+}{\eta_-} - i\pi\right)^2 & z < \frac{1}{4} \end{cases}$$

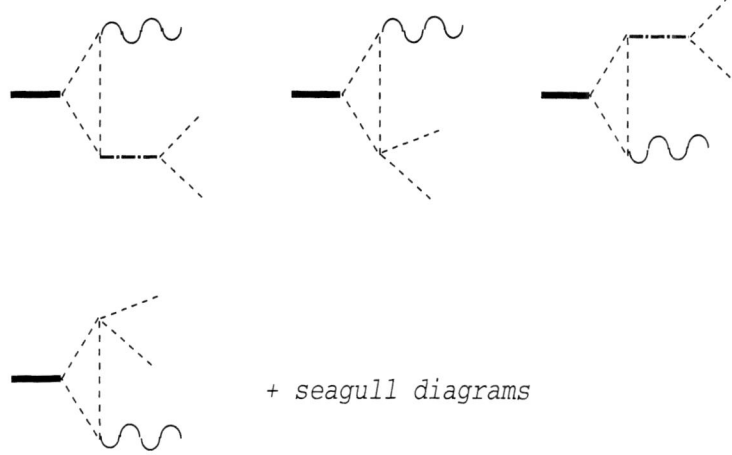

FIGURE 1. *Contributions to $\phi \to \pi^0\pi^0\gamma$ in the LSM. Dashed lines denote pseudoscalar mesons (kaons in the loops and neutral pions in the final state) while dot-dashed lines denote (intermediate) isoscalar scalar mesons (sigma and $f_0(980)$).*

with

$$g(z) = \begin{cases} (4z-1)^{\frac{1}{2}} \arcsin\left[\frac{1}{2\sqrt{z}}\right] & z > \frac{1}{4} \\ \frac{1}{2}(1-4z)^{\frac{1}{2}}(\ln\frac{\eta_+}{\eta_-} - i\pi) & z < \frac{1}{4} \end{cases}$$

$$\eta_\pm = \frac{1}{2}[1 \pm (1-4z)^{\frac{1}{2}}], \quad a = \frac{M_\phi^2}{m_{K^+}^2}, \quad b = \frac{m_{\pi\pi}^2}{m_{K^+}^2}. \tag{5}$$

The F factor appearing in Eq.(3) contains the information on the coupling constants.

$$F = 2(g_{KK\pi\pi} - \frac{g_{\sigma\pi\pi} g_{\sigma KK}}{m_{\pi\pi}^2 - m_\sigma^2 + i\Gamma_\sigma m_\sigma} - \frac{g_{f\pi\pi} g_{fKK}}{m_{\pi\pi}^2 - m_f^2 + i\Gamma_f m_f}) \tag{6}$$

The three and four-meson couplings are given by the model [8] as

$$g_{\sigma KK} = -\frac{m_\sigma^2 - m_K^2}{2f_K}(\cos\phi - \sqrt{2}\sin\phi) \quad ; \quad g_{\sigma\pi\pi} = -\frac{m_\sigma^2 - m_\pi^2}{2f_\pi}\cos\phi$$

$$g_{fKK} = -\frac{m_f^2 - m_K^2}{2f_K}(\sin\phi + \sqrt{2}\cos\phi) \quad ; \quad g_{f\pi\pi} = -\frac{m_\sigma^2 - m_\pi^2}{2f_\pi}\sin\phi \tag{7}$$

$$g_{KK\pi\pi} = -\frac{m_\sigma^2 - m_K^2}{4f_K f_\pi}.$$

Integrating over phase space we obtain the energy spectrum as

$$\frac{d\Gamma}{dm_{\pi\pi}} = \frac{\alpha_{em}}{4\pi}\frac{m_{\pi\pi}}{M_\phi}(\frac{g_{\phi KK}}{4\pi})^2(\frac{1}{4\pi})^2(\frac{M_\phi}{m_K})^4|L(m_{\pi\pi}^2)|^2|F|^2(1-\frac{m_{\pi\pi}^2}{M_\phi^2})^3\sqrt{1-\frac{4m_\pi^2}{m_{\pi\pi}^2}} \quad (8)$$

The above results for the energy spectrum depend on the scalar mixing angle ϕ and scalar meson masses. In particular is highly sensitive to the chosen value for the mixing angle. It is worth to remark that for a sigma meson mass above 600 MeV the theoretical predictions for the energy spectrum yield a desastrous result as compared with the experimental results. As a check, we have verified that our results reduces to those of χPT [10] in the heavy (and non-mixed) isoscalar scalar fields limit.

In Fig. 2 we show both theoretical and experimental results. We use $\phi = -9^0$, $m_f = 980\ MeV$, $m_\sigma = 430\ MeV$, and $\Gamma_f = 70\ MeV$ in the numerical evaluations. We also use the sigma width as dictated by the model:

$$\Gamma_\sigma = \frac{3m_\sigma^3}{32\pi f_\pi^2}((1-\frac{m_\pi^2}{m_\sigma^2})cos(\phi))^2\sqrt{1-4\frac{m_\pi^2}{m_\sigma^2}}. \quad (9)$$

In table 1 we show the theoretical predictions arising from Eq(8) for the total and partial (i.e. integrated over a limited region of the energy spectrum) Branching Ratios and compared with results from the Novosibirsk groups. Within experimental errors, agreement is satisfactory.

Table I

$m_{\pi^0\pi^0}(MeV)$	$BR(CMD2)(\times 10^{-4})$	$BR(SND)(\times 10^{-4})$	$BR_{TH}(\times 10^{-4})$
> 550	$1.06 \pm 0.09 \pm 0.06$		1.08
> 700	$0.92 \pm 0.08 \pm 0.06$	$1.00 \pm 0.07 \pm 0.12$	0.91
> 900	$0.57 \pm 0.06 \pm 0.04$	$0.50 \pm 0.06 \pm 0.06$	0.56
$total(> 2m_\pi)$	$1.08 \pm 0.17 \pm 0.09$	$1.14 \pm 0.10 \pm 0.12$	1.18

Summarizing, the VEPP-2M SND and CMD2 experimental results for the $\phi \to \pi^0\pi^0\gamma$ results are consistent with a mostly $\bar{s}s$ $f_0(980)$ meson provided we take into account the effects of the $U_A(1)$ breaking in the scalar sector. This process gives also support to the existence of a scalar meson resonance (σ) in the 400-600 MeV region. The process under consideration is highly sensitive to the scalar mixing angle and experimental results for this process require $\phi \approx -9^0$ which is consistent with other estimates [8].

ACKNOWLEDGMENTS

This work was supported by Conacyt-Mexico under project I27604-E.

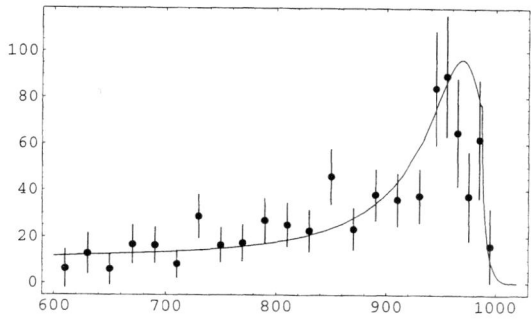

FIGURE 2. $dB(\phi \to \pi^0\pi^0\gamma)/dm_{\pi\pi} \times 10^{-8} MeV^{-1}$ as a function of the dipion invariant mass. The experimental points are taken from the SND Coll. V.M. Aulchenko et.al. [7].

REFERENCES

1. Particle Data Group, *Eur. Phys. Jour.* **C3** (1998).
2. for a recent account on this we refer to Proceedings of the Workshop on Hadron Spectroscopy, Frascati, Italy, March (1999). Frascati Physics Series Vol XV (1999), pags 75-262; T. Bressani, A. Feliciello, A. Filipi Eds.
3. See references in [9].
4. J. Wenstein and N. Isgur, *Phys Rev.* **D27**, 588 (1983); T. Barnes *Phys. Lett.* **B165**, 434 (1985).
5. N.N. Achasov, S.A. Denayin and G. N. Shestakov *Phys. Lett.* **B108**, 134 (1982); *Z.Phys.* **C16**, 55 (1982); E. P. Shabalin, *Sov. J. Nucl. Phys.* **46**, 485 (1987); N.N. Achasov and G. N. Shestakov *Z.Phys.* **C41**, 309 (1988); N.N. Achasov and V.N. Ivanchenko, *Nucl. Phys.* **B315**, 465 (1989): N.N. Achasov hep-ph/9803292.
6. N. N.achasov, hep-ph/9910540, hep-ph/9904223 ; N.N. Achasov, V.V. Gubin, hep-ph/9904439 ;
7. R.R Akhmetsin et.al. *Phys. Lett.* **B462** (1999) 380; V.M. Aulchenko et.al. *Phys. Lett.* **B440** (1998) 442, hep-ex/9807016; R.R. Akhmetsin et.al. *Phys. Lett.* **B462** (1999) 371; M. N. achasov et.al. hep-ex/9809013.
8. M. Napsuciale hep-ph/9803396 unpublished;
9. J.L. Lucio and M. Napsuciale *Phys.Lett.* **B454** (1999) 365.
10. A. Bramon et.al. *Phys. Lett.* **B289** (1992) 97.
11. A. Bramon et. al. *Phys. Lett.* **B283** (1992) 416.

Observation of the Centrally Produced $\phi\phi$ System at 800 GeV/c

M.A.Reyes[a], M.C.Berisso[b], D.C.Christian[c], J.Felix[f], A.Gara[d],
E.E.Gottschalk[c], G.Gutiérrez[c], E.P.Hartouni[e], B.C.Knapp[d],
M.N.Kreisler[b], S.Lee[b], K.Markianos[b], G.Moreno[f], M.H.L.S.Wang[b,e],
A.Wehman[c], D.Wesson[b]

[a] Universidad Michoacana de San Nicolás de Hidalgo, Morelia, Michoacán, México
[b] University of Massachussetts, Amherst, Massachussetts, USA
[c] Fermilab, Batavia, Illinois, USA
[d] Columbia University, Nevis Laboratory, New York, USA
[e] Lawrence Livermore National Laboratory, Livermore, California, USA
[f] Universidad de Guanajuato, León, Guanajuato, México

Abstract. We present evidence of the observation of the centrally produced $\phi\phi$ system at 800 GeV/c in the reaction $pp \to p_{slow}(\phi\phi)p_{fast}$. A clear bump is observed over low ϕK^+K^- and $K^+K^-K^+K^-$ background. Statistics are comparable to those previously obtained in $\pi^- p$ interactions, where three 2^{++} waves were found to comprise all cross section.

INTRODUCTION

QCD predicts the existence of gluon bound states –glueballs– and bound states of quarks and gluons –hybrids–, which should be differentiated from the conventional $q\bar{q}$ and qqq states through their unusual properties (supressed or enlarged production, too broad or too narrow decay widths), or forbidden quantum numbers. However, there is no direct evidence that any of these states has already been observed [1], leading to a missing part of QCD. One system with unusual properties is that composed of two ϕ mesons, highly produced in channels where the reaction should be suppressed due to the Okubo-Zweig-Iizuka (OZI) rule [2].

The first observation of $\phi\phi$ production was made using the BNL–MPS Spectrometer in the OZI suppressed reaction,

$$\pi^- p \to \phi\phi n \tag{1}$$

at 22.6 GeV/c [3]. This reaction was suppressed by about a factor of 10 relative to the OZI allowed reaction [4]

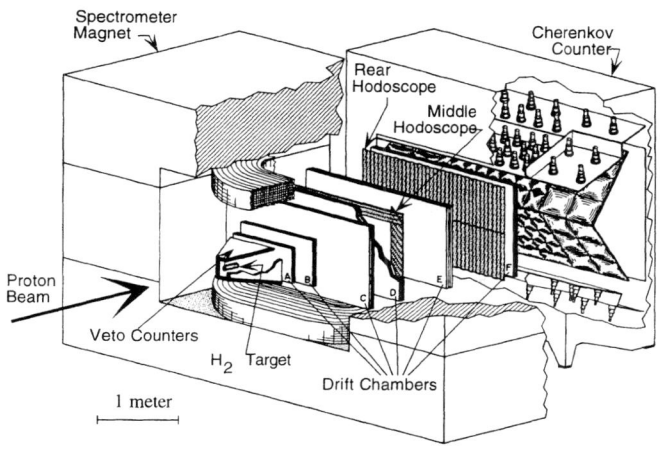

FIGURE 1. E690 multiparticle spectrometer.

$$\pi^- p \to \phi K^+ K^- n, \quad (2)$$

whereas typical OZI allowed to OZI suppressed ratios for single ϕ production [5] are on the order of 50! These unusual properties make this a useful channel for understanding gluon dynamics, and a candidate for bound gluon state observation.

With respect to resonant structures in this channel, BNL–MPS is the only experiment performing a thorough Partial Wave Analysis (PWA). In their 1985 analysis with 3652 events, they showed that three 2^{++} waves were necessary to comprise all of the cross section [6]. This result prevailed in further analysis with higher statistics.

In Central Production, the CERN Ω Spectrometer was used to search for this reaction [7,8]. The low statistics make these data not suitable to perform a PWA.

We present here evidence of the observation of the centrally produced $\phi\phi$ system in the doubly diffractive reaction,

$$pp \to p_{slow}(\phi\phi)p_{fast}, \quad \phi \to K^+ K^- \quad (3)$$

using 90% of the 5×10^9 events recorded by Fermilab E690 during the Fermilab's 1991 fixed target run. Within the experimental context described above, a thorough analysis of this data seems to be the only possible way to confirm BNL results.

E690 SPECTROMETER

The E690 apparatus consisted of a high rate, open geometry multiparticle spectrometer (Fig.1) used to measure the target system (T) in $pp \to p_{fast}(T)$ reactions,

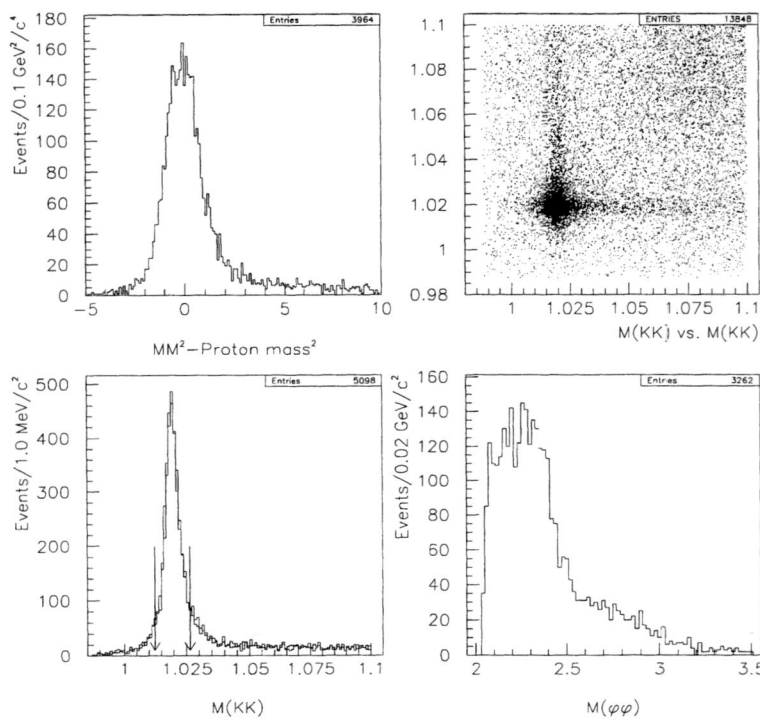

FIGURE 2. Missing mass squared minus proton mass squared for events in reaction (3) (upper left). K^+K^- invariant mass, first vs. second pair (upper right). K^+K^- invariant mass, when the other pair lies in the ϕ-mass band (lower left). $\phi\phi$ invariant mass (lower right).

and a beam spectrometer system used to measure the incident 800 GeV/c beam and scattered proton. A liquid hydrogen target was located just upstream of the multiparticle spectrometer. The 96 cell Cherenkov counter located at the downstream end of the main spectrometer magnet used Freon 114 as a radiator and had a pion threshold of 2.57 GeV/c. The E690 apparatus has been described elsewhere [9].

DATA SELECTION AND RESULTS

Final state (3) was selected by requiring a primary vertex in the LH_2 target with two positive and two negative tracks, an incoming beam track and a fast forward proton. At least one of the four charged tracks was required to be identified by the Cherenkov counter as an ambiguous K/p, the other were required to be compatible with a K identity. A minimum gap of 1.5 rapidity units was required between

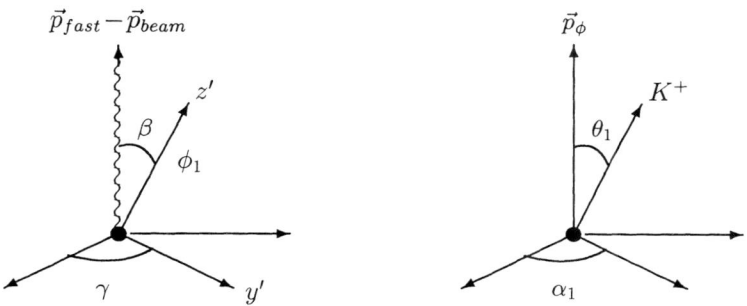

FIGURE 3. The six angles used to specify the spin and angular momentum of the $\phi\phi$ system.

each charged track and the slow proton. No direct measurement was made of the slow proton, but a kinematical cut of $p_z < 250$ MeV/c or $\arctan(p_t/p_z) > 30$ for the missing momentum was used to require that it was outside the acceptance of the detector. The missing mass squared (MM^2) minus proton mass squared shown in Fig.2 has a clear peak around zero for events in reaction (3). 24556 events were selected with $m(K^+K^-) < 1.1$ GeV/c^2 and $-10 < MM^2 - m_p^2 < 10$ GeV^2/c^4.

Fig. 2 shows the results obtained after selection. The two upper plots show the missing mass squared minus proton mass squared for events in reaction (3), and the first versus second pair mass when $-2 < MM^2 - m_p^2 < 2$ GeV^2/c^4. Only one combination per event enters this plot. We can see that $\phi\phi$ events are predominantly produced over $\phi K^+ K^-$ and $K^+K^-K^+K^-$ events, these being the only background source. The lower left plot shows the K^+K^- invariant mass, when the other pair lies in the ϕ-mass band of $1.0124 < m(K^+K^-) < 1.0264$ GeV/c^2, and with $-2 < MM^2 - m_p^2 < 2$ GeV^2/c^4. Finally, the lower right plot shows the $\phi\phi$ invariant mass with all cuts. The invariant mass plot shows the particular cross section seen by BNL–MPS, with a high bump between 2–2.5 GeV/c^2.

Some hints about PWA.

The $\phi\phi$ mass plot shows that statistics are comparable to those obtained by BNL–MPS in 1985, and, therefore, a PWA is feasible. Work in this direction is in progress, and the way to proceed is the following.

There are six angles to specify the spin and angular momentum of the $\phi\phi$ system, each ϕ decaying into a K^+K^- pair. Three of them are defined as the Gottfried-Jackson (GJ) angles of one of the ϕ mesons, in the rest frame of the $\phi\phi$ system, with the z-axis in the direction of $\vec{p}_{fast} - \vec{p}_{beam}$, and the y-axis perpendicular to the plane formed by $\vec{p}_{fast} - \vec{p}_{beam}$ and $\vec{p}_{slow} - \vec{p}_{tgt}$, in this frame. The rest are defined as the GJ angles for the K^+ of the first ϕ, in the ϕ rest frame, with the z'-axis in the direction of $\vec{p}_\phi$, and the y'-axis in the $\hat{z}$-$\hat{z}'$ plane (Fig.3).

With respect to these reference frames, the basis vectors for a given allowable combination of the total angular momentum of the $\phi\phi$ system J, orbital angular momentum L, $M = |J_z|$, Parity P, and exchange naturality η, are given by [10]

$$G^{J^P LSM^\eta}(\gamma, \beta, \alpha_1, \alpha_2, \theta_1, \theta_2) = \text{Real}\left[\frac{(1-i) - \eta(1+i)}{2} \sum_{\mu,\lambda} C(1,1,S|\mu,-\lambda) \times\right.$$

$$\left. C(L,S,J|0, \mu - \lambda)e^{-iM\gamma}e^{i\mu\alpha_1} e^{i\lambda\alpha_2}d^J_{M,\mu-\lambda}(\beta)\ d^1_{\mu,0}(\theta_1)\ d^1_{\lambda,0}(\theta_2)\right] \quad (4)$$

$I = 0$ and $C = +$ for the $\phi\phi$ system, and Bose statistics require that $L+S=$ an even number.

CONCLUSIONS

In conclusion, we report on observation of the centrally produced $\phi\phi$ system. Data is very clean, the only background being low ϕK^+K^- and $K^+K^-K^+K^-$ production. Statistics are enough to perform a Partial Wave Analysis, which is underway.

Acknowledgements

This work was funded in part by the Department of Energy under Contracts No. DE-AC02-76CHO3000 and No. DE-AS05-87ER40356, the National Science Foundation under Grants No. PHY89-21320 and No. PHY90-14879, and CONACyT de México under Grants No. 1061-E9201 and No. 3793-E9401.

REFERENCES

1. C.Caso et al., Euro.Phys.Jour. **C3**, 1 1998.
2. S.Okubo, Phys.Lett. **5**, 165 (1963); G.Zweig, CERN Report No. TH-401 and TH-412, 1964 (unpublished); J.Iizuka, Prog.Theor.Phys., Suppl. **37-38**, 21 (1966).
3. A.Etkin et al., Phys.Rev.Lett. **40**, 422 (1978).
4. A.Etkin et al., Phys.Rev.Lett. **41**, 784 (1978).
5. D.S.Ayres et al., Phys.Rev.Lett. **32**, 1463 (1974).
6. A.Etkin et al., Phys.Rev.Lett. **49**, 1620 (1982); A.Etkin et al., Phys.Lett. **B165**, 217 (1985); **B201**, 568 (1988).
7. T.A.Armstrong et al., Phys.Lett. **B166**, 245 (1986); **B221**, 221 (1989).
8. D.Barberis et al., Phys.Lett. **B432**, 436 (1998).
9. The E690 spectrometer previously used in BNL E766 is described in J.Uribe et al., Phys.Rev. **D49**, 4373 (1994). The beam chambers are described in D.C.Christian et al., Nucl.Instum.Methods Phys.Res., Sect **A345**, 62 (1994).
10. R.S.Longacre, Proceedings of the *VII International Conference on Experimental Meson Spectroscopy*, American Inst. Phys. **113**, 0051 (1984).

Final State Interference Effects in Hadronic Charm Meson Decays [1]

G. Herrera and M. I. Martínez

Centro de Investigación y de Estudios Avanzados del IPN.
Apdo. Postal 14740. México 07360, D. F. MEXICO.

Abstract. In this work two final-state interference effects in hadronic charm meson decays are analyzed. These effects can be observed in the Dalitz plot of the decay under study. In particular, the effects of Bose-Einstein correlations (BEC) and resonance production in the decay $D^+ \to K^-\pi^+\pi^+$ are studied. Previous experimental results show —besides resonances— some other interesting effects in the Dalitz plot of this decay which may be associated to final-state interactions like the ones presented here.

INTRODUCTION

The particles produced in the decay of unstable particles are not totally independent of each other. They interact to some degree by means of different phenomena. Examples of these are the Bose-Einstein Correlations (BEC) and the production and decay of resonances.

BEC effect was firstly observed by Goldhaber, Goldhaber, Lee and Pais (GGLP) [1] in 1960 in $p - \bar{p}$ annihilations. They found that the angular distribution of like-sign pion pairs was not identical to that of unlike-sign pairs; which was not predicted by the model they used. They were able to account for this difference by explicitly symmetrizing —since they are identical bosons— the wavefunction correspondig to like pairs.

However, this "identical particle" effect is not exclusive of bosons; fermions present a similar one as well. In a more practical way, this effect can be understood as follows. Whenever two (or more) identical particles, say bosons (fermions) are created in any process, they tend to occupy the same (not the same) phase space region and they are said to be "correlated" ("anti-correlated"). In the case of bosons the effect is called BEC. This is a pure quantum effect.

Here we analyze the effect of BEC and resonance production in the phase space of the decay $D^+ \to K^-\pi^+\pi^+$. These effects are perfectly visible in the Dalitz plot of the decay so we present the corresponding Dalitz plot generated by MC

---

[1] Work supported by CONACyT under contract J28391E

simulation. The effect of BEC in the Dalitz plot has already been studied before in charm meson decays [2–5], and in e^+e^- annihilations [6,7] by several authors so we are able to compare our plots.

BOSE-EINSTEIN CORRELATIONS

The original correlation function R_{BE}^G corresponding to a pair of identical pions found by GGLP [1] was

$$R_{BE}^G = 1 + e^{-\beta Q^2}, \qquad Q^2 \equiv -(p_1 - p_2)^2$$

in which they consider a Gaussian-shaped volume source. This is nothing more than the probability of producing two identical pions with 4-momenta p_1 and p_2.

In the early seventies Kopylov and Podgoretskiĭ proposed a different approach to the study of correlations between identical pions. They found the correlation function for nucleons emitted by highly excited nuclei to be [8]

$$R_{BE}^K = 1 + \frac{1}{1+(q_0\tau)^2}\left[\frac{2 J_1(qR)}{qR}\right]^2,$$

where q is the 4-momentum difference and R, τ are the space-time characteristics of the nucleus.

However this parametrization is not Lorentz invariant and does not reproduce the experimental fact of the correlation function being almost independent of the energy difference between the pions [7]. In [7] another parametrization, in terms of the components of the 4-momentum difference (see Figure 1), is proposed. This parametrization does fulfill the previous requirements, that is

$$R_{BE} = 1 + \lambda e^{-\beta q_T^2 - \gamma(q_L^2 - q_0^2)}. \tag{1}$$

This parametrization was implemented in the simulation presented here.

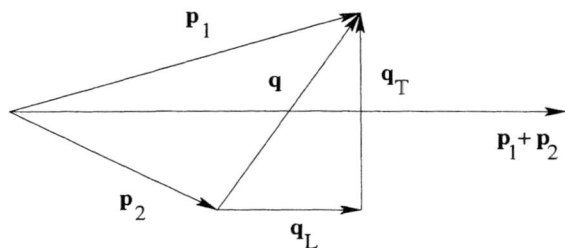

FIGURE 1. Definition of the variables involved in the "extended" GGLP parametrization.

RESONANCE PRODUCTION AND DECAY

The decay $D^+ \to K^-\pi^+\pi^+$ is dominated by the non-resonant mode whose branching ratio is approximately 95% [9]. The resonant $\bar{K}_0^*(892)$ mode whose branching ratio is approximately 21% [9] was implemented in the simulation.

We included the effect of resonance production and decay by using the correlation function found by Bowler [10]. This correlation function for the production of two identical pions one of which is the daugther of a resonance of mas m_0 and width Γ is

$$\frac{1}{(M_{23}^2 + m_0^2\,\Gamma^2)(M_{13}^2 + m_0^2\,\Gamma^2)}$$
$$\times \left[M_{23}^2 + M_{13}^2 + 2\,m_0^2\,\Gamma^2 + 2\,|\rho(Q)|^2 \left(M_{23}^2\, M_{13}^2 + m_0^2\,\Gamma^2 \right) \right]$$

with

$$M_{ij}^2 = \left(m_0^2 - m_{ij}^2 - \frac{\Gamma^2}{4} \right)^2$$

Here the subindices 1 and 2 denote the two identical pions while the subindex 3 denotes the second particle from the resonance. The quantity $\rho(Q)$ is the Fourier transform of the source distribution which corresponds, in this case, to the exponential in (1) [1,5,7].

MONTE CARLO SIMULATION

We implemented the effects to be studied in the simulation by assigning a weight —proportional to the corresponding model— to the plots for each event that was generated.

Figure 2 shows the result of a simulation in which only BEC has been included by means of the correlation function (1). This correlation function contains three parameters that allow manipulation of the Dalitz plot shape. First, the parameter λ is called the coherence parameter since it is a measure of the coherence of the source which produces the identical particles, and it can take on values between 0 and 1. The more coherent the source the smaller λ. We chosed $\lambda = 1$ in order to obtain the maximum effect of BEC on the Dalitz plot.

Since (1) is related to the Fourier transform of the source distribution, β and γ are the Fourier conjugate variables of the spatial and temporal dimensions of the source respectively. The value of β we found suitable corresponds to a transverse spatial dimension of 0.126 fm. For γ we found a corresponding energetic dimension instead of the temporal one using the uncertainty principle so that the value $\gamma = 4$ corresponds to an energy very close to the mass of the K^- meson.

By 'a suitable value' we mean one which produces a Dalitz plot comparable to that obtained experimentally such as that in figure 2(a) in ref. [3].

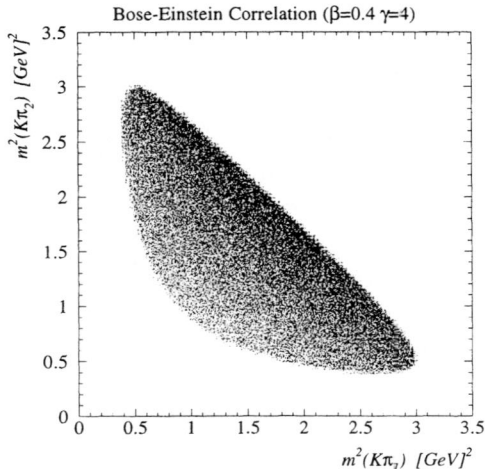

FIGURE 2. Computer generated Dalitz plot for the decay $D^+ \to K^-\pi^+\pi^+$ where only Bose-Einstein correlations are shown.

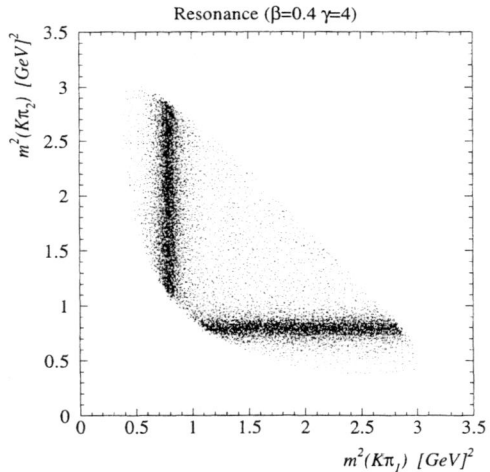

FIGURE 3. Computer generated Dalitz plot for the decay $D^+ \to K^-\pi^+\pi^+$ where only resonance effects are shown.

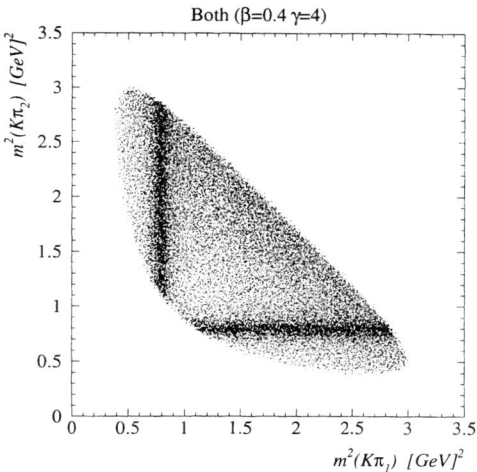

FIGURE 4. Computer generated Dalitz plot for the decay $D^+ \to K^-\pi^+\pi^+$ where both effects are shown.

Figure 3 is a simulation in which only resonance production and decay effects are shown while figure 4 includes both effects. It can be seen from this last figure that the effect of the resonance diminishes that of BEC which is reasonable since the production of the resonance decreases the correlation between the pions.

CONCLUSIONS

Even though we tried different values of β and γ our Dalitz plot for the decay $D^+ \to K^-\pi^+\pi^+$ does not reproduce experimental results (figure 2(a) in [3]). However we must take into consideration that the latter obviously includes all kind of final state interactions such as electromagnetic interactions and production and decay of other resonances. So the next step in this work should be the implementation of such effects in the simulation in order to get a more realistic one and a more accurate Dalitz plot.

We can say that final state interactions —like the ones studied here— in a decay process alter the shape of the corresponding Dalitz plot and therefore experimental fits must consider a non-constant matrix element to analize data which includes not only resonance effects but BEC as well.

REFERENCES

1. Goldhaber G., Goldhaber S., Lee W. and Pais A., *Phys. Rev.* **120**, 300 (1960).
2. MARK III Collab., Adler J. et al., *Phys. Lett.* B **196**, 107 (1987).

3. E691 Collab., Anjos J. C., et al., *Phys. Rev.* D **48**, 56 (1993).
4. E687 Collab., Frabetti P. L. et al., *Phys. Lett.* B **331**, 217 (1994).
5. Cuautle E. and Herrera G., *Phys. Lett.* B **434**, 153 (1998).
6. Aihara H. et al., *Phys. Rev.* D **31**, 996 (1985).
7. Avery P. et al., *Phys. Rev.* D **32**, 2294 (1985).
8. Kopylov G. I. and Podgoretskiĭ G. I., *Sov. J. Nucl. Phys.* **15**, 219 (1972).
9. Particle Data Group, Caso C. et al., *Eur. Phys. J.* C **3**, 187 (1998).
10. Bowler M. G., *Z. Phys.* C **46**, 305 (1990).

Λ^0 Polarization in $pp \to p\{\Lambda^0 K^+\}$ at 800 GeV/c

J. Félix[1], M.C. Berisso[2], D.C. Christian[3], A. Gara[4,a], E.E. Gottschalk[3], G. Gutierrez[3], E.P. Hartouni[5], B.C. Knapp[4], M.N. Kreisler[2,5], S. Lee[2,b], K. Markianos[2,c], G. Moreno[1], M.A. Reyes[6], M. Sosa[1], M.H.L.S. Wang[2,5], A. Wehmann[3], D. Wesson[2,d].[1]

[1] *Universidad de Guanajuato, León Guanajuato México;* [2] *University of Massachusetts, Amherst Massachusetts USA;* [3] *Fermilab, Batavia Illinois USA;* [4] *Columbia University, Nevis Labs, New York USA;* [5] *Lawrence Livermore National Laboratory, Livermore California USA;* [6] *Universidad Michoacana de SNH, Morelia Mich. México.*

Abstract. We present some preliminary measurements of Λ^0 polarization, in the final state $pp \to p\{\Lambda^0 K^+\}$ created at 800 GeV/c, as a function of P_T, x_F, and the diffractive mass of the object $\{\Lambda^0 K^+\}$.

INTRODUCTION

Contrary to naïve arguments, the baryon Λ^0 is produced with non-zero polarization in high energy pp reactions [1]. Its polarization depends on both the normalized Λ^0 longitudinal momentum with respect to the beam momentum, in the cm of the event, (x_F) and the Λ^0 transverse momentum with respect to the beam momentum $(P_T)^2$. In spite of the apparent simplicity of this phenomenon, no satisfactory explanation exists so far. All the models created to explain this phenomenon lack of predictive power and are based on hypotheses created solely to explain it[3]. Almost all measurements have been done using inclusive reactions; there are a few measurements in exclusive reactions [2,5,7]; those can provide more useful information on this phenomenon.

---

[1]) Preset address: a) IBM Watson Laboratory, Yorktown Heights, NY USA; b) SKY Computers, Inc., Chelmsford, MA USA; c) University of Washington, Seattle, Washington USA; d) OAO Corporation, Athens, Georgia USA.

THE EXPERIMENT AND DATA SELECTION

In this paper we present preliminary measurements of Λ^0 polarization performed in a big statistics sample, from the FNAL-690 experiment, with the particular final state

$$pp \to p\Lambda^0 K^+, \qquad (1)$$

created at 800 GeV/c. The experiment is described in detail elsewhere [4,5]; here we mention only relevant details to the present analysis: The E690 recorded 5×10^9 $pp \to p_{fast} X$ events during the Fermilab's 1991 fixed target run at 800 GeV/c. The sample was selected requiring a primary vertex inside the liquid hydrogen target, one incoming beam track at 800 GeV/c, four tracks in the final state, two well separated vertices -the primary where the reaction occurred and the secondary where the Λ^0 decayed-, the second vertex unambiguously identified with a Λ^0, the K^+ and the Λ^0 track assigned to the primary vertex, conservation of electric charge, of baryon number, of energy, and conservation of momentum –the difference of the total energy and the longitudinal momentum of the initial state minus the corresponding of the final state ($\Delta(E-Pl)$) was required to be inside $(-0.030, 0.024)$ GeV/c; and the squared event transversal momentum (SP_T) to be less than 0.005 $GeV/c-$.

Λ^0 POLARIZATION

We analyzed 24 487 events with the foregoing characteristics. The P_T distribution is shown in Fig. 1 and the x_F distribution in Fig. 2. The M_X distribution -the mass of the diffracted object ($\Lambda^0 K^+$)- is shown in Fig. 3.

A Monte Carlo study of the spectrometer acceptance for the states represented by $Eq.1$ was carried out, faithfully simulating all aspects of the spectrometer and of particles production -including the polarization of Λ^0-. In general, the Monte Carlo satisfactorily reproduced data distributions. Just to show this agreement, in the above three figures, Monte Carlo distributions were superimposed; data show error bars. The agreement is very good.

We investigated the sample for polarization using the angular distribution of the p from the Λ^0 decay ($\Lambda^0 \to p\pi^-$) in the center of mass of Λ^0. The quantization axis, the reference axis to measure the angular distribution, was the normal to the production plane $\mathbf{n} = \frac{\mathbf{P}_{beam} \times \mathbf{P}_\Lambda}{|\mathbf{P}_{beam} \times \mathbf{P}_\Lambda|}$. The polarization was extracted fitting to a straight line the angular distribution splitted into bins of x_F, P_T, and $M_X{}^2$. In each case, we integrated over the other two variables. An example of $cos\theta$ distribution is given in Fig. 4; also in this case the Monte Carlo distribution is superimposed. In general the acceptance corrections are small and the polarization from the $cos\theta$ distribution corrected and uncorrected by acceptance are statistically equivalent. We present the results with out correction by acceptance.

We tested the programs used to extract the polarization determining the polarization from a Monte Carlo sample, following exactly the same procedure used to extract the polarization of the sample; in all cases, the polarization was consistent with zero.

RESULTS AND DISCUSSION

The Λ^0 polarization as a function of x_F is shown in Fig. 5; in this case the sign is the opposite observed in inclusive reactions. As a function of P_T is shown in Fig. 6; in this case the polarization was multiplied by (-1) to compare with inclusive measurements, for these are done mainly in the $x_F > 0$ region. This experiment works in the $x_F < -0.3$ region. And as a function of M_X is shown in Fig. 7. Here we also multiplied the data by (-1) to compare with measurements done in the $x_F > 0$ region.

The Λ^0 polarization follows the same trend in exclusive reactions as well as in inclusive reactions, except for a sign. In the whole sample, the polarization as a function of x_F is different from inclusive measurements in a sign. The minus sign appears because the sample ($Eq.1$) is collected in the $x_F < -0.3$ region. As a function of M_X the polarization is a linear function. For $M_X > 2.2\ GeV$ the experiment overlap with that of Ref. 7; in this region the measurements agree. As M_X increases, the polarization decreases linearly. In general, in the threshold $M_X \sim 1.61\ GeV$ the polarization is almost $\sim +0.67$, decreases linearly to zero at $M_X \sim 2.2\ GeV$ and continues linearly up to ~ -0.67 at $M_X \sim 3.25\ GeV$. The striking behaviour of Λ^0 polarization, as a function of P_T, is the positive polarization observed in the $P_T < 0.65\ GeV/c$ region, or negative, since we inverted the sign to compare. The polarization is zero at $P_T \sim 0.0$, decreases as a function of P_T, reaches a minimum and increases up to zero at $P_T \sim 0.65$; this value, where the polarization gets zero, must depend on the average of x_F. Beyond $P_T \sim 0.65$ the polarization increases almost linearly with P_T. With more statistics we can trace this behaviour of Λ^0 polarization as a function of x_F and P_T simultaneously.

Summarizing, we have measured the Λ^0 polarization in the exclusive final state represented by the $Eq.1$. We found that this polarization is a function of x_F, P_T, and M_X. However, it is not a linear function as it is in inclusive measurements and high multiplicity exclusive final states. As a function of M_X is a linear one; it is $+0.67$ at $M_X \sim 1.61\ GeV$ and decreases linearly to zero at $M_X \sim 2.2\ GeV$ and goes linearly to -0.67 at $M_X \sim 3.2\ GeV$.

ACKNOWLEDGEMENTS

We acknowledge the assistance of the technical staff at the AGS at Brookhaven National Laboratory and the superb efforts by the staffs at the University of Massachusetts, Columbia University, and Fermilab. This work was supported in part by National Science Foundation Grants No. PHY90-14879 and No. PHY89-21320, by the Department of Energy Contracts No. DE-AC02-76CHO3000, No. DE-AS05-

87ER40356 and No. W-7405-ENG-48, and by CoNaCyT of México under Grants 1061-E9201, 458100-5-3793E, and 458100-5-4009PE.

REFERENCES

1. G. Bunce et al., Phys. Rev. Lett. **36**, 1113 (1976); G. Bunce et al., Phys. Lett. **B 86**, 386 (1979); B. Lundberg et al., Phys. Rev. **D 40**, 3557 (1989); K. Heller et al., Phys. Rev. Lett. **41**, 607 (1978); K. Heller et al., Phys. Lett. **B 68**, 480 (1977); F. Lomanno et al., Phys. Rev. Lett. **43**, 1905 (1979); S. Erhan et al., Phys. Lett. **B 82**, 301 (1979); F. Abe et al., Phys. Rev. Lett. **50**, 1102 (1983); K. Raychaudhuri et al., Phys. Lett. **B 90**, 319 (1980); A. M. Smith et al., Phys. Lett. **B 185**, 209 (1987); K. Heller et al., Phys. Lett. **B 68**, 480 (1977); K. Heller et al., Phys. Rev. Lett. **51**, 2025 (1983).
2. J. Félix et al., Phys. Rev. Lett. **76**, 22 (1996); J. Félix et al., Phys. Rev. Lett. **82**, 5213 (1999); J. Félix, Ph.D. thesis, Universidad de Guanajuato, México, 1994.
3. T. A. DeGrand et al., Phys. Rev. **D 24**, 2419 (1981); B. Andersson et al., Phys. Lett. **85B**, 417 (1979); J. Szweed et al., Phys. Lett. **105B**, 403 (1981); K. J. M. Moriarty et al., Lett. Nuovo Cimento, **17** 366 (1976); S. M. Troshin and N. E. Tyurin Sov. J. Nucl. Phys. **38**(4), Oct. 1983; J. Soffer and N.E. Törnqvist Phys. Rev. Lett. **68**, 907 (1992); Y. Hama and T. Kodama Phys. Rev. D 48, 3116 (1993); R. Barni et al. Phys. Lett. B 296 (1992) 251-255; W. G. D. Dharmaratna and G. R. Goldstein Phys. Rev. D **53** 1073 (1996); W. G. D. Dharmaratna and G. R. Goldstein Phys. Rev. D **41** 1731 (1990); S. M. Troshin and N. E. Tyurin Phys. Rev. D **55** 1265 (1997); L. Zuo-Tang and C. Boros Phys. Rev. Lett. **79** 3608 (1997).
4. J. Uribe et al., Phys. Rev. **D 49**, 4373 (1994); E. P. Hartouni et al., Phys. Rev. Lett. **72**, 1322 (1994); E. E. Gottschalk et al., Phys. Rev. **D 53**, 4756 (1996); D. C. Christian et al., Nucl. Instr. and Meth. **A345**, 62 (1994); B. C. Knapp and W. Sippach, IEEE Trans. on Nucl. Sci. **NS-27**, 578 (1980); E. P. Hartouni et al., ibid. **NS-36**, 1480 (1989); B. C. Knapp, Nucl. Instrum. Methods A **289**, 561 (1990).
5. S. Lee, Ph. D. Thesis, University of Massachusetts, Amherst, 1994.
6. K. Heller et al., Phys. Lett. 68 **B** 480(1977).
7. T. Henkes et al., Phys. Lett. **B 283**, (1992) 155.

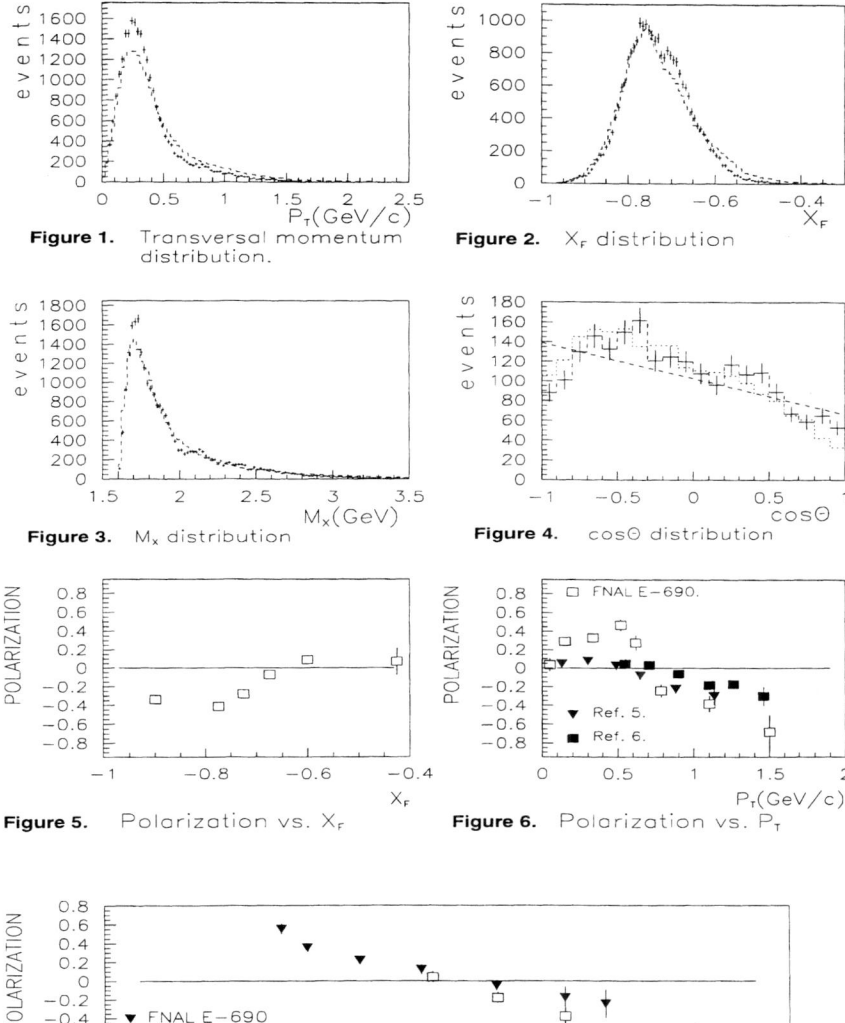

Figure 1. Transversal momentum distribution.

Figure 2. X_F distribution

Figure 3. M_x distribution

Figure 4. $\cos\Theta$ distribution

Figure 5. Polarization vs. X_F

Figure 6. Polarization vs. P_T

Figure 7. Polarization vs. M_x

Strong interaction corrections to the neutron beta decay and high precision.

A. García[a], J.L. García–Luna.[a,b]

[a]*Departamento de Física, Centro de Investigación y de Estudios Avanzados del IPN, Apartado Postal 14-740, 07000 México, Distrito Federal, Mexico.*

[b]*Universidad de Guadalajara, Av. Juarez 975 S.J., Guadalajara, Jal.*

Abstract. We present, in the neutron beta decay, expressions for the decay rate and the electron asymmetry that contain the theoretical effects at the 10^{-4} level. This accuracy is better than the current experimental precision that experiments allow. We consider the effects of the second class current and the radiative corrections. We compare the values of the CKM matrix element $|V_{ud}|$ from this decay with the values both from superallowed fermi beta decays and the unitary of the CKM matrix.

At present, some observables of neutron beta decay are well measured, namely, the decay rate R and the electron asymmetry α_e. We can with the neutron beta decay test to high precision the standard model or study other structures beyond it. The goal of the present analysis is to find the expressions for these observables that contain the theoretical uncertanties at the 10^{-4} level. This accuracy is better than the current experimental precision that modern experiments allow. For this aim it is necessary to study the effects of strong interactions, of the radiative corrections, and of the recoil of the proton.

In the case of neutron beta decay the standard model reproduces the effective hamiltonian of the $V - A$ theory. In this theory the information of the strong interactions is parametrized by mean of form factors. R without radiative corrections, R^0, but taking into acount the proton recoil is given by [1]

$$R^0 = \frac{m_e^5 G_F^2}{2\pi^3}|Vud|^2 \left[1.6335 f_1^2 + 2.3775 \times 10^{-6} f_1 f_2 + 1.1625 \times 10^{-3} f_1 f_3 + 1.5836 \times 10^{-6} f_2^2 + 4.8325 \times 10^{-7} f_3^2 + 3.1835 \times 10^{-6} f_1 \lambda_{f_1} + 4.9014 g_1^2 - 1.0161 \times 10^{-2} g_1 g_2 - 6.2017 \times 10^{-6} g_2^2 + 1.1453 \times 10^{-5} g_1 \lambda_{g_1}\right], \quad (1)$$

f_1 and g_1 are the vector and the axial-vector form factors, f_2 and g_2 are known as weak magnetism and weak electricity form factors, respectively, and f_3 and g_3 are the scalar and pseudo-scalar form factors. $\frac{m_e^5 G_F^2}{2\pi^3} = 0.11613 \times 10^{-3} s^{-1}$ and $|V_{ud}|$ is the CKM matrix element.

The terms λ_{f_1} and λ_{g_1} approximately have the value 2.45 and they are the corrections due to the four-momentum transfer q^2 dependence of f_1 and g_1. We do not exhibit the contributions of $g_1 g_3$, g_3^2, λ_{f_2}, λ_{g_2} and, λ_{g_3} since their coefficients are at the 10^{-8} level, i.e. well below the current experimental precision.

To determine the form factors it is necessary to make some assumptions. If we assume $SU(2)$ symmetry and use the CVC hypothesis, we get $f_1(0) = 1$ and $f_2(0) = 1.8529$. Because there are no second class currents $f_3(0) = 0$ and $g_2(0) = 0$. With the PCAC hypothesis $g_3(0) \approx 114$. The form factor $g_1(0)$ remains as a free parameter.

Because we want to make an analysis to high precision, it is convenient to estimate how the form factors behave under $SU(2)$ symmetry breaking (isospin breaking) to the 10^{-4} level. Isospin symmetry is a very good symmetry, its breaking is known to be at most the 3% level. The corrections due to this breaking to f_2 and f_3 will contribute to R^0 below 10^{-4} as can be seen in Eq.(1). The change of f_1 away from CVC has been estimated in Ref. [2] to be 5×10^{-5} and its contribution to R^0 is seen to be 1.9×10^{-6}. The reason why f_1 is changed so little is the validity of the Ademollo-Gatto theorem [3]. The contribution of g_3 can be seen to be ignorable also.

The corrections to g_2 owing to $SU(2)$ breaking are dominate by the effects of confinement, as given in Ref. [4] to be $g_2 = \frac{\Delta m}{2\omega}$. Δm is the mass difference between the quark u and d, ω the energy of each quark inside the nucleon. So, $g_2 = 0.006$. If we substute this value in the term $1.0161 \times 10^{-2} g_1 g_2$ of Eq.(1), with $g_1 \approx 1.25$ we would obtain 7.6×10^{-5} which is below the current experimental precision.

Under the above considerations, the expression for the transition rate at the 10^{-4} level becomes

$$R^0 = \frac{m_e^5 G_F^2}{2\pi^3} |Vud|^2 (1.6335) \left(1 + 3.0005\lambda^2\right), \qquad (2)$$

where we have factorized the 1.6335. $\lambda = g_1/f_1$ and is still a free parameter.

The electron asymmetry without radiative corrections α_e^0 is defined as follows:

$$\alpha_e^0 = 2 \frac{N \downarrow - N \uparrow}{N \downarrow + N \uparrow}, \qquad (3)$$

where $N \downarrow (N \uparrow)$ is the number of electrons that have $\theta_e < \frac{1}{2}\pi (\theta_e > \frac{1}{2}\pi)$, θ_e is the angle between the electron momentum and the spin of the neutron.

α_e^0 has already been calculated [1] making the same assumption for the form factors that we did to derive Eq.(2) and taking into acount the proton recoil. We obtain

$$\alpha_e^0 = \frac{-0.20916 \times 10^{-3} + 0.2766\lambda - 0.2775\lambda^2}{0.1897 + 0.5692\lambda^2}. \qquad (4)$$

This completes our discussion of the effects of the strong interaction and the recoil of the proton. We next discuss the effects of radiative corrections. The radiative corrections have been separated according to their properties into model-independent radiative corrections (MIRC) and model-dependent radiative corrections (MDRC). While some authors [1,5,6] indicate that the additivity of different orders, of the Fermi function, in α arises in a natural way other authors assume the factorization of these [7,8]. We found that the usual factorization assumptions in this corrections is no longer acceptable at the level of precision we studied. One must use the additivity property of the radiative corrections, other wise a perceptible bias is unnecessarily introduced. Both MIRC and MDRC produce effects that must be incorporated into R and α_e.

Let us now collect all the radiative corrections [9]. The complete result with contributions up to 10^{-4} level is

$$R^{RC} = \frac{m_e^5 G_F^2}{2\pi^3}|V_{ud}|^2 \left(1+3\lambda^2\right) \left\{ \frac{1}{m_e^5} \int \left(f_\alpha + \frac{\alpha(M_P)}{2\pi} g(E) + f_{\alpha^2} + \delta_{\alpha^2} + \right.\right.$$
$$\left.\left. + \frac{\alpha}{2\pi} \left[\ln \frac{M_P}{M_A} + 2C\right] + \frac{\alpha(M_P)}{2\pi} A_g + S(M_P, M_Z) \right) ds \right\}. \tag{5}$$

where $ds = El(E_m - E)^2 dE$ is the phase space factor, $f_\alpha(f_{\alpha^2})$ is the term that represents the Fermi function $F(E)$ to $O(\alpha)[O(\alpha^2)]$. They arise from the Coulomb correction in a natural way [1,5,6]. $g(E)$ and δ_{α^2} are the contributions of the real and virtual photons, whose explicit form is given in Ref. [7,9]. The terms $\ln \frac{M_W}{M_A} + 2C$ represent the asymptotic and non-asymptotic corrections with a photon, $C = 0.881 \pm 0.030$ and the mass M_A is taken within the interval $630 MeV \leq M_A \leq 2520 MeV$ and is the only source of uncertainty out of control in the radiative corrections. $A_g = -0.34$ represents the perturbative QCD corrections of order $O(\alpha_s)$. $S(M_P, M_Z)$ comes from the Z_0 boson interchange and contains an up-date of the renormalization group corrections [9,10]. Notice that the recoil of the proton in the radiative corrections contributes a negligible amount

The more compact numerical expression of R is the following

$$R = 0.1897|V_{ud}|^2(1+3.0005\lambda^2)(1+0.0738 \pm 0.0008) \times 10^{-3} s^{-1}. \tag{6}$$

The radiative corrections to α_e follow the same lines as for R. They are given explicitly in [1]. At the 10^{-4} they must be taken into account. As dicussed in [1], the result is $\alpha_e = \alpha_e^0(1-0.0012)$. Including α_e^0 of Eq.(4) explicitly, we have the compact numerical expression

$$\alpha_e = \frac{-0.2089 \times 10^{-3} + 0.2763\lambda - 0.2772\lambda^2}{0.1897 + 0.5692\lambda^2}. \tag{7}$$

Eqs.(6) and (8) contain all the effects up to the 10^{-4} level in the formulas for the two best measured observables in neutron beta decay. We now turn application of this two formulas. $|V_{ud}|$ is the most precisely determined entry of the CKM matrix. Its value can be determined from three sources, namely, the unitarity of the CKM, superallow fermi transitions (SFT), and neutron beta decay. Here we shall obtain the value (or values) that current data of neutron beta decay give for $|V_{ud}|$ and compare it (or them) with the former two.

In SFT nine different decay modes have been measured. Currently, the accepted value of $|V_{ud}|$ is 0.9735 ± 0.0005 [11]. The unitarity of the CKM matrix gives $|V_{ud}| = 0.9756 \pm 0.0005$. This value is obtained from $|V_{ud}|^2 + |V_{us}|^2 + |V_{ub}|^2 = 1$ using $|V_{us}| = 0.2196 \pm 0.0023$ and $|V_{ub}| = 0.0032 \pm 0.0008$ [11]. To determined $|V_{ud}|$ from neutron beta decay we use the experimental values of R and α_e. The latter can be replaced by λ, if its experimental value is quoted directly. Since this is the case, we shall use λ and not α_e. There are two experimental values of λ, namely, $\lambda_A = -1.2740 \pm 0.0030$ [12] and $\lambda_{LYB} = -1.2624 \pm 0.0024$ [13]. Notice that these two values differ from one another by more than three standard deviations. They are statistically incompatible and, accordingly, should not be averaged, but dealt with separately. The decay rate R has been measured many times and a consistent unique value is currently available, $R = (1.1278 \pm 0.0024) \times 10^{-3} s^{-1}$ [11]. $|V_{ud}|$ can now be determined using Eq.(6). We obtain

$$|V_{ud}|^n_{\lambda_{LYB}} = 0.9786 \pm 0.0019 \pm 0.0004, \tag{8}$$

$$|V_{ud}|^n_{\lambda_A} = 0.9712 \pm 0.0022 \pm 0.0004, \tag{9}$$

these two values differ from one another by more than three standard deviations, as was to be expected. The first error quoted in Eqs.(8) and (9) corresponds to the errors in λ and R, the second one arises from the theoretical uncertainty in R of Eq.(6).

Both of these two values of Eqs.(8) and (9) violate the unitarity of the CKM matrix, the first from above, the second from below. Also the first disagrees rather strongly with the $|V_{ud}|$ of SFT; however, the second seems to agree with this latter.

It is amusing to consider the average of the two values of λ, as proposed in [11], $\lambda_{aver} = -1.2670 \pm 0.0019$. In this case one obtains

$$|V_{ud}|^n_{\lambda_{averag}} = 0.9756 \pm 0.0016 \pm 0.0004. \tag{10}$$

This value is perfectly compatible with unitarity. But, it should be clear that this is a delusion.

In conclusion, we have obtained formulas for the decay rate and electron spin-asymmetry of neutron beta decay, R and α_e, that contain all the relevant corrections up to the 10^{-4} level. We have considered the effects of strong interactions, of the radiative corrections, and of the recoil of the proton. At this level, the formulas depend only on three unknown parameters, namely, $|V_{ud}|$, λ, and M_A. The latter may be taken as systematic bias and incorporated into errors bars. λ may be determined experimentally (measuring it through α_e). Thus, only $|V_{ud}|$ remains as a real free parameter in this decay.

We discussed in detail the relevance of many contributions. The breaking of isospin, which affects CVC, does not contribute appreciably, the second class curent corrections are also quite negligible. However, apart from the model dependent M_A

in the radiative corrections, we found that the usual factorization assumptions in this corrections is no longer acceptable at the level of precision we studied. One must use the additivity property of the radiative corrections, otherwise a perceptible bias is unnecessarily introduced.

The main use of the precision neutron data is, as we have implied above, an independent determination of $|V_{ud}|$. At present, the experimental situation is not clear and one cannot yet obtain a unique $|V_{ud}|$ in this decay. However, one may expect that this situation will soon be improved.

The $|V_{ud}|$ for this decay will be useful in several ways. Compared to the $|V_{ud}|$ from SFT it may help to control better the theoretical uncertanties in those decays. Compared to the unitarity $|V_{ud}|$ it may help either to detect the existence of new physics (or physics beyond the standard model) or to put severe constraints on its existence.

In addition to improving the experimental measurements in neutron beta decay, let us emphasize that more theoretical efforts are requiered. In particular, it should be very interesting to determine M_A with more theoretical accuracy. This is a difficult problem, but by now it should be clear that its solution would keep all effects at the 10^{-4} level well under control.

ACKNOWLEDGMENTS.

The authors acknowledge Partial Support of CONACyT (México). The work of J.L.G.L. was Partial Support by Universidad de Guadalajara (Mexico) is also acknowledged.

REFERENCES

1. A. García and P. Kielanowski, Hyperon Beta Decay, Lecture Notes in Physics **222** (Springer–Verlag), Berlin, Heidelberg, New York, Tokyo, 1985.
2. N. Paver and Riazuddin, Phys. Lett. **B260**, 421 (1991), A. Bramon, A. Grau, and G. Panchieri, Phys. Lett. **344**, 234 (1995).
3. R. Behrends and A. Sirlin, Phys. Rev. Lett.**4**, 186(1964), M. Ademollo and R. Gatto, Phys. Rev. Lett.**13**, 264(1964).
4. A. Halprin, et al, Phys. Rev. **D14**, 2343(1976).
5. A. Sirlin, Phys. Rev. **164**, 1767(1967).
6. T. A. Halpern, Phys. Rev. **C1**, 1928(1970), T. A. Halpern and B. Chern, Phys. Rev. **C175**, 1314(1968).
7. A. Sirlin, Rev. Mod. Phys. **50**, 573(1978).
8. D. H. Wilkinson, Nucl. Phys. **A377**, 474(1982).
9. A. García, J.L. García-Luna unpublished
10. W.J. Marciano and A. Sirlin, Phys. Rev. Lett. **56**,22 (1986).
11. C. Caso et al, Particle Data Group. E. Phys. J. **C3**, 1(1998).
12. H. Abele et al, Phys.Lett. **B407**, 212(1997).
13. P. Liaud, et al, Nucl. Phys. **A612**, 53(1997), B. Yerozolimsky, et al, Phys.Lett. **B412**, 240(1997); P. Bopp, et. al., Phys. Rev. Lett. **56**, 919 (1986).

Single spin asymmetries and the Thomas precession mechanism[†]

G. Domínguez-Zacarías and G. Herrera-Corral

Physics Department - CINVESTAV
PO. Box 14-740
07000 Mexico D.F. MEXICO.

Abstract. We study the asymmetries of inclusive π^+, π^0 and π^- production in the interaction of a polarized with a non polarized proton, in the frame of a two component model. Our model is consistent with the experimental data available.

INTRODUCTION

In the naive quark model the proton consists of three valence quarks (uud), the spin structure however seems to be more complex. The spin structure of the proton has been studied by several experiments (EMC,SMC at CERN; E142/E143 at SLAC and FNAL-E704). The spin-dependent structure function $g_1(x)$, for the proton was determined in 1988 for the first time and a disagreement with the *Ellis-Jaffe* sum rule [1] was found. This result implies that the spin of the proton may not be completely accounted by the constituents quarks. As a result many experiments were planned and different measurements are now available. The production of π^a [1)] in the interaction of polarized with unpolarized protons has been studied. The asymmetry production of π^a when the spin of the proton comes up or down may be a manifestation of the spin structure of the proton.

THE TWO MODEL COMPONENT

The spin asymmetry is defined as:

$$A_N(X_F, P_T) = \frac{\sigma \uparrow - \sigma \downarrow}{\sigma \uparrow + \sigma \downarrow} \quad (1)$$

Where $\sigma \uparrow$ and $\sigma \downarrow$ represent the differential cross sections for π^a production when the proton comes with the spin up or down respectively.

[1)] Where a denotes the three pion charges $(+, -, 0)$
[†] This work has been supported by CONACYT under contract 29273E

In a two components model:

$$\frac{d\sigma^{\pi^a}}{dX_F dP_T} = \frac{d\sigma^{\pi^a, REC}}{dX_F dP_T} + \frac{d\sigma^{\pi^a, QCD}}{dX_F dP_T}, \qquad (2)$$

where REC and QCD indicate that pions originate in a recombination (REC) and/or fragmentation (QCD) process. The QCD component in equation (1) can be obtained from $e^+ + e^- \rightarrow q\bar{q}$ reactions, where the quarks in the final state hadronize in a similar way as they do in $g+g \rightarrow q\bar{q}$ or $q+\bar{q} \rightarrow q\bar{q}$. In a fragmentation process quarks give origin to hadrons(mostly π-mesons) [5,6]. According with the distribution obtained in e^+e^- annihilation at center of mass energies between 3.6 and 5.2 GeV [7], we use

$$\frac{d\sigma}{dX_F} = A e^{-BX_F}, \qquad (3)$$

for the distribution of π that originate in the fragmentation of quarks. In the recombination process, the production of π^+ ($u\bar{d}$) and π^- ($\bar{u}d$), ocurrs when a $\bar{d}$ ($\bar{u}$) quarks from the sea of the proton joints a valence quark $u(d)$. In the process the sea quaks are accelerated [2,3]. In hyperon polarization models, the Thomas precession mechanism seems to be responsible of spin alignment.

We use a recombination model where Thomas precession is present plus a QCD correction to study the asymmetries. We make use of the correlation beetwen the polarized proton ($\uparrow$) and the u quark in agreement with experimental measurements [4]. To produce a π^+, the valence $u(\uparrow)$ quark in the polarized proton recombines with a $\bar{d}$ from the sea, the $\bar{d}$ quark must have a spin down. To produce a π^- meson, the valence d ($\downarrow$) quark from the polarized proton $(\uparrow)$ recombines with a $\bar{u}$ quark from the sea. In the π^0 case, the valence u quark from the polarized proton combine with $\bar{u}$ quark from the sea. See Figure 1.

To first order in ω_T [2] the asymmetries for π^+, π^- and π^0 production are given by :

$$A_N^{\pi^+}(X_F, P_T) = \frac{4(1-2\xi)}{5(1+2\xi)^2 M^2 [\frac{1}{2} + \frac{M^4}{X_F P} A \exp(-BX_F)]} P_{T_{\pi^+}} \qquad (4)$$

$$A_N^{\pi^-}(X_F, P_T) = -\frac{4(1-2\xi)}{5(1+2\xi)^2 M^2 [\frac{1}{2} + \frac{M^4}{X_F P} A \exp(-BX_F)]} P_{T_{\pi^-}} \qquad (5)$$

$$A_N^{\pi^0}(X_F, P_T) = \frac{8(1-2\xi)[\frac{2}{M^6} - \frac{1}{M_1^6}]}{5(1+2\xi)^2 [\frac{2}{M^4} + \frac{1}{M_1^4} + \frac{A}{(2X_F P)^2} \exp(-BX_F)]} P_{T_{\pi^0}}, \qquad (6)$$

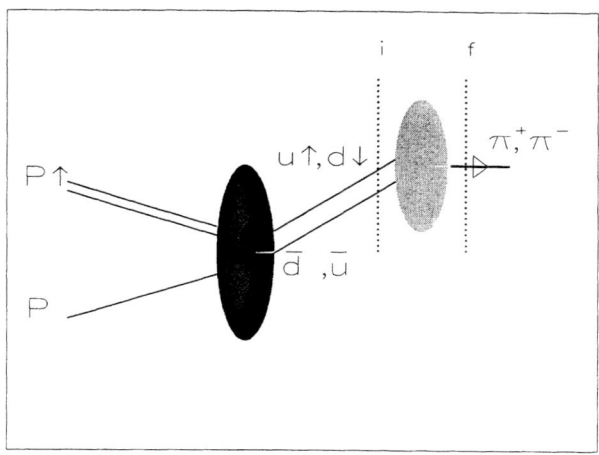

FIGURE 1. Amplitude for the reaction $P \uparrow P \to \pi^{\pm} + X$ in a recombination picture.

where,

$$M^2 = \left[\frac{P_v^2 + m_v^2}{\xi} + \frac{P_s^2 + m_s^2}{1-\xi} - P_{T_{\pi^a}}^2 - m_{\pi^a}^2\right], \quad (7)$$

and the subindices v and s denote the valence and sea quark respectively. The parameter ξ is parametrized [2] as

$$\xi = \frac{1}{2}(1 - X_F) + 0.1 X_F. \quad (8)$$

Figure 2 shows the spin asymmetries from equations (4) to (7) together with experimental data [8]. The absolute magnitude of the asymmetry for π^+ and π^- are similar, and reproduce experimental results qualitatively. The two different lines represent our predictions (equation 4 and 5) with two different values of B. In the π^+ case the experimental data are between our two predictions. In the π^- case only one experimental point is between these predictions, but describe satisfactorily these data, considering that there are reasons to believe that recombination of π^- must be slightly different than for π^+.

In Figure 3 the asymmetries for π^+ and π^- in two ranges of P_T are shown. The model reproduces the experimental data in these ranges. In Figure 4 we show the asymmetry for π^o [9]. Our model is in agreement with experimental data in the differents ranges of P_T too. Thomas precession seems to be responsible of spin alignment and the flavor decomposition of spin observed in nucleuons at SMC and

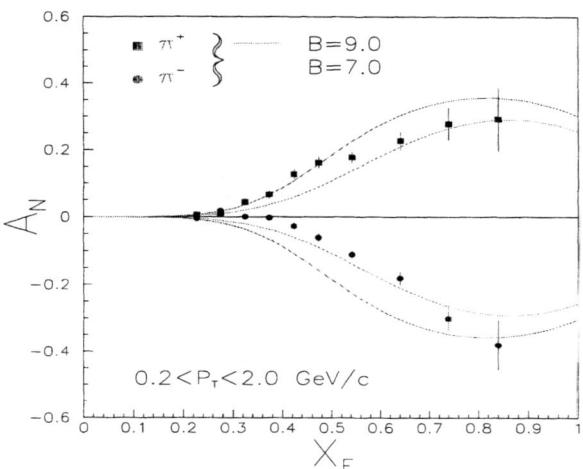

FIGURE 2. The asymmetry A_N for π^+ and π^- production versus X_F and the two components model for two values of B.

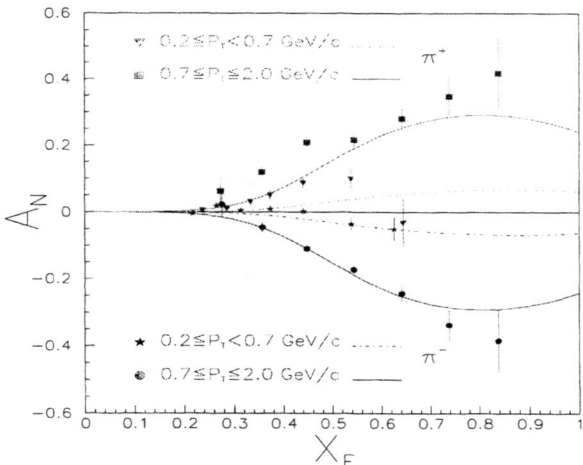

FIGURE 3. A_N versus X_F for π^+ and π^- production in two P_T ranges and the two components model prediction with $B = 9.0$.

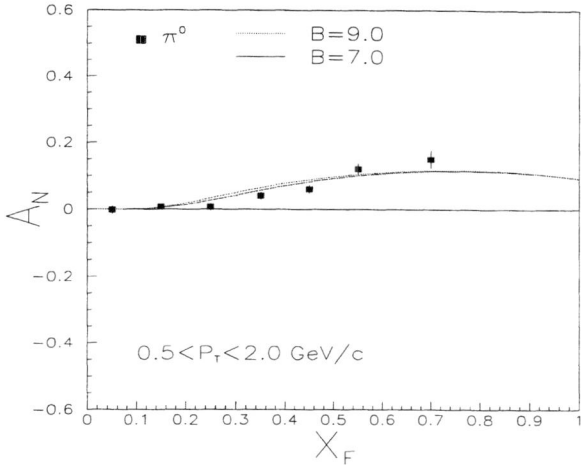

FIGURE 4. A_N versus X_F for π^0 production and the two components model prediction.

HERMES is shown to be consistent with the experimentally observed asymmetry in the framework of a two component model.

REFERENCES

1. J.D. Bjorken *Phys. Rev D* **148**, 1467 (1966).
   J. Ellis and R.L. Jaffe *Phys. Rev D* **9**, 1444 (1974); **10** 1669 (1974).
   Ashman J. el al. *European Muon Collaboration, Phys. Lett. B* **206**, 364 (1988).
   V.W. Hughes, V. Papavassiliou, R. Piegaia, K.P. Schuler and G. Baum, *Phys. Lett. B* **212**, 511 (1988).
2. DeGrand T.A. and Miettinen H.I.,*Phys. Rev. D* **23** (1981) 1227; **24** (1981) 2419.
3. Bonner B. E. et al. *Phy. Rev. D* **38** (1988) 729. Saroff S. et al. *Phys. Rev. Lett.*;**64** (1990) 995.
4. K. Ackerstaff et al. *Hermes Collaboration, Phys. Lett. B* **464**, 123 (1999).
5. Halzen F. and Martin D. Alan *Quarks and Leptons* **An Introductory Course in Modern Particle Physics**. Wiley, New York 48 (1984).
6. Taizo Muta *Foundations of Quantum Chromodynamics*, 1987
7. DASP Collab., *Nucl. Phys. B* **148** (1979) 189.
8. D.L. Adams et al. *FNAL E704 Collaborations, Phys. Lett. B* **345**, 479 1995.
9. D.L. Adams et al. *FNAL E704 Collaborations, Phys. Lett. B* **261**, 201 1991

LIST OF PARTICIPANTS

Joao dos Anjos	CBPF	janjos@lafex.cbpf.br
Erika Alvarez	IF-UNAM	erika@ft.ifisicacu.unam.mx
Moises Araiza	IF-BUAP	moises@sirio.ifuap.buap.mx
Jorge I. Aranda	IFM-UMSNH	jaranda@ginette.ifm.umich.mx
Juan Carlos Arteaga	Cinvestav	JuanC.Arteaga@fis.Cinvestav.mx
Alejandro Ayala	ICN-UNAM	ayala@nuclecu.unam.mx
Rene Asomoza	Cinvestav	rasomoza@mail.Cinvestav.mx
Mateo Barkovich	ICN-UNAM	mateobar@hotmail.com
Enrique Barradas	FCFM-BUAP	barradas@fcfm.buap.mx
Juan Barranco	UDLA	fa097798@mail..udlap.mx
Argelia Bernal	UDLA	fa095715@mail.udlap.mx
Jaime Besprosvany	IF-UNAM	bespro@ft.ifisicacu.unam.mx
Antonio Bouzas	Cinvestav-UM	abouzas@kin.cieamer.conacyt.mx
Alejandro Cabo	ICIMAF	cabo@cidet.icmf.inf.cu
German A. Calderon	Cinvestav	gacalde@fis.Cinvestav.mx
Regnier A. Cano	UADY	
Karla Cantun	UADY	kbca@monika.cieamer.conacyt.mx
Marcos Cardoso	UNESP	mcr@ift.unesp.br
Alexandra Carreño	ICN-UNAM	alex@nuclecu.unam.mx
Guillermo Contreras	Cinvestav-UM	jgcn@moni.cieamer.conacyt.mx
Maritza de Coss	Cinvestav-UM	mdecoss@jade.cieamer.conacyt.mx
Elizabetta Crescio	INFN/Turin	crescio@to.infn.it
Lorenzo Diaz Cruz	IF-BUAP	ldiaz@sirio.ifuap.buap.mx
Juan Carlos D'Olivo	ICN-UNAM	dolivo@nuclecu.unam.mx
Galileo Dominguez	Cinvestav	gdz@fis.Cinvestav.mx
Eckhard Elsen	DESY	elsen@mail.desy.de
Jurgen Engelfried	IF-UASLP	jurgen@ifisica..uaslp.mx
Catalina Espinoza	IF-UNAM	catalina@ft.ifisicacu.unam.mx
Julian Felix Valdez	IFUG	felix@ifug.ugto.mx
Arturo Fernandez	FCFM-BUAP	afernand@fcfm.buap.mx
Alain Flores	FCFM-BUAP	afflores@fcfm.buap.mx
Jorge Flores	CCF-UNAM	jfv@servidor.unam.mx
Julio C. Flores	Cinvestav	julio@fis.Cinvestav.mx
Tomas Flores	UADY	
Ricardo Gaitan	FES-UNAM	Ricardo.Gaitan@fis.Cinvestav.mx
Luis A. Gallegos	IFUG	gallegos@ifug2.ugto.mx
Carlos Garcia Canal	U. Nal-La Plata	garcia@venus.fisica.unlp.edu.ar
Augusto Garcia	Cinvestav	augarcia@fis.Cinvestav.mx
Gerardo Garcia	IFUG	ggarcia@ifug2.ugto.mx
Israel Garcia	UADY	
Jose Luis Garcia	Cinvestav	jlgarcia@fis.Cinvestav.mx
Juan Jose Godina	Cinvestav	jj@fis.Cinvestav.mx
Virendra Gupta	Cinvestav-UM	virendra@kin.cieamer.conacyt.mx

Albino Hernandez	ESFM-IPN	albino@esfm.ipn.mx
Jaime Hernandez	IF-BUAP	jhernand@sirio.ifuap.buap.mx
Javier Hernandez	FCFM-BUAP	javierh@phyun0.ucr.edu
Gerardo Herrera	Cinvestav	gherrera@fis.Cinvestav.mx
Wilbert Herrera	UADY	
Rodrigo Huerta	Cinvestav-UM	rhuerta@kin.cieamer.conacyt.mx
Jacobo Konigsberg	U. Florida	konigsberg@fnald.fnal.gov
Francisco Larios	Cinvestav-UM	flarios@kin.cieamer.conacyt.mx
Leon Lederman	Illinois I. Tech	lederman@fnal.gov
Idelfonso Leon	Cinvestav	Idelfonso.Leon@fis.Cinvestav.mx
Ling-Fong Li	Carnegie-Mellon U	lfli@cmuhep2.phys.cmu.edu
Dennys A. Lopez	IF-BUAP	armando@sirio.ifuap.buap.mx
Gabriel Lopez	Cinvestav	glopez@fis.Cinvestav.mx
Lao-tse Lopez	FCM-BUAP	lao-tse@fcfm.buap.mx
Axel de la Macorra	IF-UNAM	macorra@ft.ifisicacu.unam.mx
Jose E. Madriz	IF-UMSNH	edgar@ginette.ifm.umich.mx
G. Maniocchi	INFN/GP	
Roberto Martinez	U Nal Bogota	romart@ciencias.unal.edu.co
Mario I. Martinez	Cinvestav	mim@fis.Cinvestav.mx
Jose A. Mendez	Cinvestav-UM	jmendez@kin.cieamer.conacyt.mx
George Mikenberg	Weizmann Inst.	G.Mikenberg@cern,ch
Joaquin Miranda	UADY	
Alfonso Mondragon	IF-UNAM	amondra@ft.ifisicacu.unam.mx
Benjamin Morales	IF-UNAM	bamr@ft.ifisicacu.unam.mx
Antonio Morelos	IF-UASLP	morelos@dec1.ifisica.uaslp.mx
Enrique Moreno	IA-UNAM	emoreno@astroscu.unam.mx
Laura Muñoz S.	IFUG	lamusa@ifug3.ugto.mx
Gabriela Murguia	IF-UNAM	gabriela@ft.ifisicacu.unam.mx
Mauro Napsuciale	IFUG	mauro@ifug2.ugto.mx
Jorge Luis Navarro	U del Atlantico	jnavarro@mailcity.com
Alejandro Ordaz	UADY	
Alfredo del Oso	IF-UNAM	alfredo@ft.ifisicacu.unam.mx
Rodrigo Pelayo	Cinvestav	rpelayo@fis.Cinvestav.mx
Monica Pepe	INFN/PG	monica.pepe@PG.infn.it
Miguel A. Perez	Cinvestav	mperez@fis.Cinvestav.mx
William Ponce	U Antioquia	wponce@fisica.udea.edu.co
Juan Ponciano	U N La Plata	ponciano@venus.fisica.unlp.edu.ar
Carlos A. Ramirez	U I Santander	cramirez@hemeroteca.icfes.gov.co
Fernando Ramirez	FC-UAEM	framirez@servm.fc.uaem.mx
Alfredo Raya	IFM-UMSNH	raya@ginette.ifm.umich.mx
Alejandro Reyes	IF-UNAM	Alejandro.Reyes@cern.ch
Marco A. Reyes	IFM-UMSNH	marco@ifm1.ifm.umich.mx
Tania Rivera	IF-UNAM	tania@fenix.ifisicacu.unam.mx
Ezequiel Rodriguez	IF-UNAM	ezequiel@ft.ifisicacu.unam.mx
Simon Rodriguez	IFUG	srodi@ifug1.ugto.mx

Alberto de Roeck	DESY/CERN	deroeck@mail.desy.de
Eduardo Rosado	UADY	eduardo@monika.cieamer.conacyt.mx
Andrey Rostovstev	ITEP	rostov@iris1.itep.ru
Marti Ruiz Altaba	IF-UNAM	marti@ft.ifisicacu.unam.mx
Alvaro de Rujula	CERN	Alvaro.Derujula@cern.ch
Alberto Sanchez	Cinvestav	Alberto.Sanchez@fis.Cinvestav.mx
Gabriel Sanchez	Cinvestav-UM	gsanchez@kin.cieamer.conacyt.mx
Sahu Sarira	ICN-UNAM	sarira@nuclecu.unam.mx
Francisco Sastre	Cinvestav-UM	sastre@kin.cieamer.conacyt.mx
Jurgen Schukraft	CERN	schukraft@cern.ch
Jordi Sod	IF-UNAM	jordi@mail.internet.com.mx
Ulises Solis	ICN-UNAM	solis@nuclecu.unam.mx
Rainer Stamen	U Dortmund	stamen@Physik.Uni-Dortmund.DE
Gilberto Tavares	Cinvestav	gtv@fis.Cinvestav.mx
Eduardo S. Tututi	ECFM-UMSNH	tututi@zeus.ccu.umich.mx
German Valencia	State Iowa U	valencia@iastate.edu
Andrea Vargas	Cinvestav	andrea@fis.Cinvestav.mx
Eugenia Vargas	UADY	
Mauricio Vargas	U Nal Bogota	mvargas@ciencias.ciencias.uanl.edu.co
Roberto Vega	SMU	vega@pascal.physics.smu.edu
Luis M. Villase or	IFM-UMSNH	villasen@zeus.ccu.umich.mx
Ramona Vogt	LBL/UC Davis	vogt@mail-nsdth.lbl.gov
Axel Weber	IFM-UMSNH	axel@io.ifm.umich.mx
Jose Wudka	U Cal Riverside	wudka@phyun5.ucr.edu
Cristo M. Yee	U A Sinaloa	e027018@alumno.uasnet.mx
Miguel Zambrano	UADY	
Ivan V. Zuloaga	U Nal Bogota	iazulo@ciencias.ciencias.uanl.edu.co
Arnulfo Zepeda	Cinvestav	zepeda@fis.Cinvestav.mx

Maru Rodriguez	Organizing Committee
Monika Kullova	Organizing Committee
Lupita Aguilar	Organizing Committee

Author Index

A

Anjos, J. C., 172, 267
Asomoza, R., 247

B

Berisso, M. C., 370, 381
Besprosvany, J., 289
Bonvicini, V., 360

C

Cabo, A., 294
Cerello, P., 360
Christian, D. C., 370, 381
Contreras, J. G., 276
Crescio, E., 360
Cuautle, E., 172

D

de la Macorra, A., 320
De Roeck, A., 122
Diaz-Cruz, J. L., 299
Domínguez-Zacarías, G., 391

E

Engelfried, J., 102

F

Félix, J., 259, 370, 381
Flores, J., 252

G

Gaitán, R., 337
Gara, A., 370, 381
García, A., 386

García, E., 337
García Canal, C. A., 199
García-Luna, J. L., 386
Giubellino, P., 360
Gottschalk, E. E., 370, 381
Gutiérrez, G., 370, 381

H

Hartouni, E. P., 370, 381
Hernández-Galeana, A., 337
Hernández-Montoya, R., 360
Herrera, G., 250, 375, 391

K

Knapp, B. C., 370, 381
Kolojvari, A., 360
Konigsberg, J., 69, 271
Kreisler, M. N., 370, 381

L

Larios, F., 346
Lee, S., 370, 381
Li, L.-F., 16
Lucio, J. L., 365

M

Markianos, K., 370, 381
Martínez, M. I., 375
Martinez, R., 326
Mazza, G., 360
Mondragón, A., 310
Montano, L., 360
Montero, J. C., 315
Morelos, A., 255
Moreno, G., 259, 370, 381

N

Napsuciale, M., 365
Nissinen, J., 360
Nouais, D., 360

P

Pérez, M. A., 342, 346
Piccinelli, G., 320
Pleitez, V., 315
Ponce, W. A., 332

R

Rashevsky, A., 360
Reyes, M. A., 370, 381
Rigol, M., 294
Rivera-Rebolledo, J. M., 337
Rivetti, A., 360
Rodriguez, J.-A., 326
Rodriguez, M. C., 315
Rodríguez-Jáuregui, E., 310
Rostovtsev, A., 351

S

Sahu, S., 355
Sánchez-Hernández, A., 76, 263
Sassot, R., 199

Schukraft, J., 3
Sistema Tele Yucatán Canal 13, 285
Sosa, M., 381

T

Tavares-Velasco, G., 342, 346
Toscano, J. J., 346
Tosello, F., 360

V

Vacchi, A., 360
Valencia, G., 45
Vargas, M., 326
Vogt, R., 152

W

Wang, M. H. L. S., 370, 381
Weber, A., 305
Wehman, A., 370, 381
Wesson, D., 370, 381
Wudka, J., 81

Z

Zepeda, A., 280
Zuluaga, I. D., 326